T0211919

Integral Equation Methods for Electromagnetic and Elastic Waves

© Springer Nature Switzerland AG 2022
Reprint of original edition © Morgan & Claypool 2009

Integral Equation Methods for Electromagnetic and Elastic Waves
Weng Cho Chew, Mei Song Tong and Bin Hu

ISBN-13: 978-3-031-00579-4   paperback

ISBN-13: 978-3-031-01707-0   ebook

DOI10.1007/978-3-031-01707-0

A Publication in the Springer series
LECTURES ON COMPUTATIONAL ELECTROMAGNETICS #12

Lecture #12

Series Editor: Constantine A. Balanis, Arizona State University

Series ISSN: 1932-1252   print
Series ISSN: 1932-1716   electronic

First Edition
10 9 8 7 6 5 4 3 2 1

Integral Equation Methods for Electromagnetic and Elastic Waves

Weng Cho Chew
University of Hong Kong and
University of Illinois at Urbana-Champaign

Mei Song Tong
University of Illinois at Urbana-Champaign

Bin Hu
Intel Research

SYNTHESIS LECTURES ON COMPUTATIONAL ELECTROMAGNETICS #12

ABSTRACT

Integral Equation Methods for Electromagnetic and Elastic Waves is an outgrowth of several years of work. There have been no recent books on integral equation methods. There are books written on integral equations, but either they have been around for a while, or they were written by mathematicians. Much of the knowledge in integral equation methods still resides in journal papers. With this book, important relevant knowledge for integral equations are consolidated in one place and researchers need only read the pertinent chapters in this book to gain important knowledge needed for integral equation research. Also, learning the fundamentals of linear elastic wave theory does not require a quantum leap for electromagnetic practitioners.

Integral equation methods have been around for several decades, and their introduction to electromagnetics has been due to the seminal works of Richmond and Harrington in the 1960s. There was a surge in the interest in this topic in the 1980s (notably the work of Wilton and his coworkers) due to increased computing power. The interest in this area was on the wane when it was demonstrated that differential equation methods, with their sparse matrices, can solve many problems more efficiently than integral equation methods. Recently, due to the advent of fast algorithms, there has been a revival in integral equation methods in electromagnetics. Much of our work in recent years has been in fast algorithms for integral equations, which prompted our interest in integral equation methods. While previously, only tens of thousands of unknowns could be solved by integral equation methods, now, tens of millions of unknowns can be solved with fast algorithms. This has prompted new enthusiasm in integral equation methods.

KEYWORDS

Integral Equations, Computational Electromagnetics, Electromagnetic Waves, Linear Vector Spaces, Energy Conservation Theorem, Low-Frequency Problems, Dyadic Green's Function, Fast Inhomogeneous Plane Wave Algorithm, Elastic Waves

Dedication

We dedicate this book
to our predecessors
in this field.

Contents

Preface

This monograph is an outgrowth of several years of work. It began with a suggestion by Joel Claypool from Morgan & Claypool Publishers and Constantine Balanis of Arizona State University on writing a monograph on our recent contributions to computational electromagnetics (CEM). Unfortunately, we had just written a book on CEM a few years ago, and it was hard to write a new book without repeating many of the materials written in *Fast and Efficient Algorithms in Computational Electromagnetics* (FEACEM).

In the midst of contemplating on a topic, it dawned upon me that there was no recent books on integral equation methods. There had been books written on integral equations, but either they have been around for a while, such as books edited by Mittra, and by Miller, Medgyesi-Mitschang and Newman, or they were written by mathematicians, such as the book by Colton and Kress.

Much of the knowledge in integral equation methods still resides in journal papers. Whenever I have to bring researchers to speed in integral equation methods, I will refer them to study some scientific papers but not a book. So my thinking was that if we could consolidate all important knowledge in this field in one book, then researchers just need to read the pertinent chapters in this book to gain important knowledge needed for integral equation research.

Integral equation methods have been around for several decades, and their introduction to electromagnetics has been due to the seminal works of Richmond and Harrington. After the initial works of Richmond and Harrington in the 1960s, there was a surge in the interest in this topic in the 1980s (notably the work of Wilton and his coworkers) due to the increased power of computers. The interest in this area was on the wane when it was demonstrated that differential equation methods, with its sparse matrices, can solve many problems more efficiently than integral equations. However, in recent decades (1990s), due to the advent of fast algorithms, there was a revival in integral equation methods in electromagnetics.

Much of our work in recent years has been in fast algorithms for integral equations, which prompted our interest in integral equation methods. While previously, only tens of thousands of unknowns can be solved by integral equation methods, now, tens of millions of unknowns can be solved with fast algorithms. This has prompted new enthusiasm in integral equation methods. This enthusiasm had also prompted our writing the book FEACEM, which assumed that readers would have the basic knowledge of integral equation methods. This book would help to fill in that knowledge.

With this book, important relevant knowledge for integral equations are consolidated in one place. Much of the knowledge in this book exists in the literature. By rewriting this knowledge, we hope to *reincarnate* it. As is in most human knowledge, when it was first discovered, only few could understand it. But as the community of scholars came together to digest this new knowledge, regurgitate it, it could often be articulated in a more lucid and succinct form. We hope to have achieved a certain level of this in this book.

Only one chapter on this book is on fast algorithm, namely, Chapter 7, because the topic of this chapter has not been reported in FEACEM. This chapter is also the outgrowth of BH's PhD dissertation on fast inhomogeneous plane wave algorithm (FIPWA) for free space and layered media.

Since many researchers in electromagnetics can easily learn the knowledge related to elastic waves, as many of our researchers do, we decide to include Chapter 8 in this book. It is written primarily for practitioners of electromagnetics—Learning linear elastic wave theory does not require a quantum leap for electromagnetic practitioners. Because we have recently applied some of our expertise in solving the elastic wave problems, including the application of fast algorithm, we decide to include this chapter.

In this book, Chapters 1 to 6 were primarily written by WCC, while Chapter 7 was written by BH, and MST wrote Chapter 8, the last Chapter. We mutually helped with the proofreading of the chapters, and in particular MST helped extensively with the proofreading, editing, as well as the indexing of the book.

As for the rest of the book, Chapter 1 introduces some concepts in CEM, and tries to contrast them with concepts in computational fluid dynamics (CFD), which itself is a vibrant field that has been studied for decades.

Chapter 2 is on the fundamentals of electromagnetics. We relate some fundamental concepts to concepts in linear vector space theory. It is *de rigueur* that students of CEM be aware of these relationships. After all, a linear integral equation eventually becomes a matrix equation by projecting it into a subspace in the linear vector space.

Chapter 3 gives a preliminary introduction to integral equation methods without being bogged down by details. It emphasizes mainly on a heuristic and physical understanding of integral equation methods. For many practitioners, such a level of understanding is sufficient.

Chapter 4 is not for the faint hearted as it delves into more details than Chapter 3. It discusses integral equations for impenetrable bodies as well as penetrable bodies. It deals with a very important issue of uniqueness of many of these integral equation methods. It introduces Gedanken experiments to prove the nonuniqueness of these solution methods, and discusses remedies for them.

Chapter 5 is on low-frequency methods in integral equations. Integral equation methods, as well as differential equation methods, are plagued with instabilities when the frequency is extremely low, or the wavelength is very long compared to the structure being probed. This is because we are switching from the regime of wave physics to the regime of circuit physics (discussed in Chapter 1). This transition presage a change in solution method when one goes into the low-frequency regime, which I have nicknamed "twilight zone!"

Chapter 6 reports on how one formulates the integral equations in the layered media. The derivation of the dyadic Green's function for layered media is often a complex subject. It involves complicated Sommerfeld integrals, which often are poorly convergent. We propose a new formulation, which harks back to the technique discussed in *Waves and Fields in Inhomogeneous Media*, as oppose to the Michalski-Zheng formulation, which is quite popular in the literature.

Integral equation methods are in general quite complex compared to differential equation methods. To convert integral equations to matrix equations, one has to project the Green's operator onto a subspace by numerical quadrature. This numerical quadrature involves singular integrals, and can be quite challenging to perform accurately. We regret not being able to discuss these numerical quadrature techniques in this book. These often belong to the realm of applied mathematics on numerical quadrature of singular integrals. Much has been written about it, and the monograph does not allow enough space to discuss this topic.

However, we hope that we have discussed enough topics in this short monograph that can pique your interest in this very interesting and challenging field of integral equation methods. Hopefully, many of you will pursue further research in this area, and help contribute to new knowledge to this field!

Weng Cho CHEW
June 2008, Hong Kong

Acknowledgements

Many in our research group gave feedback and helped proofread the manuscript at various stages. Thanks are due Mao-Kun LI, Yuan LIU, Andrew Hesford, Lin SUN, Zhiguo QIAN, Jie XIONG, Clayton Davis, William Tucker, and Bo HE. During my stay as a visiting professor at Nanyang Technological University, I received useful feedback from Eng Leong TAN on some of the chapters. Peter Monk gave useful feedback on part of Chapter 2.

We are grateful to our research sponsors for their supports in the last decade, notably, AFOSR, SAIC-DEMACO, Army-CERL, Raytheon-UTD, GM, SRC, Intel, IBM, Mentor-Graphics, and Schlumberger. WCC also thanks supports from the Founder Professorship and Y.T. Lo Endowed Chair Professorship, as well as supports from the IBM Visiting Professorship at Brown University, the Distinguished Visitor Program at the Center for Computational Sciences, AFRL, Dayton, the Cheng Tsang Man Visiting Professorship at NTU, Singapore, and the MIT Visiting Scientist Program. Supports from Arje Nachman and Mike White of the government agencies are gratefully acknowledged. We appreciate interactions with partners and colleagues in the industry, notably, Steve Cotten, Ben Dolgin, Val Badrov, HP Hsu, Jae Song, Hennning Braunisch, Kemal Aygun, Alaeddin Aydiner, Kaladhar Radhakrishnan, Alina Deutsch, Barry Rubin, Jason Morsey, Li-Jun Jiang, Roberto Suaya, Tarek Habashy, and John Bruning. Also, discussions with university partners on various projects such as Ben Steinberg, Steve Dvorak, Eric Michielsen, and John Volakis have been useful. Last but not least, constant synergism with members of the EM Laboratory at Illinois, Shun-Lien Chuang, Andreas Cangellaris, Jianming Jin, Jose Schutt-Aine, and Jennifer Bernhard are enlightening. This book is also dedicated to the memory of Y.T. Lo, Don Dudley, and Jin Au Kong, who have been WCC's mentors and friends.

CHAPTER 1

Introduction to Computational Electromagnetics

1.1 Mathematical Modeling—A Historical Perspective

In the beginning, mathematics played an important role in human science and technology advancements by providing arithmetic, geometry, and algebra to enable scientists and technologists to solve complex problems. For example, precalculus was used by Kepler to arrive at laws for predicting celestial events in the 1600s. In the late 1600s and early 1700s, Leibniz and Newton developed early calculus for a better description of physical laws of mechanics [1]. Later, the development of differential equations enabled the better description of many physical phenomena via the use of mathematics. The most notable of these were the Navier-Stokes equations for understanding the physics of fluids in the 1800s [2–4].

Computational fluid dynamics (CFD) [5] has a longer history than computational electromagnetics (CEM) (Richardson 1910 [6], Courant, Friedrichs, and Lewy 1928 [7], Lax 1954 [8], Lax and Wendroff 1960 [9], MacCormack 1969 [10]). The early works are mainly those of mathematicians looking for numerical means to solve partial differential equation (PDE), with the knowledge that many such equations have no closed form solution. Hence, the reason for the early development of CFD is simple: many fluids problems are analytically intractable. This stems from the nonlinearity inherent in the Navier-Stokes equations. Early fluids calculations were done numerically, albeit laboriously, with the help of human "computers". In fact, the design of the first electronic computers was driven by the need to perform laborious fluids calculations (Eniac 1945, Illiac I 1952 [11]).

In contrast, most electromagnetic media are linear. Hence, the governing equations for electromagnetics are, for the most part, linear. The linearity of electromagnetic equations allows for the application of many analytic techniques for solving these equations.

The electromagnetic equations are:

$$\nabla \times \mathbf{E} = -\frac{\partial \mathbf{B}}{\partial t} \tag{1.1}$$

$$\nabla \times \mathbf{H} = \frac{\partial \mathbf{D}}{\partial t} + \mathbf{J} \tag{1.2}$$

$$\nabla \cdot \mathbf{D} = \rho \tag{1.3}$$

$$\nabla \cdot \mathbf{B} = 0 \tag{1.4}$$

These equations are attributable to the works of Faraday (Eq. (1.1), Faraday's law 1843 [12]), Ampere (Eq. (1.2), Ampere's law 1823 [13]), Coulomb (Eq. (1.3), Coulomb's law 1785 [14]; it was enunciated in a divergence form via Gauss' law), and Gauss (Eq. (1.4), Gauss' law 1841 [15]). In 1864, Maxwell [16] completed the electromagnetic theory by adding the concept of displacement current ($\partial \mathbf{D}/\partial t$) in Eq. (1.2). Hence, this equation is also known as generalized Ampere's law. The set of equations above is also known as Maxwell's equations. The original Maxwell's equations were expressed in some 20 equations. It was Heaviside [17] who distilled Maxwell's equations into the above four equations. Hence, they are sometimes known as the Maxwell-Heaviside equations. These equations have been proven to be valid from subatomic length scale to intergalactic length scale.

Because of their linearity, one useful technique for solving electromagnetic equations is the principle of linear superposition. The principle of linear superposition further allows for the use of the Fourier transform technique. Hence, one can Fourier transform the above equations in the time variable and arrive at

$$\nabla \times \mathbf{E} = i\omega \mathbf{B} \tag{1.5}$$

$$\nabla \times \mathbf{H} = -i\omega \mathbf{D} + \mathbf{J} \tag{1.6}$$

$$\nabla \cdot \mathbf{D} = \rho \tag{1.7}$$

$$\nabla \cdot \mathbf{B} = 0 \tag{1.8}$$

The preceding equations are the frequency-domain version of the electromagnetic equations. When the region is source free, that is, $\rho = 0$ and $\mathbf{J} = 0$, as in vacuum, $\mathbf{B} = \mu_0 \mathbf{H}$ and $\mathbf{D} = \epsilon_0 \mathbf{E}$, the simplest homogeneous solution to the above equations is a plane wave of the form $\mathbf{a}\exp(i\mathbf{k} \cdot \mathbf{r})$ where $\mathbf{a}$ is orthogonal to $\mathbf{k}$ and $|\mathbf{k}| = \omega\sqrt{\mu_0\epsilon_0}$. Hence, Maxwell's work bridged the gap between optics and electromagnetics because scientists once thought that the electromagnetic phenomenon was different from the optical phenomenon.

Also, for electrodynamic problems, the last two of the above equations can be derived from the first two with appropriate initial conditions. Hence, the last two equations can be ignored for dynamic problems [18].

The use of Fourier transform technique brings about the concept of convolution. The concept of convolution leads to the development of the Green's function technique [19]. From the concept of Green's functions, integral equation can be derived and solved [18].

As a result, electromagneticists have a whole gamut of analytic tools to play with. Many engineering problems and theoretical modeling can be solved by the application of these tools, offering physical insight and shedding light on many design problems. The use of intensive computations to arrive at critical data can be postponed until recently. Hence, CEM concepts did not develop until 1960s [20, 21]. Although laborious calculations were done in the early days, they were not key to the development of many technologies.

1.2 Some Things Do Not Happen in CEM Frequently— Nonlinearity

Nonlinearity is not completely absent in electromagnetics. Though most nonlinear phenomena are weak, they are usually amplified by short wavelengths, or high frequencies. When the wavelength is short, any anomalies will be amplified and manifested as a cumulative phase phenomenon. For instance, the Kerr effect, which is a 10^{-4} effect, is responsible for the formation of solitons at optical frequencies [22].

Nonlinear effects do not occur frequently in electromagnetics. A nonlinear phenomenon has the ability to generate high frequency fields from low frequency ones (Just imagine $f(x) = x^2$. If $x(t) = \sin \omega t$, $f[x(t)] = \sin^2 \omega t$ has a frequency component twice as high as that of $x(t)$). Nonlinearity can give rise to shock front and chaos. These phenomena do not occur in linear electromagnetics.

Nonlinearity in electromagnetics can be a blessing as well as a curse. The successful propagation of a soliton in an optical fiber was heralded as a breakthrough in optics for optical communications whereby a pulse can propagate distortion-free in an optical fiber for miles needing no rejuvenation. However, the nonlinear phenomenon in an optical fiber is besieged with other woes that make the propagation of a pure soliton difficult. After two decades of research, no commercially viable communication system using solitons has been developed. On the contrary, optical fibers are engineered to work in the linear regime where dispersion management is used to control pulse shapes for long distance propagation. Dispersion management uses alternating sections of optical fibers with positive and negative dispersion where pulse expansion is followed by pulse compression to maintain the integrity of the signal [23].

1.3 The Morphing of Electromagnetic Physics

Electromagnetic field is like a chameleon—its physics changes as the frequency changes from static to optical frequencies [24]. Electromagnetics is described by fields—a field is a function of 3D space and time. These fields satisfy their respective PDEs that govern their physics. Hence, electromagnetic physics changes depending on the regime. The morphing of the physics in electromagnetics can be easily understood from a simple concept: the comparison of derivatives. PDEs that govern electromagnetics have many derivative operators in them— depending on the regime of the physics, different parts of the partial derivative operators become important.

Another important concept is that of the boundary condition. The boundary condition dictates the manner in which a field behaves around an object. In other words, a field has to contort around an object to satisfy the boundary condition. Hence, the variation of a field around an object is primarily determined by the geometrical shape of the object.

1.3.1 Comparison of Derivatives

Much physics of electromagnetics can be understood by the boundary condition and the comparison of derivatives.[1] Because the boundary condition shapes the field, the derivative of the field is inversely proportional to the size of the object. If the typical size of the object is of length L, then

$$\frac{\partial}{\partial x} \sim \frac{1}{L}, \qquad \frac{\partial^2}{\partial x^2} \sim \frac{1}{L^2} \tag{1.9}$$

From the above, we can see that around small structures, the space derivative of the field is large.

Because the Maxwell-Heaviside equations contain primarily first derivatives in space and time, it is the comparison of the importance of these derivatives that delineates the different regimes in electromagnetics. For a time-harmonic field, with angular frequency ω, we can define an associated plane wave with wavelength λ such that $c/\lambda = \omega/(2\pi)$.

$$\frac{\partial}{\partial t} \sim \omega \sim \frac{c}{\lambda} \leftrightarrow \frac{\partial}{\partial x} \sim \frac{1}{L} \tag{1.10}$$

The comparison of time derivative (which is proportional to λ^{-1}) to the space derivative (which is proportional to L^{-1}) delineates different regimes in electromagnetics.

Consequently, electromagnetic physics can be divided into the following regimes:

- Low frequency, $L \ll \lambda$.

- Mid frequency, $L \approx \lambda$.

- High frequency, $L \gg \lambda$.

When the object size is much less than the wavelength, the physics of the field around it is dominated by the fact that the time derivative is unimportant compared to the space derivative. The field physics is close to that of static field.

In the mid-frequency regime, the space derivative is equally important compared to the time derivative, but the size of the object, and hence the boundary condition, still controls the nature of the solution.

In the high-frequency regime, the boundary condition plays a lesser role compared to the equations in shaping the solution. Hence, the solution resembles the homogeneous solutions of electromagnetic equations, which are plane waves. Plane waves are mathematical representation of rays, and hence, ray physics becomes more important in this regime.

In the above category, we are assuming that the point of observation is of the same order as the size of the object L. If one were to observe the field at different distances d from the object, the physics of the field will change according to the above argument. Hence, we have:

- Near field, $d \ll \lambda$.

- Mid field, $d \approx \lambda$.

- Far field, $d \gg \lambda$.

---

[1] This technique is often used in fluids to arrive at numbers such as the Reynolds number [2].

In the near-field regime, electrostatic physics is important. In the mid-field regime, wave physics is important, whereas in the far-field regime, ray physics is important.[2] We will elaborate on the above regimes in more detail in the following discussions.

1.3.2 Low-Frequency Regime

When the frequency is low or the wavelength is long, the governing equation for the electromagnetic field is the static equation, which is Laplace's equation or Poisson's equation. Laplace's equation is an elliptic equation where the solution is smooth without the existence of singularities (or shocks) [25]. Furthermore, solution of Laplace's equation is nonoscillatory, implying that it cannot be used to encode spatial information needed for long distance communication. Because the solution is smooth, multigrid [26–28] and wavelets [29, 30] methods can be used to accelerate the solution in this regime with great success.

Moreover, at static, the magnetic field and the electric field are decoupled from each other. Hence in this regime, electromagnetic equations reduce to the equation of electrostatics and equation of magnetostatics that are completely decoupled from each other [31–33]. The equation of electrostatics also governs the working principle of capacitors, whereas the equation of magnetostatics governs the working principle of inductors. Furthermore, when the frequency is very low, the current decomposes itself into a divergence-free component and a curl-free component. This is known as the Helmholtz decomposition, and it can be gleaned from the following equations, which are reduced from electromagnetic equations when the frequency is vanishingly small or $\omega \to 0$.

$$\left. \begin{array}{c} \nabla \times \mathbf{H} = \mathbf{J} \\ \nabla \cdot \mathbf{B} = 0 \end{array} \right\} \text{magnetostatic} \tag{1.11}$$

$$\left. \begin{array}{c} \nabla \times \mathbf{E} = 0 \\ \nabla \cdot \mathbf{D} = \rho = \lim \frac{\nabla \cdot \mathbf{J}}{i\omega} \end{array} \right\} \text{electrostatic} \tag{1.12}$$

The top part of Eq. (1.11) is Ampere's law. It says that a current $\mathbf{J}$ produces a magnetic field $\mathbf{H}$. However, since $\nabla \cdot (\nabla \times \mathbf{H}) = 0$, it follows that this current satisfies

$$\nabla \cdot \mathbf{J} = 0 \tag{1.13}$$

or is divergence-free.

The last equation in (1.12) follows from the continuity equation

$$\nabla \cdot \mathbf{J} + \frac{\partial \rho}{\partial t} = 0 \ \text{ or } \ \nabla \cdot \mathbf{J} = i\omega\rho \tag{1.14}$$

It implies that the charge ρ that produces the electric field $\mathbf{E}$ in (1.12) comes from the nondivergence free part of $\mathbf{J}$. Hence, in the limit when $\omega \to 0$,

$$\mathbf{J} = \mathbf{J}_{irr} + \mathbf{J}_{sol} \tag{1.15}$$

namely, the current $\mathbf{J}$ decomposes itself into an irrotational component $\mathbf{J}_{irr}$ (which is curl-free but not divergence-free) and solenoidal component $\mathbf{J}_{sol}$ (which is divergence-free but not

---

[2]It is to be noted that in optics, the ray physics regime can be further refined into the Fresnel zone and the Fraunhoffer zone [34].

curl-free). The irrotational current models the electroquasistatic fields, whereas the soleinodal current models the magnetoquasistatic fields. Again, as noted earlier, this is known as Helmholtz decomposition. In order for the last equation in (1.5) to produce a finite charge ρ, it is seen that $\mathbf{J}_{irr} \sim O(\omega)$ when $\omega \to 0$.

Many CEM methods, which are developed for when ω is nonzero, break down as $\omega \to 0$. This is because those methods, although capable of capturing the physics of the electromagnetic field when ω is not small, fail to capture the physics correctly when $\omega \to 0$.

1.3.3 Mid Frequency Regime

The mid-frequency regime is where the electric field and magnetic field are tightly coupled together. It is also the regime where electromagnetic fields become electromagnetic waves because of this tight coupling. The field, instead of being smooth as in solutions to Laplace's equation, becomes oscillatory as is typical of a wave phenomenon. The oscillatory nature of the field can be used to encode information, and this information can travel uncorrupted over vast distances such as across galaxies. Hence, electromagnetic waves are extremely important for long-distance communications.

In the mid-frequency regime, electromagnetic fields can be described by hyperbolic-type PDEs. If a wave contains broadband signals all the way from static to high frequencies, it can propagate singularities such as a step discontinuity. This is the characteristic of hyperbolic PDEs.

The oscillatory nature of the field in this regime also precludes methods such as multigrid and wavelets from working well in this regime because the solution is not infinitely smooth [24, 30].

1.3.4 High-Frequency Regime

The high-frequency regime is where the wavelength becomes extremely short compared to the dimension of interest. In this regime, the wave solution becomes plane-wave like. Plane waves behave like rays. Even though electromagnetic field is still described by the wave equation, a hyperbolic equation, asymptotic methods using the short wavelength as a small parameter can be used to extract the salient feature of the solution. Alternatively, we can think of rays as particle-like. This is what happens to light rays.

Consequently, light rays can be described by the transport equation and the eikonal equation. These equations can be derived from the wave equation under the high-frequency and short-wavelength approximation [35].

The high-frequency regime also allows for the beam-forming of electromagnetic waves. For instance, electromagnetic field can be focused as in optics [36–45]. Hence, antennas with high directivity such as a reflector antenna can be designed at microwave frequencies. At lower frequencies, such an antenna will have to be unusually large to increase the directivity of electromagnetic wave.

1.3.5 Quantum Regime

In addition to the above, one has to be mindful of another regime when the frequency is high. A quantum of electromagnetic energy is proportional of hf, and if hf is not small ($h = 6.624 \times 10^{-34} Js$ is Planck's constant and f is the frequency in Hertz, we then have the quantum regime. In this regime, electromagnetics has to be described by quantum physics. The electromagnetic field, which is an observable, has a quantum operator analogue. This is a property not derivable from the wave equation, but is a result of the quantum nature of light. Hence, in the optical regime when light wave interacts with matter, it is important to consider quantum mechanics to fully model its physics [46].

Because of the minute value of Planck's constant, the quantization of electromagnetic energy is unimportant unless the frequency is very high, as in optical frequencies. Because of the quantization of electromagnetic energy, light waves interact more richly with matter, giving rise to a larger variety of physical phenomena such as nonlinear phenomena.

1.4 Matched Asymptotics

Matched asymptotics was first used in fluids to solve problems perturbatively by dividing into regions near the boundary and away from the boundary, because the physics of fluids is different near a boundary surface and away from a boundary surface. One way to apply matched asymptotics in electromagnetics is to use the near-field/far-field dichotomy. The yardstick for near and far field is then the wavelength. Another way is to exploit the difference in the solution when it is close to a singularity and far away from a singularity.

In electromagnetics, matched asymptotics can be used to piece together solutions in different regions in space [47–49]. For instance, the scattering solution of a small sphere is first obtained by Rayleigh using matched asymptotics: the near field is used to match the boundary condition on sphere, whereas the far field is obtained by using radiation from a dipole field. Kirchhoff obtained the approximate solution of the capacitance of two circular disks [50] by breaking the solution into region close to the edge of the disk (where the field is singular) and away from the edge where the solution is smooth. The work was followed by other workers [51–54]. Approximate formulas for the resonant frequencies of microstrip disks and approximate guidance condition for microstrip lines can be similarly obtained [55,56].

1.5 Why CEM?

Because of the linearity of Maxwell-Heaviside equations, analytic tools have helped carried the development of electromagnetic technology a long way. Theoretical modeling has often played an important role in the development of science and technology. For a long time, theoretical modeling in electromagnetics can be performed with the abundance of analytic tools available. The analytic solutions often correspond to realistic problem in the real world.

Hence, early theoretical modeling in electromagnetics is replete with examples of mathematical virtuosity of the practitioners. Examples of these are the Rayleigh solution to the waveguide problem [57], solution of Rayleigh scattering by small particles [34,58], Sommerfeld solutions to the half-plane problem [59] and the half-space problem [60], Mie series solution to the scattering by spheres [61,62] etc. Because electromagnetics was predated by the theory of sound and fluids [63], much of the mathematics developed for sound and fluids can be used in electromagnetics with some embellishment to account for the vector nature of the field. Many of the scattering solutions by canonical shapes is documented in Bowman *et al.* [64].

Later, when science and technology called for solutions to more complex problems, theoretical modeling was performed with the help of perturbation and asymptotic methods. In asymptotic methods, the wavelength is usually used as a small parameter for a perturbation expansion. This is the field of physical optics, geometrical optics, geometrical theory of diffraction, etc [36–45]. The solution is usually ansatz based. For example, when the frequency is high, or the wavelength short, the wavenumber $k = 2\pi/\lambda$ is large, the scattering solution from a complex object may be assumed to be dominated by a few salient features as a consequence of reflection and diffraction. Hence, the approximate solution may be a superposition of solutions of the form

$$\psi^{sca}(\mathbf{r}) = \frac{e^{ikr}}{r^{\alpha}} \left[a_0 + a_1 \frac{1}{kr} + a_2 \frac{1}{(kr)^2} + a_3 \frac{1}{(kr)^3} + \cdots \right] \tag{1.16}$$

The constants α and a_i's are obtained by matching solution with canonical solutions, such as the Sommerfeld half-plane solution, and the Mie series solution after Watson's transformation [65].

However, the leaps and bounds progress of computer technology opens up new possibilities for solving the electromagnetic equations more accurately for theoretical modeling in many applications. This is especially so when the structure is complex. Instead of being driven by qualitative engineering where physical intuition is needed, a quantitative number can be arrived at using CEM that can be used as a design guide.

Moreover, the use of CEM can reduce our dependency on experiments, which are usually by cut-and-try method. Experiments are usually expensive, and if design decisions can be made by virtual experiment or virtual prototyping using simulation and modeling techniques based on CEM, design cost can be substantially reduced. CEM preserves a large part of the physics of the electromagnetic equations when they are solved, and hence, it can represent a high-fidelity virtual experiment quite close to the real world.

Asymptotic and approximate methods serve the design needs at high frequencies. However, they become less reliable when the frequency is low or when the wavelength is long. For this regime, CEM methods are more reliable than approximate methods. One good example is the scattering from an aircraft by mega-Hertz electromagnetic waves, where the wavelength is not short enough to allow for short-wavelength approximation [66].

1.6 Time Domain versus Frequency Domain

In many physical problems, such as fluids, the equations are inherently nonlinear. Hence, they do not lend themselves to Fourier transform methods. For many applications, however, the electromagnetic equations are linear. Hence one can choose to solve the electromagnetic equations in the frequency domain other than the time domain, as noted before. The frequency-domain equations are obtained by Fourier transforming the electromagnetic equations in time.

Alternatively, one can achieve the simplicity of frequency domain by assuming a pure time-harmonic solution that is completely sinusoidal where the time dependence at any given point in space is of the form $A\cos(\omega t + \phi)$. Since one can write

$$A\cos(\omega t + \phi) = \frac{1}{2}\left(Ae^{-i\phi}e^{-i\omega t} + \text{C.C.}\right) \tag{1.17}$$

where C.C. stands for "complex conjugate", it is simpler to work with complex time-harmonic signals of the form $e^{-i\omega t}$ with complex amplitudes. Such signals are called phasors [34]. To recover the real signal, one just adds the complex conjugate part to the complex signal, or simply, takes the real part of the complex signal.

For linear equations, a pure time-harmonic solution can exist as an independent solution, whereas for nonlinear equations, the incipient of a time-harmonic signal will generate higher harmonics, causing the mixing of harmonics (see Section 1.2).

The advantages of solving the electromagnetic equations in the frequency domain is the removal of time dependence of the electromagnetic equations, and hence the resultant equations are simpler (all $\partial/\partial t$ can be replaced with $-i\omega$). Another advantage is that each frequency component solution can be solved for independently of the other frequency component—hence, they can be solved in an embarrassingly parallel fashion with no communication overhead at all. This multitude of time-harmonic solutions for different frequencies can then be superposed to obtain the time domain solution [67].

On the other hand, the time-domain solution has to be sought by considering the inter-dependence of time steps. Although in parallel computing, interprocessor communication can be reduced via the domain-decomposition method, some communications are still required [68, 69].

1.7 Differential Equation versus Integral Equation

For a nonlinear problem, it is more expedient to solve the equations in their differential forms. However, when a problem is linear, the principle of linear superposition allows us to first derive the point source response of the differential equation. This is the Green's function [19], or the fundamental solution of the differential equation. With the Green's function, one can derive an integral equation whose solution solves the original problem. We shall illustrate this in the frequency domain for a simple scalar wave equation [18].

Given the wave equation in the time domain

$$\nabla^2\phi(\mathbf{r}, t) - \frac{1}{v^2}\frac{\partial^2}{\partial t^2}\phi(\mathbf{r}, t) = s(\mathbf{r}, t) \tag{1.18}$$

By assuming a complex time-harmonic signal with $e^{-i\omega t}$ dependence, the wave equation becomes the Helmholtz equation as follows with an arbitrary source on the right hand side:

$$\nabla^2\phi(\mathbf{r}, \omega) + k^2\phi(\mathbf{r}, \omega) = s(\mathbf{r}, \omega), \quad k = \omega/v \tag{1.19}$$

The Green's function is the solution when the right-hand side is replaced by a point source,

$$\nabla^2 g(\mathbf{r}, \mathbf{r}') + k^2 g(\mathbf{r}, \mathbf{r}') = -\delta(\mathbf{r} - \mathbf{r}') \tag{1.20}$$

where the dependence of $g(\mathbf{r}, \mathbf{r}')$ on ω is implied. By the principle of linear superposition, the solution to the original Helmholtz equation can be expressed as

$$\phi(\mathbf{r}, \omega) = -\int_V d\mathbf{r}' g(\mathbf{r}, \mathbf{r}') s(\mathbf{r}', \omega) \tag{1.21}$$

For an impenetrable scatterer, the Green's function can be used to derive an integral equation of scattering

$$-\phi_{inc}(\mathbf{r}) = \int_S dS' g(\mathbf{r}, \mathbf{r}') s(\mathbf{r}'), \quad \mathbf{r} \in S \tag{1.22}$$

where $\phi_{inc}(\mathbf{r})$ is the incident field, $g(\mathbf{r}, \mathbf{r}')$ is the free-space Green's function given by

$$g(\mathbf{r}, \mathbf{r}') = \frac{e^{ik|\mathbf{r}-\mathbf{r}'|}}{4\pi|\mathbf{r} - \mathbf{r}'|} \tag{1.23}$$

and $s(\mathbf{r}')$ is some unknown surface source residing on the surface of the scatterer. Physically, the above equation says that the surface source $s(\mathbf{r}')$ generates a field that cancels the incident field within the scatterer. To solve the integral equation implies finding the unknown $s(\mathbf{r}')$.

The advantage of solving an integral equation is that often times, the unknown $s(\mathbf{r}')$ resides on a 2D manifold or surface [70].[3] Whereas if a differential equation is solved, the unknown is the field $\phi(\mathbf{r})$ that permeates the whole of a 3D space. Consequently, it often requires more unknowns to solve a differential equation compared to an integral equation. More precisely, the unknown count for $s(\mathbf{r}')$ on a 2D surface scales as L^2 where L is the typical length of the scatterer, whereas the unknown count for the field $\phi(\mathbf{r})$ scales volumetrically as L^3. Hence, when the problem size is very large, the cruelty of scaling law will prevail, implying that differential equations usually require more unknowns to solve compared to integral equation.

Another advantage of solving integral equation is that the Green's function is an exact propagator that propagates a field from point A to point B. Hence, there is no grid dispersion error of the kind that exists in numerical differential equation solvers where the field is propagated from point A to point B via a numerical grid. For wave problems, the grid dispersion error causes an error in the phase velocity of the wave, causing cumulative phase error that becomes intolerable. Hence, the grid dispersion error becomes a more severe problem with increased problem size. The grid discretization density has to increase with increased problem size, prompting the unknown scaling to increase with more than L^3 [68, 71].

On the other hand, numerical differential equation solvers are easier to implement compared to integral equation solvers. In integral equation solvers, a proper way to evaluate integrals with singular integrand is needed. For an N unknown problem, the differential equation solver also solves an equivalent sparse matrix system with $O(N)$ storage greatly reducing the storage requirements, but the matrix system associated with integral equation is usually a dense matrix system requiring $O(N^2)$ storage and more than $O(N^2)$ central processing unit (CPU) time to solve. However, the recent advent of fast integral equation solver has removed these bottlenecks [66, 70, 72–75]. Previously, only dense matrix systems with tens of thousands of unknowns can be solved, but now, dense matrix systems with tens of millions of unknowns can be solved [76, 77].

---

[3]The pros and cons of differential equation and integral equation solvers were discussed in greater detail in this reference. We are repeating them here for convenience.

1.8 Nondissipative Nature of Electromagnetic Field

Electromagnetic waves in vacuum and many media are nondissipative. That explains the intergalactic distance over which electromagnetic waves can travel. The popular Yee scheme [20] in CEM is nondissipative albeit dispersive. The nondissipative nature of a numerical scheme preserves a fundamental property of the electromagnetic field. However, many numerical schemes in CFD have been borrowed to solve electromagnetic problems [5, 9, 78–80]. These numerical schemes are dissipative, and hence, one has to proceed with caution when applying these numerical schemes to solve CEM problems.

The nondissipative nature of the electromagnetic field also implies that two electromagnetic systems can affect each other even though they may be very far apart. Even short length scale phenomenon can propagate over long distances such as light rays.

1.9 Conclusions

Differential equation solvers are used almost exclusively in many fields because of the nonlinearity of the relevant equations. On the other hand, because of the linearity of the Maxwell-Heaviside equations, analytic methods, such as the time-domain method and the integral equation method, can be used to arrive at alternative methods to solve the electromagnetic equations such as integral equation solvers. Traditional integral equation solvers are inefficient, but with the advent of fast solvers, they are more efficient than before, and have become even more efficient than differential equation solvers in many applications.

There has been an interest in applying differential equation techniques to solve electromagnetic equations. When applied to an aircraft scattering problem, these solvers use resources that scale more pathological compared to fast integral equation solvers. One way to curb the scaling law in resource usage is to combine differential equation solvers with integral equation solvers where the computational domain is truncated by a fast integral equation solver ([70] Chapter 13, [81],). If memory and CPU resource are not an issue, differential equation techniques can be used to obtain broadband simulation data in electromagnetics. If memory alone is not an issue, fast time-domain integral equation solver can be used to obtain broadband data in electromagnetics ([70] Chapter 18, [82]).

The morphing of electromagnetic physics from static to long wavelengths, to mid wavelengths, and then to short wavelengths implies that different techniques need to be used in these different regimes. Many complex problems remain in electromagnetics that call for more sophisticated computational algorithms. One of these problems is the simulation of complex structures inside a computer at the level of motherboard, package, and chip, as shown in Figure 1.1. It entails multiscale features that cannot be easily tackled when using conventional methods.

A computer at the finest level consists of computer chips that contain tens of millions of transistors. Here, X and Y lines in a multilayer fashion route electrical signals among these transistors (Figure 1.1(c)). A chip forms a data processing unit that has thousands of electrical input and output wires. A chip communicates outside itself including receiving power supply via the package (Figure 1.1(b)). Different processing units together form a system, and different systems, including a power supply system, form a computer as seen at

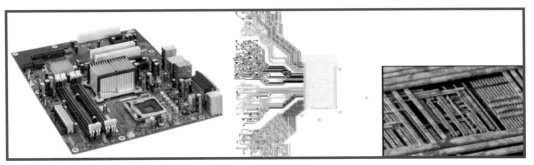

(a) Board level complexity.   (b) Package level complexity.   (c) Chip level complexity.

Figure 1.1: A complex multiscale structure coming from the real world as found inside a computer at the level of (a) motherboard, (b) package, and (c) chip. (Courtesy of Intel.)

the motherboard level (Figure 1.1(a)) just inside the computer chassis.

Appendix: Complexity of an Algorithm

In almost all modern numerical methods of solving electromagnetic problems, an object (or region, or domain) is first decomposed into many small subobjects (or subregions, or subdomain) which are much smaller than a wavelength. Then a matrix equation is derived that accounts for the interaction of these subobjects (or subregions, or subdomains). Each subobject is usually given one or more unknowns. The number of subobjects (or subregions, or subdomains) represents the degree of freedom of an object, and we often say that an object has been discretized in a numerical method. The number of unknowns to be solved for represents the degree of freedom of the system.

The two most important measures of the efficiency of a numerical method are: (1) the memory requirements, and (2) the computational time. Ideally, both of them should be small. For example, the storage of the unknowns in the system, and the pertinent matrix elements requires the usage of the random access memory (RAM) of a computer for efficiency. The solving of the matrix equation requires computational resource such as CPU time. The memory usage may scale as $O(N^\alpha)$ and the CPU time may scale as $O(N^\beta)$ where α and β are numbers larger than one. These scaling properties of the memory and computation time with respect to the unknown number N are respectively known as the memory complexity and the computational complexity.

At this point, it is prudent to define the meaning of $O(N^\beta)$ (read as "order N^β") and similar expressions. When

$$R \sim O(N^\beta), \quad N \to \infty, \tag{A-1}$$

one implies that there exists a constant C such that

$$R \to CN^\beta, \quad N \to \infty, \tag{A-2}$$

where R is the resource usage: it can be memory or CPU time. In this case, we say that the asymptotic complexity of the algorithm is $O(N^\beta)$ or just N^β. For large-scale computing, where N is usually very large, the term "complexity" refers to "asymptotic complexity". The complexity of an algorithm also determines the scaling property of the resource usage by the algorithm. For instance, if $\beta = 2$, the resource usage quadruples every time the number of unknowns, N, doubles.

When R refers to CPU time, the constant C in Equation (A-2) depends on the speed of the hardware, and the detail implementation of the algorithm. However, the complexity of an algorithm does not depend on the constant C.

Depending on the algorithm used, α can be as large as two and β can be as large as three. The intent of an efficient numerical algorithm is to make these numbers as close to one as possible.

When α and β are large, the resource needed can easily swamp the largest supercomputers we have. For instance, Gaussian elimination for inverting a matrix is of complexity N^3. The CPU time can be easily of the order of years when N is large. This is known as the "cruelty of computational complexity" in computer science parlance.

Recently, the advent of fast algorithms in reducing α and β when solving these equations, as well as the leaps and bounds progress in computer technology, which helps to reduce the constant C in Equation (A-2), make many previously complex unsolvable problems more accessible to scientists and engineers using a moderately sized computer.

Alternatively, some workers have designed algorithms to solve problems using the hard drive as well as the RAM as storage space. These are usually known as out-of-core solvers. Even then, if the memory and computational complexity of the algorithm is bad, the hard-drive memory can easily become insufficient.

Parallel computers can be used to speed up solutions to large problems, but they do not reduce the computational complexity of an algorithm. If a problem takes CPU time proportional to N^β, namely,

$$T \to CN^\beta, \quad N \to \infty, \tag{A-3}$$

an ideal parallel algorithm to solve the same problem with P processors has CPU time that scales as

$$T \to \frac{1}{P}CN^\beta, \quad N \to \infty. \tag{A-4}$$

Such an ideal algorithm is said to have a perfect speed up, and 100% parallel efficiency. However, because of the communication cost needed to pass data between processors, the latency in computer memory access as problem size becomes large, 100% parallel efficiency is generally not attained. In this case, the CPU time scales as

$$T \to \frac{e}{100P}CN^\beta, \quad N \to \infty, \tag{A-5}$$

where e is the parallel efficiency in percentage. As can be seen from above, the use of parallel computers does not reduce the computational complexity of the algorithm, but it helps to reduce the proportionality constant in the resource formula.

Bibliography

[1] D. E. Smith, *History of Mathematics*, New York: Dover Publication, 1923.

[2] A. R. Choudhuri, *The Physics of Fluids and Plasma*, Cambridge UK: Cambridge University Press, 1998.

[3] C. L. M. H. Navier, "Mémoire sur les lois du mouvement des fluides," *Mém. Acad. Sci. Inst. France*, vol. 6, pp. 389–440, 1822.

[4] G. G. Stokes, "On the theories of the internal friction of fluids in motion, and of the equilibrium and motion of elastic solids," *Trans. Cambridge Philos. Soc.*, vol. 8, pp. 287–319, 1845.

[5] J. C. Tannehill, D. A. Anderson, and R. H. Pletcher, *Computational Fluid Mechanics and Heat Transfer*, First edition, New York: Hemisphere Publ., 1984. Second edition, Philadelphia: Taylor & Francis, 1997.

[6] L. F. Richardson, "The approximate arithmetical solution by finite differences of physical problems involving differential equations, with an application to th stresses in a masonry dam," *Philos. Trans. R. Soc. London, Ser. A*, vol. 210, pp. 307–357, 1901.

[7] R. Courant, K. O. Friedrichs, and H. Lewy, "Über die partiellen differenzengleichungen der mathematischen physik," *Math. Ann.*, vol. 100, pp. 32–74, 1928. (Translated to: "On the partial difference equations of mathematical physics," *IBM J. Res. Dev.*, vol. 11, pp. 215–234, 1967.

[8] P. D. Lax, "Weak solutions of nonlinear hyperbolic equations and their numerical computation," *Commun. Pure Appl. Math.*, vol. 7, pp. 159–193, 1954.

[9] P. D. Lax and B. Wendroff, "Systems of conservation laws," *Commun. Pure Appl. Math.*, vol. 13, pp. 217–237, 1960.

[10] R. W. MacCormack, "The effect of viscosity in hypervelocity impact cratering," *AIAA Paper 69-354*, Cincinnati, OH, 1969.

[11] P. E. Ceruzzi, *A History of Modern Computing*, Cambridge, MA: MIT Press, 1998.

[12] M. Faraday, "On static electrical inductive action," *Philos. Mag.*, 1843. M. Faraday, Experimental Researches in Electricity and Magnetism. vol. 1, London: Taylor & Francis, 1839.; vol. 2, London: Richard & John E. Taylor, 1844; vol. 3, London: Taylor & Francis, 1855. Reprinted by Dover in 1965. Also see M. Faraday, "Remarks on Static Induction," Proc. R. Inst., Feb. 12, 1858.

[13] A. M. Ampère, "Mémoire sur la théorie des phénomènes électrodynamiques," *Mem. Acad. R. Sci. Inst. Fr.*, vol. 6, pp. 228–232, 1823.

[14] C. S. Gillmore, *Charles Augustin Coulomb: Physics and Engineering in Eighteenth Century Frrance*, Princeton, NJ: Princeton University Press, 1971.

[15] C. F. Gauss, "General theory of terrestial magnetism," *Scientific Memoirs*, vol. 2, ed. London: R & J.E. Taylor, pp. 184–251, 1841.

[16] J. C. Maxwell, *A Treatise of Electricity and Magnetism,* 2 vols, Clarendon Press, Oxford, 1873. Also, see P. M. Harman (ed.), *The Scientific Letters and Papers of James Clerk Maxwell, Vol. II, 1862–1873*, Cambridge, U.K.: Cambridge University Press, 1995.

[17] O. Heaviside, "On electromagnetic waves, especially in relation to the vorticity of the impressed forces, and the forced vibration of electromagnetic systems," *Philos. Mag.*, 25, pp. 130–156, 1888. Also, see P. J. Nahin, "Oliver Heaviside," *Sci. Am.*, pp. 122–129, Jun. 1990.

[18] W. C. Chew, *Waves and Fields in Inhomogeneous Media*, New York: Van Nostrand Reinhold, 1990, reprinted by Piscataway, New Jersey: IEEE Press, 1995.

[19] G. Green, An Essay on the Application of Mathematical Analysis to the Theories of Electricity and Magnetism, T. Wheelhouse, Nottingham, 1828. Also, see L. Challis and F. Sheard, "The green of Green functions," *Phys. Today*, pp. 41–46, Dec. 2003.

[20] K. S. Yee, "Numerical solution of initial boundary value problems involving Maxwell's equation in isotropic media," *IEEE Trans. Antennas Propag.*, vol. 14, pp. 302–307, 1966.

[21] R. F. Harrington, *Field Computation by Moment Method,* Malabar, FL: Krieger Publ., 1982. (Original publication, 1968.)

[22] Y. R. Shen, *The Principles of Nonlinear Optics*, New York: John Wiley & Sons, 1991.

[23] G. P. Agrawal, *Fiber-Optic Communication Systems,* New York: John Wiley & Sons, 1997.

[24] W. C. Chew, "Computational Electromagnetics—the Physics of Smooth versus Oscillatory Fields," *Philos. Trans. R. Soc. London, Ser. A, Math., Phys. Eng. Sci. Theme Issue Short Wave Scattering*, vol. 362, no. 1816, pp. 579–602, Mar. 15, 2004.

[25] R. J. LeVeque, *Numerical Methods for Conservation Laws*, ETH Lectures in Mathematics Series, Basel: Birkhuser Verlag, 1990.

[26] R. P. Federonko, "A relaxation method for solving elliptic equation," *USSR Comput. Math. Phys.*, vol. 1, pp. 1092–1096, 1962.

[27] A. Brandt, "Multilevel adaptive solutions to boundary value problems," *Math. Comput.*, vol. 31, pp. 333–390, 1977.

[28] O. Axelsson and V. A. Barker, *Finite Element Solution of Boundary Value Problems: Theory and Computation*, New York: Academic Press, 1984.

[29] S. Jaffard, "Wavelet methods for fast resolution of elliptic problems," *SIAM J. Numer. Anal.*, vol. 29, no. 4, pp. 965–986, 1992.

[30] R. L. Wagner and W. C. Chew, "A study of wavelets for the solution of electromagnetic integral equations," *IEEE Trans. Antennas Propag.*, vol. 43, no. 8, pp. 802–810, 1995.

[31] D. R. Wilton and A. W. Glisson, "On improving the electric field integral equation at low frequencies," *URSI Radio Sci. Meet. Dig.*, pp. 24, Los Angeles, CA, Jun. 1981.

[32] J. S. Zhao and W. C. Chew, "Integral Equation Solution of Maxwell's Equations from Zero Frequency to Microwave Frequencies," *IEEE Trans. Antennas Propag., James R. Wait Memorial Special Issue*, vol. 48. no. 10, pp. 1635–1645, Oct. 2000.

[33] W. C. Chew, B. Hu, Y. C. Pan and L. J. Jiang, "Fast Algorithm for Layered Medium," *C. R. Phys.*, vol. 6, pp. 604–617, 2005.

[34] J. A. Kong, *Electromagnetic Wave Theory*, New York: John Wiley & Sons, 1986.

[35] M. Born and E. Wolf, *Principles of Optics*, New York: Pergammon Press, 1980.

[36] B. D. Seckler and J. B. Keller, "Geometrical theory of diffraction in inhomogeneous media," *J. Acoust. Soc. Am.*, vol. 31, pp. 192–205, 1959.

[37] J. Boersma, "Ray-optical analysis of reflection in an open-ended parallel-plane waveguide: II-TE case," *Proc. IEEE*, vol. 62, pp. 1475–1481, 1974.

[38] H. Bremmer, "The WKB approximation as the first term of a geometric-optical series," *Commun. Pure Appl. Math.*, vol. 4, pp. 105, 1951.

[39] D. S. Jones and M. Kline, "Asymptotic expansion of multiple integrals and the method of stationary phase," *J. Math. Phys.*, vol. 37, pp. 1–28, 1958.

[40] V. A. Fock, *Electromagnetic Diffraction and Propagation Problems*, New York: Pergamon Press, 1965.

[41] R. C. Hansen, (ed.), *Geometric Theory of Diffraction*, Piscataway, New Jersey: IEEE Press, 1981.

[42] R. G. Kouyoumjian and P. H. Pathak, "A uniform geometrical theory of diffraction for an edge in a perfectly conducting surface," *Proc. IEEE*, vol. 62, pp. 1448–1461, 1974.

[43] S. W. Lee and G. A. Deschamps, "A uniform asymptotic theory of electromagnetic diffraction by a curved wedge," *IEEE Trans. Ant. Progat.*, vol. 24, pp. 25–35, 1976.

[44] P. H. Pathak, "An asymptotic analysis of the scattering of plane waves by a smooth convex cylinder," *Radio Sci.*, vol. 14, pp. 419, 1979.

[45] S. W. Lee, H. Ling, and R. C. Chou, "Ray tube integration in shooting and bouncing ray method," *Microwave Opt. Tech. Lett.*, vol. 1, pp. 285–289, Oct. 1988.

[46] H. Haken, *Light: Waves, Photons, Atoms*, North Holland, Amsterdam, 1981.

[47] M. D. Van Dyke, *Perturbation Methods in Fluid Mechanics*, New York: Academic Press, 1975.

[48] J. Kevorkian, J. D. Cole, *Perturbation Methods in Applied Mathematics*, New York: Spring Verlag, 1985.

[49] A. H. Nayfeh, *Perturbation Methods*, New York: Wiley-Interscience, 1973.

[50] G. Kirchhoff, "Zur Theorie des Kondensators," *Monatsb. Deutsch. Akad. Wiss.*, Berlin, pp. 144, 1877.

[51] S. Shaw, "Circular-disk viscometer and related electrostatic problems," *Phys. Fluids*, vol. 13, no. 8, pp. 1935, 1970.

[52] W. C. Chew and J. A. Kong, "Asymptotic formula for the capacitance of two oppositely charged discs," *Math. Proc. Cambridge Philos. Soc.*, vol. 89, pp. 373–384, 1981.

[53] W. C. Chew and J. A. Kong, "Microstrip capacitance for a circular disk through matched asymptotic expansions," *SIAM J. Appl. Math.*, vol. 42, no. 2, pp. 302–317, Apr. 1982.

[54] S. Y. Poh, W. C. Chew, and J. A. Kong, "Approximate formulas for line capacitance and characteristic impedance of microstrip line," *IEEE Trans. Microwave Theory Tech.*, vol. MTT-29, no. 2, pp. 135–142, Feb. 1981.

[55] W. C. Chew and J. A. Kong, "Asymptotic eigenequations and analytic formulas for the dispersion characteristics of open, wide microstrip lines," *IEEE Trans. Microwave Theory Tech.*, vol. MTT-29, no. 9, pp. 933–941, Sept. 1981.

[56] W. C. Chew and J. A. Kong, "Asymptotic formula for the resonant frequencies of a circular microstrip antenna," *J. Appl. Phys.*, vol. 52, no. 8, pp. 5365–5369, Aug. 1981.

[57] J. W. Strutt Rayleigh (Lord Rayleigh), "On the passage of electric waves through tubes, or the vibra cylinder," *Phi. Mag.*, vol. 43, pp. 125–132, 1897.

[58] E. A. Ash and E. G. S. Paige (eds.), *Rayleigh Wave Theory and Application*, Berlin: Spring-Verlag, 1985.

[59] A. Sommerfeld, "Mathematische Theorie der Diffraction," *Math. Ann.*, vol. 47, no. s319, pp. 317–374, 1896.

[60] A. Sommerfeld, "Uber die Ausbreitung der Wellen in der drahtlosen Telegraphie," *Ann. Phys.*, vol. 28, pp. 665–737, 1909.

[61] G. Mie, "Beiträge zur optik trüber medien speziel kolloidaler metallösungen," *Ann. Phys. (Leipzig)*, vol. 25, pp. 377, 1908.

[62] P. Debye, "Der Lichtdruck auf Kugeln von beliebigem Material," *Ann. Phys.*, vol. 4, no. 30, pp. 57, 1909.

[63] J. W. Strutt Rayleigh (Lord Rayleigh), *Theory of Sound*, New York: Dover Publ., 1976. (Originally published 1877.)

[64] J. J. Bowman, T. B. A. Senior, and P. L. E. Uslenghi, *Electromagnetic and Acoustic Scattering by Simple Shapes*, Amsterdam: North-Holland, 1969.

[65] G. N. WATSON, "The Diffraction of Electric Waves by the Earth," *Proc. R. Soc. London. Ser. A*, vol. 95, 1918, pp. 83–99.

[66] J. M. Song, C. C. Lu, W. C. Chew, and S. W. Lee, "Fast Illinois solver code (FISC)," *IEEE Antennas Propag. Mag.*, vol. 40, no. 3, pp. 27–33, 1998.

[67] J. M. Song and W. C. Chew, "Broadband time-domain calculations using FISC," *IEEE Antennas Propag. Soc. Int. Symp. Proc.*, San Antonio, Texas, vol.3, pp. 552–555, Jun. 2002.

[68] A. Taflove, *Computational Electrodynamics: The Finite Difference Time Domain Method*, Norwood, MA: Artech House, 1995.

[69] W. Hoefer, V. Tripathi, and T. Cwik, "Parallel and Distributed Computing Techniques for Electromagnetic Modeling," in *IEEE APS/URSI Joint Symposia*, University of Washington, 1994.

[70] W. C. Chew, J. M. Jin, E. Michielssen, and J. M. Song, eds., *Fast and Efficient Algorithms in Computational Electromagnetics*, Berlin: Artech House, 2001.

[71] R. Lee and A. C. Cangellaris, "A study of discretization error in the finite element approximation of wave solution," *IEEE Trans. Antennas Propag.*, vol. 40, no. 5, pp. 542-549, 1992.

[72] V. Rokhlin, "Rapid solution of integral equations of scattering theory in two dimensions," *J. Comput. Phys.*, vol. 36, no. 2, pp. 414–439, 1990.

[73] C. C. Lu and W. C. Chew, "A multilevel algorithm for solving boundaryvalue scattering," *Microwave Opt. Tech. Lett.*, vol. 7, no. 10, pp. 466-470, Jul. 1994.

[74] J. M. Song and W. C. Chew, "Multilevel fastmultipole algorithm for solving combined field integral equations of electromagnetic scattering," *Microwave Opt. Tech. Lett.*, vol. 10, no. 1, pp. 14-19, Sept. 1995.

[75] B. Dembart and E. Yip, "A 3D fast multipole method for electromagnetics with multiple levels," *Eleventh Ann. Rev. Prog. Appl. Comput. Electromag.*, vol. 1, pp. 621-628, 1995.

[76] S. Velamparambil, W. C. Chew, and J. M. Song, "10 million unknowns, is it that large," *IEEE Antennas Propag. Mag.*, vol.45, no.2, pp. 43–58, Apr. 2003.

[77] M. L. Hastriter, "A Study of MLFMA for Large-Scale Scattering Problems," *Ph.D. Thesis*, Dept. ECE, U. Illinois, Jun. 2003.

[78] V. Shankar, "A gigaflop performance algorithm for solving Maxwell's Equations of electromagnetics," *AIAA Paper 911578*, Honolulu, Jun. 1991.

[79] J. S. Shang, and D. Gaitonde, "Scattered Electromagnetic Field of a Reentry Vehicle," *AIAA Paper 940231*, Reno, Nevada, Jan. 1994.

[80] J. S. Shang, "Computational Electromagnetics," *ACM Comput. Surv.*, vol. 28, no. 1, pp. 97–99, 1996.

[81] J. M. Jin and V. V. Liepa, "Application of hybrid finite element method to electromagnetic scattering from coated cylinders," *IEEE Trans. Antennas Propag.*, vol. 36, pp. 50–54, Jan. 1988.

[82] A. A. Ergin, B. Shanker, and E. Michielssen, "Fast evaluation of transient wave fields using diagonal translation operators," *J. Comput. Phys.*, vol. 146, pp. 157–180, 1998.

CHAPTER 2

Linear Vector Space, Reciprocity, and Energy Conservation

2.1 Introduction

Linearity allows us to use the Fourier transform technique to transform and simplify the electromagnetic equations. We express all field and current functions of time using Fourier transform. For example,

$$\mathbf{F}(\mathbf{r},t) = \frac{1}{2\pi} \int_{-\infty}^{\infty} \mathbf{F}(\mathbf{r},\omega) e^{-i\omega t} d\omega \tag{2.1}$$

Substituting the above into the electromagnetic equations for the fields and currents, the equations become independent of time in the Fourier space, but the fields and currents are now functions of frequency and they are complex valued. Furthermore, $\partial \mathbf{F}(\mathbf{r},t)/\partial t$ Fourier transforms to $-i\omega \mathbf{F}(\mathbf{r},\omega)$. Consequently, the electromagnetic equations in the Fourier space become

$$\nabla \times \mathbf{E}(\mathbf{r},\omega) = i\omega \mathbf{B}(\mathbf{r},\omega) - \mathbf{M}(\mathbf{r},\omega) \tag{2.2}$$
$$\nabla \times \mathbf{H}(\mathbf{r},\omega) = -i\omega \mathbf{D}(\mathbf{r},\omega) + \mathbf{J}(\mathbf{r},\omega) \tag{2.3}$$
$$\nabla \cdot \mathbf{D}(\mathbf{r},\omega) = \rho_e(\mathbf{r},\omega) \tag{2.4}$$
$$\nabla \cdot \mathbf{B}(\mathbf{r},\omega) = \rho_m(\mathbf{r},\omega) \tag{2.5}$$

Each of these fields and currents can then be considered a representation of a time-harmonic signal. These complex fields and currents in the Fourier space are generally referred to as being in the frequency domain. The magnetic current $\mathbf{M}(\mathbf{r},\omega)$, and the magnetic charge $\rho_m(\mathbf{r},\omega)$ are regarded as fictitious, and are added to symmetrize the above equations.

In the electromagnetic equations, the second two equations are derivable from the first two equations for time varying fields and current. Hence, there are actually only two equations with four unknown quantities $\mathbf{E}$, $\mathbf{H}$, $\mathbf{D}$, and $\mathbf{B}$, assuming that the driving sources of the system $\mathbf{J}$ and $\mathbf{M}$ are known. The constitutive relations, which describe the material media, have to be invoked to render the above equations solvable [1]. For free space, the constitutive relations are $\mathbf{B} = \mu_0\mathbf{H}$ and $\mathbf{D} = \epsilon_0\mathbf{E}$, thus reducing the number of unknowns from four to two.

If the material media is nonlinear for which there is a nonlinear relation between $\mathbf{D}$ and $\mathbf{E}$ for instance, then a purely time-harmonic $\mathbf{E}$ will produce a $\mathbf{D}$ with more than one harmonics because of the second and higher order harmonic generation. Then a pure time-harmonic (monochromatic) signal cannot exist in such a system.

As mentioned in the Chapter 1, linearity permits the definition of the Green's function and the expression of solutions by invoking the principle of linear superposition. Consequently, the linearity of electromagnetic equations makes their solutions amenable to a large number of mathematical techniques, even though many of these techniques are restricted to linear problems only, such as linear vector space techniques. Because there are a number of mathematical concepts that are important for understanding computational electromangetics (CEM) and integral equation solvers, these concepts will be reviewed here.

This chapter is necessary also because a large part of electromagnetic theory can be understood from the concepts of linear vector spaces. The introduction of linear vector space concepts to students of computational electromagnetics is mandatory (*de rigueur*) because such concepts are very useful in understanding wave physics and computational electromagnetics (CEM). But many students of electromagnetics are not introduced to such concepts.

In this chapter, additionally, we will look at the reciprocity theorem and the energy conservation theorem from the viewpoint of a linear vector space [2]. It can be shown that these properties are intimately related to the symmetry and Hermitian properties of the linear operators in electromagnetics.

Some of the concepts expressed here are buried in, or may be derived from, the works of Reed and Simon [3], Felsen and Marcuvitz [4], Hanson and Yakovlev [5], or Collin [6], but the points expressed here have not been explicitly stated before.

2.2 Linear Vector Spaces

For linear electromagnetics, the computation is done in a linear vector space with linear operators. Examples of linear vector spaces are found in linear algebra, where one usually deals with vectors in N dimensions. Solving linear systems with N unknowns forms the underlying gist of CEM. How then do we arrive at such systems from the electromagnetic equations?

The electromagnetic equations are defined in a continuum space. The continuum space entails four physical space-time dimensions described by the x, y, z, and t variables. Because these space-time variables are infinitely divisible, electromagnetic fields and sources are described by infinite degree of freedom. One can think of them as an equivalent matrix system with uncountably infinite number of variables. An example of such infinite dimensional space with uncountably infinite degree of freedom is the Hilbert space.

However, it is cumbersome if not impossible to perform computations in this infinite dimensional space. Hence, a systematic way is needed to approximate this space with a discrete, finite dimensional space. One way to arrive at a discrete, finite system is with the subspace projection method. This is the underlying principle behind the finite element method (FEM) and the method of moments (MOM). Another method is to use finite-difference approxima-tion of space time in a finite domain of space, or to use Nyström method to approximate an integral equation. In Nyström method, integrals are approximated with quadrature rules, so that a continuum integral is replaced with a discrete summation.

Most of the linear vector space concepts are similar to those in vector space in linear algebra, except that the vectors can be infinite dimensional. These vectors can be countably infinite, or uncountably infinite. A finite summation in linear algebra will be replaced by an infinite summation, or an integral, which may or may not converge.

A linear vector space V consists of a set of infinitely many vectors that can be added and scaled. The added vectors are members of the space V, and so are the scaled vectors. A set of vectors, $\{f_n, n = 1, \ldots, N\}$, are linearly independent if

$$\sum_{n=1}^{N} a_n f_n = 0 \qquad (2.6)$$

implies that $a_n = 0, n = 1, \ldots, N$. This set of vectors spans the vector space V if every element f in the vector space can be written as

$$f = \sum_{n=1}^{N} a_n f_n \qquad (2.7)$$

The above concept can be extended to the case when N is infinite, or when the vectors are uncountably infinite. In the case of uncountably infinite set, an integral replaces the summation above.

For this linear vector space, we can define linear operators (linear transformations) that transform a vector from linear vector space V to another vector in linear vector space W. For example we can write

$$f = \mathbb{L}v \qquad (2.8)$$

where f is a new vector generated when the operator $\mathbb{L}$ operates on v. Here, f can be in the same space as v lives in or it can be outside the space that v is in. The space of functions that $\mathbb{L}$ can operate on is known as the *domain* of $\mathbb{L}$ or $D(\mathbb{L})$. The space spanned by vectors that $\mathbb{L}v$ can generate for all $v \in D(\mathbb{L})$ is known as the *range* of $\mathbb{L}$ or $R(\mathbb{L})$.

The domain or range of $\mathbb{L}$ can change depending on the form of $\mathbb{L}$. For instance, if $\mathbb{L}$ contains derivatives, they may render the action of $\mathbb{L}$ on some functions undefined, eliminating these functions from the domain and range of $\mathbb{L}$ in the classical sense. However, one can expand the domain and range to contain distributions.

In electromagnetics, $\mathbb{L}$ can be a differential operator, an integral operator or a mixture thereof. We can think of (2.8) as the elevation of linear algebra concept where $\mathbb{L}$ is analogous to a matrix operator and v, and f are simple vectors in matrix algebra. Instead of being indexed by discrete indices in matrix algebra, (2.8) can represent functions whose "indices" are a continuum, and hence of infinite dimension. For example, an explicit form of (2.8) in one dimension is

$$f(x) = \int_a^b dx' L(x, x')v(x'), \qquad a < x < b \qquad (2.9)$$

This is the analogy of

$$\mathbf{f} = \overline{\mathbf{L}} \cdot \mathbf{v} \tag{2.10}$$

in matrix algebra.

Also, in electromagnetics, one works with fields in the Fourier space or the frequency domain, which allows for complex numbers. Hence, in frequency-domain electromagnetics, a vector space embodies complex-valued vectors as elements. As such, the inner products in electromagnetics have to be defined so that they can account for the possibilities of different physical interpretations. We will first outline the inner products in electromagnetics and compare them with the inner products commonly used in mathematics and quantum physics.

2.3   Inner Products for Electromagnetics

To use any subspace projection method, we need to define an inner product in the linear vector space. Before defining an inner product, one should be cognizant that a function $f(x)$ is actually the analogy of a component of a vector f_j in linear algebra, where the former is "indexed" by an indenumerable variable x, whereas the latter is indexed by a denumerable integer. This analogy can be extended to multidimensional functions in space. By drawing upon this analogy, the inner product in an infinite dimensional space can be defined.

In electromagnetics, the inner product between two functions is defined as [7–10]

$$\langle f, g \rangle = \int f(\mathbf{r}) g(\mathbf{r}) d\mathbf{r}, \tag{2.11}$$

where the integral is assumed to converge. This inner product involves elements $f(\mathbf{r})$ and $g(\mathbf{r})$, which either live in one, two, or three dimensions.

The above is analogous to

$$\mathbf{f}^t \cdot \mathbf{g} = \sum_{i=1}^{N} f_i g_i \tag{2.12}$$

in matrix algebra, and hence, we call this the inner product between two functions or infinite dimensional vectors. The difference between an inner product of a finite dimensional vector and an infinite dimensional vector is that the latter inner product may not exist because the integral may diverge.

The inner product in Eq. (2.11) is physically related to the reaction between two field quantities [11]. We shall call the above the reaction inner product. In the literature, it has been called the pseudo inner product [5] or the bilinear form [12]. In subsequent sections, unless otherwise stated, we will use $\langle , \rangle$ to imply an electromagnetic or reaction inner product.

The electromagnetic reaction inner product has its usefulness: it can be shown that for a large class of media, called reciprocal media, the electromagnetic equations are analogous to a complex symmetric system in linear algebra.

In CEM, when the reaction inner product is used, it is to find the matrix representation of electromagnetic operators. If the electromagnetic operator is symmetric, the matrix system obtained is complex symmetric, resulting in storage reduction. Because the reaction inner product is given the physical meaning of a reaction, such a physical concept can be used as a sanity check in many experimental measurements and engineering designs [11].

When we need to define an inner product for the energy, we will explicitly denote one of the vectors with a complex conjugation, namely,

$$\langle f^*, g \rangle = \int f^*(\mathbf{r}) g(\mathbf{r}) d\mathbf{r}, \tag{2.13}$$

where $f^*(\mathbf{r})$ means the complex conjugation of $f(\mathbf{r})$. Because the above has the physical meaning of complex energy or power [13], we shall call the above the energy inner product. In this paper, f and g can be n-vectors (column vectors with n elements). In this case, the products in the integrands of the right-hand sides of (2.11) and (2.13) are replaced by dot products.

2.3.1   Comparison with Mathematics

The above inner product is different from that defined in the mathematical literature [12, 14], where the inner product is invariably defined as

$$\langle f, g \rangle_H = \int f^*(\mathbf{r}) g(\mathbf{r}) d\mathbf{r}. \tag{2.14}$$

where the subscript H emphasizes that the conjugation of f is implied. The above inner product has its elegance because it can be used to define the norm of a function as

$$||f||^2 = \int f^*(\mathbf{r}) f(\mathbf{r}) d\mathbf{r}. \tag{2.15}$$

With this definition, the norm of a function $||f||^2 > 0$ if $f \neq 0$. A norm can be used as a metric to measure the "distance" between two vectors.

A linear vector space with a defined norm is known as a normed vector space. Different norms can be used to distinguish different normed linear vector spaces. For example, the set of all functions f such that the integral in (2.15) exists and is of finite value is known as the L_2 space. More complicated norms can be defined to allow for different normed spaces. For instance, a norm can be defined such that

$$||f||_{\mathbb{A}}^2 = \int \int f^*(\mathbf{r}) \mathbb{A}(\mathbf{r}, \mathbf{r}') f(\mathbf{r}') d\mathbf{r} d\mathbf{r}', \tag{2.16}$$

where f and $\mathbb{A}$ are chosen so that the above is always a bounded positive number. When $||f||_{\mathbb{A}}^2$ is positive for all candidate f, then $\mathbb{A}$ is termed a positive operator. When $||f||_{\mathbb{A}}^2$ is bounded, for all f such that $||f|| = 1$, then $\mathbb{A}$ is termed bounded.

2.3.2   Comparison with Quantum Mechanics

The electromagnetic inner products are different from the inner product of quantum mechanics [15, 16]. In quantum mechanics, one studies quantum systems, which have (experimentally) measurable quantities called observables and state vectors describing the state of a quantum system. The observables are measurable quantities such as momentum and position, and they are represented by operators in a quantum system. The state vector is represented by a function $f(\mathbf{r})$. The inner product between two state vectors is defined in Dirac's bra and ket notation as

$$\langle f|g \rangle = \int f^*(\mathbf{r})g(\mathbf{r})d\mathbf{r}. \tag{2.17}$$

This is similar to the energy inner product previously defined. If $f = g$, then the integrand on the right-hand side represents a probability function, and this definition of inner product ensures the positivity of this integrand. Probability functions, which are normalized to one, cannot be destroyed. In other words, the net "energy" of a state vector, $\int |f(\mathbf{r})|^2 d\mathbf{r}$, is always constant. Hence, a quantum system is "lossless" from that perspective. This is reflected by the fact that the Hamiltonian operator that governs the time evolution of the state vector is Hermitian, with real eigenvalues.

To convert an observable operator $\mathbb{A}$ of a quantum system into something that can be measured in the laboratory, one computes the quantity

$$\langle f|\mathbb{A}|f \rangle = \int f^*(\mathbf{r})\mathbb{A}(\mathbf{r},\mathbf{r}')f(\mathbf{r}')d\mathbf{r}d\mathbf{r}', \tag{2.18}$$

which is called the expectation value of the operator $\mathbb{A}$. Because this value represents a measurable quantity, it must always be real. Hence, the operator $\mathbb{A}$ has to be self-adjoint or Hermitian to ensure its reality. In other words, the definition of quantum-mechanics inner product ensures the reality of the expectation value.

2.4   Transpose and Adjoint of an Operator

Many mathematics books define the adjoint of an operator, but not the transpose. Such definitions may be sufficient for quantum mechanics, but are not suitable for electromagnetics [10]. We define the transpose of an operator $\mathbb{G}$, denoted by $\mathbb{G}^t$, to be

$$\langle f, \mathbb{G}g \rangle = \langle g, \mathbb{G}^t f \rangle, \quad \forall g \in D(\mathbb{G}), \forall f \in R(\mathbb{G}), \tag{2.19}$$

where $D(\mathbb{G})$ and $R(\mathbb{G})$ stand for the domain and the range of $\mathbb{G}$, respectively. The definition of the transpose of a matrix is simple because we just need to swap the indices. But in an infinite-dimensional vector space, it is more convenient to define the transpose using inner products given by (2.19). Notice that the range and the domain of $\mathbb{G}^t$ are swapped compared to those of $\mathbb{G}$. An operator is symmetric if $\mathbb{G}^t = \mathbb{G}$.[1]

The adjoint of an operator $\mathbb{G}$, denoted by $\mathbb{G}^a$, is defined by

$$\langle f^*, \mathbb{G}g \rangle = \langle g^*, \mathbb{G}^a f \rangle^*, \quad \forall g \in D(\mathbb{G}), \forall f \in R(\mathbb{G}). \tag{2.20}$$

This concept is similar to the conjugate transpose of a matrix. An operator is self-adjoint if $\mathbb{G}^a = \mathbb{G}$. It is similar to the concept of a Hermitian matrix. Hence, we shall use $\mathbb{G}^\dagger$ to denote the adjoint of an operator as well, which is a common notation used to denote the conjugate transpose of a matrix in linear algebra.

---

[1]In many textbooks, this is called the "adjoint" instead of the "transpose" of an operator [6]. We prefer to call this the "transpose" because of its analogy to the "transpose" concept in linear algebra.

2.5   Matrix Representation

An operator in an infinite-dimensional space is difficult to compute. It is customary to reduce such an operator to an approximate, finite-dimensional matrix operator by finding its matrix representation.

Given an operator equation

$$\mathbb{L}f = g, \tag{2.21}$$

where f and g are functions, we can let

$$f(\mathbf{r}) \approx \sum_{n=1}^{N} f_n b_n(\mathbf{r}), \tag{2.22}$$

where $b_n(\mathbf{r})$ are linearly independent vectors or functions, f_n are constants, $D_N = \text{span}\{b_n(\mathbf{r}), n$ $1, \ldots, N\}$ is a space that approximates the original domain space of $\mathbb{L}$, or $D(\mathbb{L})$. D_N may be a subspace of $D(\mathbb{L})$ because D_N is finite-dimensional but $D(\mathbb{L})$ is infinite-dimensional. If $\mathbb{L}$ is a differential operator, then f in (2.21) is nonunique unless boundary conditions are specified. Hence, $\mathbb{L}$ has to be augmented with boundary conditions to be invertible.

We can use (2.22) in (2.21), multiply the resultant equation with

$$t_m(\mathbf{r}), m = 1, \ldots, N$$

and integrate to obtain

$$\sum_{n=1}^{N} f_n \langle t_m, \mathbb{L}b_n \rangle = \langle t_m, g \rangle, \quad m = 1, \ldots, N. \tag{2.23}$$

In order for (2.21) to have a solution, g must be in the range space $R(\mathbb{L})$ of $\mathbb{L}$. Therefore, if $R_N = \text{span}\{t_m, m = 1, \ldots, N\}$ is a good approximation to $R(\mathbb{L})$, then (2.23) is a good replacement equation to solve instead of (2.21). Moreover, (2.23) can be written in a matrix form

$$\mathbf{L} \cdot \mathbf{f} = \mathbf{g}, \tag{2.24}$$

where

$$[\mathbf{L}]_{mn} = L_{mn} = \langle t_m, \mathbb{L}b_n \rangle, \tag{2.25}$$

$$[\mathbf{f}]_n = f_n, \tag{2.26}$$

$$[\mathbf{g}]_n = \langle t_n, g \rangle = g_n. \tag{2.27}$$

In the above, $\mathbf{L}$ is the matrix representation of the operator $\mathbb{L}$, and $\mathbf{f}$ and $\mathbf{g}$ are the vector representations of the functions f and g in the subspaces spanned by $\{f_n, n = 1, \ldots, N\}$ and $\{g_n, n = 1, \ldots, N\}$, respectively.

The functions $b_n(\mathbf{r}), n = 1, \ldots, N$ are known as the expansion (or basis) functions, and together, they form a basis that spans the subspace that approximates the domain of the operator. The functions $t_n(\mathbf{r}), n = 1, \ldots, N$ are known as the testing (or weighting) functions, and together, they form a basis that spans the subspace that approximates the range of the operator. The above method is a subspace projection method, where the linear operator has been approximated by subspaces that approximate its domain and its range spaces.

The above method of finding the matrix representation of an operator is known as the method of moments (MOM) or the boundary element method (BEM) when the operator is an integral operator. It is known as the finite element method (FEM) when the operator is a differential operator; the name Galerkin's method is also used when the testing function is the same as the expansion function. Mathematically, MOM and FEM are similar methods, where an operator equation in an infinite-dimensional space is projected onto a subspace spanned by a finite set of expansion functions.

When the operator is symmetric, the range space is the same as the domain space. In this case, the matrix representation of the original operator is a symmetric matrix when the testing functions are the same as the expansion functions, and memory saving in matrix storage ensues.

Because we have approximated an infinite dimensional space with a subspace of finite dimension, the matrix representation will not be an exact representation. But it can be made increasingly accurate by increasing N if the set of expansion functions is complete when N tends to infinity. In other words, they can be used to approximate any function to arbitrary accuracy when N tends to infinity.

However, letting N tend to infinity is impractical for most computations. The trick is on how to obtain a good solution with finite computational resource, that is, with N being a finite number. Invocation of heuristics is necessary in this case. To make (2.24) a high fidelity representation of the original equation (2.21), the vector representations $\mathbf{f}$ and $\mathbf{g}$ must be good representations of the original functions f and g. For instance, if g is singular, one may invoke heuristics to infer that f is also singular, and hence, appropriately distribute the expansion functions to capture the essence of their singularities. Otherwise, an exorbitantly large number of expansion functions may be needed to approximate these singular functions well.

As previously noted, an operator $\mathbb{L}$ may operate on some function f that may produce a field g that is singular. But in CEM, we are interested in $L_{mn} = \langle t_m, \mathbb{L} b_n \rangle$ being a computable quantity. There could be choice of expansion and testing functions that may render the matrix element L_{mn} infinite or undefined. This is highly undesirable in a numerical computation. Hence, we restrict our expansion and testing functions with computable inner products when the matrix representation of the operator is sought. This is the same as requiring physical quantities to be finite, because the inner product in electromagnetics can be interpreted as either a reaction or an energy.

Oftentimes, seeking the matrix representation of an operator can also be related to the minimization of a quadratic functional by the Rayleigh-Ritz method ([10, Chapter 5]). The minimization of the quadratic functional can be shown to be equivalent to solving the original operator equation.

2.6   Compact versus Noncompact Operators

A smooth function, which is bandlimited, can be approximated well by a finite number of expansion functions, such as Fourier series or a set of polynomials, to within exponential accuracy. A compact operator can be thought of as a smoothing operator. It converts a function with high spectral components to a smooth function. In other words, the range space of a compact operator can be approximated by a finite number of expansion function accurately.

In CEM, the matrix representations of the operators are often obtained by using subdomain expansion and testing functions. These are functions whose domains have finite supports in the 2D or 3D space over which they are defined. An example of them is a pulse expansion function in one dimension. We use h to describe the typical size of the subdomain. When h is small, these functions are capable of modeling fields and sources that vary rapidly as a function of space: in other words, they can model functions with high spectral frequencies. Moreover, when h is small, many expansion functions are needed to model a function, and hence the dimension N of the matrix representation increases.

The nature of a compact operator can be understood from its matrix representation when $h \to 0$ or $N \to \infty$. When $h \to 0$, the degrees of freedom of the function increase, and more high-frequency eigenvectors can be formed. However, these higher-frequency eigenvectors will have smaller eigenvalues because of the smoothing property of the compact operator. Hence, the eigenvalue of the matrix representation, which are more numerous, become increasingly smaller and cluster around 0. In other words, a compact operator annihilates high-frequency eigenvectors. On the other hand, when its eigenvalues become increasingly larger when $h \to 0$, the operator is noncompact. A noncompact operator enhances the high frequency components of a function it acts on.

Many integral operators eliminates the high-frequency components, and hence, they are compact. Differential operators, on the other hand, amplify the high-frequency components, and hence, they are noncompact.

2.7   Extension of Bra and Ket Notations

In Hilbert space, the use of bra and ket notation is insufficient for electromagnetics as two kinds of inner products are used in electromagnetics, whereas in quantum mechanics, or mathematics, only one kind of inner product is implied. However, the notations in linear algebra are quite complete and unambiguous. We will adopt notations similar to the bra and ket, such that they do not give rise to ambiguity when used in electromagnetics.

Because the inner product

$$\langle f, g \rangle \tag{2.28}$$

is analogous to $\mathbf{f}^t \cdot \mathbf{g}$, we will liken $g\rangle$ to be analogous to a column vector in linear algebra, while $\langle f$ to be analogous to a row vector (the transpose of a column vector) in linear algebra. The comma between $\langle f$ and $g\rangle$ denoting an inner product, is analogous to the dot in linear algebra notation. We can think of $\langle f$ as the transpose of $f\rangle$. For a conjugate transpose, we will use the notation $\langle f^*$. For the lack of a better name, we will call $\langle f$ a *braem*, and $g\rangle$ a *ketem* for use in electromagnetics.

Furthermore, we will use

$$\mathbb{L}, g\rangle \tag{2.29}$$

to denote the action of the operator $\mathbb{L}$ on the vector $g\rangle$. This is analogous to $\overline{\mathbf{L}} \cdot \mathbf{g}$ in linear algebra, and hence, the comma here is still analogous to a dot product. Hence, we can think of $\mathbb{L}, g\rangle$ in this extended notation here to be analogous to $\mathbb{L}g$ in many mathematics literature.

Furthermore, for a projection of the operator onto two vectors, we denote it by

$$\langle f, \mathbb{L}, g \rangle. \tag{2.30}$$

This is analogous to $\mathbf{f}^t \cdot \overline{\mathbf{L}} \cdot \mathbf{g}$ in linear algebra where each comma is analogous to a dot product.

For the outer product between two vectors in Hilbert space, we will use the notation $f\rangle\langle g$, and this is analogous to $\mathbf{f}\mathbf{g}^t$ in linear algebra. It is to be noted that $f\rangle\langle g$ is not analogous to $f(x)g(x)$ but $f(x)g(x')$.

Dirac has used a vertical bar to denote an inner product. In his notation, a vector is denoted by $|f\rangle$, whereas $\langle f|$ invariably is analogous to conjugate transpose in linear algebra. He has used $\mathbb{L}|g\rangle$ for (2.29), and $\langle f|\mathbb{L}|g\rangle$ for (2.30) [15, 16]. Because of the implicit conjugate transpose in the notations, they are not convenient for use in electromagnetics.

2.8   Orthogonal Basis versus Nonorthogonal Basis

When a basis set is orthogonal, they can be normalized such that they are orthonormal, so that

$$\langle f_n, f_m \rangle = \delta_{nm} \tag{2.31}$$

where δ_{nm} is the Kronecker delta function. If f_n and f_m are members of an orthonormal basis set, then we can define an identity kernel in the subspace spanned by f_n such that

$$I(x, x') = \sum_{n=1}^{N} f_n(x)f_n(x') \tag{2.32}$$

We define the subspace as $S_n = \text{span}\{f_n(x), n = 1, \ldots, N, a < x < b\}$. If $g(x) \in S_n$, then the action (or operation) of $I(x, x')$ on $g(x)$ is given by

$$\int_a^b dx' I(x, x')g(x') = \sum_{n=1}^{N} f_n(x)\langle f, g \rangle = g(x) \tag{2.33}$$

We can also think of $\sum_n f_n(x)f_n(x')$ as the outer product between two vectors indexed by x and x'. Hence, it can also be expressed as

$$\mathbb{I} = \sum_{n=1}^{N} f_n\rangle\langle f_n \tag{2.34}$$

We see that $\mathbb{I}$ behaves like an integral operator in this example.

When $f_n(x)$ are nonorthogonal, but complete in a subspace, we can expand a function $g(x)$ in that subspace by

$$g(x) = \sum_{n=1}^{N} a_n f_n(x) \tag{2.35}$$

To solve for a_n, we test the above by $f_m(x), m = 1, \ldots, N$ to get

$$\langle f_m, g \rangle = \sum_{n=1}^{N} a_n \langle f_m, f_n \rangle \ , \ \ m = 1, \ldots, N \tag{2.36}$$

The above can be expressed as a matrix equation

$$\mathbf{g} = \overline{\mathbf{F}} \cdot \mathbf{a} \tag{2.37}$$

where

$$[\mathbf{g}]_n = \langle f_n, g \rangle, \quad [\overline{\mathbf{F}}]_{mn} = \langle f_m, f_n \rangle, \quad [\mathbf{a}]_n = a_n \tag{2.38}$$

The matrix $\overline{\mathbf{F}}$ is also known as the Grammian [17]. Hence, we can solve (2.37) to get

$$\mathbf{a} = \overline{\mathbf{F}}^{-1} \cdot \mathbf{g} \tag{2.39}$$

we can write (2.35) as

$$g(x) = \mathbf{f}^t(x) \cdot \mathbf{a} = \mathbf{f}^t(x) \cdot \overline{\mathbf{F}}^{-1} \cdot \mathbf{g} \tag{2.40}$$

where $[\mathbf{f}(x)]_n = f_n(x)$. Hence, we can express $\mathbf{g} = \langle \mathbf{f}, g \rangle$ in (2.38) and (2.40) becomes

$$g(x) = \mathbf{f}^t(x) \cdot \overline{\mathbf{F}}^{-1} \cdot \langle \mathbf{f}, g \rangle \tag{2.41}$$

or

$$g \rangle = \mathbf{f}^t \rangle \cdot \overline{\mathbf{F}}^{-1} \cdot \langle \mathbf{f}, g \rangle \tag{2.42}$$

We can identify an identity operator

$$I = \mathbf{f}^t \rangle \cdot \overline{\mathbf{F}}^{-1} \cdot \langle \mathbf{f} \tag{2.43}$$

When f_n's are orthonormal, $\overline{\mathbf{F}}^{-1} = \overline{\mathbf{I}}$, and the above becomes

$$\mathbb{I} = \mathbf{f}^t \rangle \cdot \langle \mathbf{f} \tag{2.44}$$

Eq. (2.44) is an alternative way of writing (2.34).

2.9 Integration by Parts

One of the most frequently used mathematical tricks in electromagnetics is integration by parts. It is based on the product rule of derivatives. Such a rule appears for scalar and vector fields in different form [18].

$$\nabla \cdot (\phi \mathbf{A}) = \nabla \phi \cdot \mathbf{A} + \phi \nabla \cdot \mathbf{A} \tag{2.45}$$

$$\nabla \cdot (\mathbf{A} \times \mathbf{B}) = \mathbf{B} \cdot \nabla \times \mathbf{A} - \mathbf{A} \cdot \nabla \times \mathbf{B} \tag{2.46}$$

$$\nabla(\phi_1 \phi_2) = \phi_2 \nabla \phi_1 + \phi_1 \nabla \phi_2 \tag{2.47}$$

$$\nabla \times (\phi \mathbf{A}) = \phi \nabla \times \mathbf{A} - \mathbf{A} \times \nabla \phi \tag{2.48}$$

It is from this rule, together with Gauss' divergence theorem, that many integral identities follow. One can integrate (2.45) over a volume V to obtain, after invoking Gauss' theorem,

$$\oint_S \hat{n} \cdot (\phi \mathbf{A}) dS = \int_V dV (\nabla \phi \cdot \mathbf{A}) + \int_V dV (\phi \nabla \cdot \mathbf{A}) \tag{2.49}$$

where S is the surface of V. If either ϕ or $\mathbf{A}$, or both are of finite support, we can pick V to be larger than the support of ϕ or $\mathbf{A}$. Then the left-hand side of (2.49) is zero because ϕ or $\mathbf{A}$, or both are zero on S, and we have

$$\int_V dV \nabla \phi \cdot \mathbf{A} = - \int_V dV (\phi \nabla \cdot \mathbf{A}) \tag{2.50}$$

Similarly, from (2.46), we have

$$\oint_S \hat{n} \cdot (\mathbf{A} \times \mathbf{B}) dS = \int_V dV \mathbf{B} \cdot \nabla \times \mathbf{A} - \int_V dV \mathbf{A} \cdot \nabla \times \mathbf{B} \tag{2.51}$$

If either $\mathbf{A}$ or $\mathbf{B}$, or both are of finite support, then, we can pick V to be larger than the support of $\mathbf{A}$ or $\mathbf{B}$, and the left-hand side of (2.51) is zero, because either $\mathbf{A}$ or $\mathbf{B}$, or both are zero on S, and we have

$$\int_V dV \mathbf{B} \cdot \nabla \times \mathbf{A} = \int_V dV \mathbf{A} \cdot \nabla \times \mathbf{B} \tag{2.52}$$

Eq. (2.47) is similar to Eq. (2.45) if we replace $\phi = \phi_1$ and $\mathbf{A} = \mathbf{a}\phi_2$ where $\mathbf{a}$ is a constant vector. Therefore, from (2.47),

$$\oint_S \hat{n}(\phi_1 \phi_2) dS = \int_V dV \phi_1 \nabla \phi_2 + \int_V dV \phi_2 \nabla \phi_1 \tag{2.53}$$

The above becomes

$$\int_V dV \phi_1 \nabla \phi_2 = - \int_V dV \phi_2 \nabla \phi_1 \tag{2.54}$$

if either ϕ_1 or ϕ_2 is of finite support.

To derive a formula from (2.48), we use the fact that

$$\int_V dV \nabla \times \mathbf{F} = \int_S dS(\hat{n} \times \mathbf{F}) \tag{2.55}$$

This can be derived by applying Gauss' divergence theorem to the function $\mathbf{C} = \mathbf{a} \times \mathbf{F}$ where $\mathbf{a}$ is an arbitrary constant, and noting that $\nabla \cdot \mathbf{C} = \nabla \cdot (\mathbf{a} \times \mathbf{F}) = -\mathbf{a} \cdot \nabla \times \mathbf{F}$ and $\hat{n} \cdot \mathbf{a} \times \mathbf{F} = -\mathbf{a} \cdot \hat{n} \times \mathbf{F}$. Using the above, we have

$$\oint dS \hat{n} \times \phi \mathbf{A} = \int_V dV \phi \nabla \times \mathbf{A} - \int_V dV \mathbf{A} \times \nabla \phi \tag{2.56}$$

or

$$\int_V dV \mathbf{A} \times \nabla \phi = \int_V dV \phi \nabla \times \mathbf{A} \tag{2.57}$$

if either ϕ or $\mathbf{A}$ is of finite support.

The above are some of the most commonly used integration-by-parts formulas in electromagnetics.

2.10   Reciprocity Theorem—A New Look

The reciprocity theorem is attributed to Lorentz [19], but variations of it have been derived. A simpler form for the Poisson's equation has been stated by George Green [20]. The reciprocity theorem can be restated from the perspective of a linear vector space: the linear operator associated with the electromagnetic equations is a symmetric operator. For the statement of the reciprocity theorem, the electromagnetic equations are augmented with a fictitious magnetic current $\mathbf{M}$ to make them symmetrical. Namely, the first two equations are

$$\nabla \times \mathbf{E}(\mathbf{r}, \omega) = i\omega \mathbf{B}(\mathbf{r}, \omega) - \mathbf{M}(\mathbf{r}, \omega), \tag{2.58}$$

$$\nabla \times \mathbf{H}(\mathbf{r}, \omega) = -i\omega \mathbf{D}(\mathbf{r}, \omega) + \mathbf{J}(\mathbf{r}, \omega). \tag{2.59}$$

For anisotropic media where $\mathbf{B} = \overline{\boldsymbol{\mu}} \cdot \mathbf{H}$, and $\mathbf{D} = \overline{\boldsymbol{\epsilon}} \cdot \mathbf{E}$, we can write electromagnetic equations as

$$\begin{bmatrix} 0 & \nabla \times \overline{\mathbf{I}} \\ \nabla \times \overline{\mathbf{I}} & 0 \end{bmatrix} \cdot \begin{bmatrix} \mathbf{E} \\ \mathbf{H} \end{bmatrix} + i\omega \begin{bmatrix} \overline{\boldsymbol{\epsilon}} & 0 \\ 0 & -\overline{\boldsymbol{\mu}} \end{bmatrix} \cdot \begin{bmatrix} \mathbf{E} \\ \mathbf{H} \end{bmatrix} = \begin{bmatrix} \mathbf{J} \\ -\mathbf{M} \end{bmatrix}. \tag{2.60}$$

where $\nabla \times \overline{\mathbf{I}} \cdot \mathbf{E} = \nabla \times \mathbf{E}$, and overbar over a boldface character implies that it is a second rank tensor. The above can be more compactly rewritten as

$$\mathbb{L} \begin{bmatrix} \mathbf{E} \\ \mathbf{H} \end{bmatrix} = \begin{bmatrix} \mathbf{J} \\ -\mathbf{M} \end{bmatrix}, \tag{2.61}$$

where

$$\mathbb{L} = \begin{bmatrix} 0 & \nabla \times \overline{\mathbf{I}} \\ \nabla \times \overline{\mathbf{I}} & 0 \end{bmatrix} + i\omega \begin{bmatrix} \overline{\boldsymbol{\epsilon}} & 0 \\ 0 & -\overline{\boldsymbol{\mu}} \end{bmatrix}. \tag{2.62}$$

We shall prove that the above operator is symmetric when $\overline{\boldsymbol{\epsilon}}$ and $\overline{\boldsymbol{\mu}}$ are symmetric tensors. Many authors associate a negative sign with one of the $\nabla \times$ operators, but that will make the $\mathbb{L}$ operator nonsymmetric [4, 21, 22].

For the fields produced by two different groups of sources, we have

$$\mathbb{L} \begin{bmatrix} \mathbf{E}_1 \\ \mathbf{H}_1 \end{bmatrix} = \begin{bmatrix} \mathbf{J}_1 \\ -\mathbf{M}_1 \end{bmatrix}, \tag{2.63}$$

$$\mathbb{L} \begin{bmatrix} \mathbf{E}_2 \\ \mathbf{H}_2 \end{bmatrix} = \begin{bmatrix} \mathbf{J}_2 \\ -\mathbf{M}_2 \end{bmatrix}. \tag{2.64}$$

Multiplying (2.63) by $[\mathbf{E}_2, \mathbf{H}_2]$ and (2.64) by $[\mathbf{E}_1, \mathbf{H}_1]$, we have

$$\left\langle [\mathbf{E}_2, \mathbf{H}_2], \mathbb{L} \begin{bmatrix} \mathbf{E}_1 \\ \mathbf{H}_1 \end{bmatrix} \right\rangle = \left\langle [\mathbf{E}_2, \mathbf{H}_2], \begin{bmatrix} \mathbf{J}_1 \\ -\mathbf{M}_1 \end{bmatrix} \right\rangle = \langle \mathbf{E}_2, \mathbf{J}_1 \rangle - \langle \mathbf{H}_2, \mathbf{M}_1 \rangle, \tag{2.65}$$

$$\left\langle [\mathbf{E}_1, \mathbf{H}_1], \mathbb{L} \begin{bmatrix} \mathbf{E}_2 \\ \mathbf{H}_2 \end{bmatrix} \right\rangle = \left\langle [\mathbf{E}_1, \mathbf{H}_1], \begin{bmatrix} \mathbf{J}_2 \\ -\mathbf{M}_2 \end{bmatrix} \right\rangle = \langle \mathbf{E}_1, \mathbf{J}_2 \rangle - \langle \mathbf{H}_1, \mathbf{M}_2 \rangle, \tag{2.66}$$

where the inner product above is defined by integrating over a volume V bounded by a surface S. Writing out the left-hand sides of (2.65) and (2.66) explicitly, we have

$$\text{LHS}(2.65) = \langle \mathbf{E}_2, \nabla \times \mathbf{H}_1 \rangle + \langle \mathbf{H}_2, \nabla \times \mathbf{E}_1 \rangle + i\omega \langle \mathbf{E}_2, \overline{\boldsymbol{\epsilon}} \cdot \mathbf{E}_1 \rangle - i\omega \langle \mathbf{H}_2, \overline{\boldsymbol{\mu}} \cdot \mathbf{H}_1 \rangle, \tag{2.67}$$

$$\text{LHS}(2.66) = \langle \mathbf{E}_1, \nabla \times \mathbf{H}_2 \rangle + \langle \mathbf{H}_1, \nabla \times \mathbf{E}_2 \rangle + i\omega \langle \mathbf{E}_1, \overline{\epsilon} \cdot \mathbf{E}_2 \rangle - i\omega \langle \mathbf{H}_1, \overline{\mu} \cdot \mathbf{H}_2 \rangle. \qquad (2.68)$$

If $\overline{\epsilon}$ and $\overline{\mu}$ are symmetric tensors, then the last two terms of (2.67) and (2.68) are equal to each other. We can apply integration by parts to (2.67) to get

$$\langle \mathbf{E}_2, \nabla \times \mathbf{H}_1 \rangle + \langle \mathbf{H}_2, \nabla \times \mathbf{E}_1 \rangle = \langle \mathbf{H}_1, \nabla \times \mathbf{E}_2 \rangle + \langle \mathbf{E}_1, \nabla \times \mathbf{H}_2 \rangle$$
$$+ \oint_S dS \hat{n} \cdot (\mathbf{E}_2 \times \mathbf{H}_1) - \oint_S dS \hat{n} \cdot (\mathbf{E}_1 \times \mathbf{H}_2), \qquad (2.69)$$

where S is the surface bounding V. Now using (2.69) in (2.67), and then subtracting (2.67) and (2.68), we have

$$\text{LHS}(2.65) - \text{LHS}(2.66) = \oint_S dS \hat{n} \cdot (\mathbf{E}_2 \times \mathbf{H}_1) - \oint_S dS \hat{n} \cdot (\mathbf{E}_1 \times \mathbf{H}_2), \qquad (2.70)$$

The surface integral can be made to vanish if the surface S is a perfect electric conductor (PEC), a perfect magnetic conductor (PMC), or a surface with a certain impedance boundary condition (IBC), implying that the right-hand side of the above is zero.

Alternatively, if the sources $\mathbf{J}_1, \mathbf{M}_1, \mathbf{J}_2, \mathbf{M}_2$ are of finite support, we can make the volume infinitely large so as to take S to infinity. Using the fact that $\mathbf{E}$ and $\mathbf{H}$ satisfy the radiation condition and degenerate to plane waves, the surface integrals in (2.69) cancel and hence vanish.

Hence, we conclude that $\text{LHS}(2.65) = \text{LHS}(2.66)$.[2] This also implies that the $\mathbb{L}$ operator defined in (2.61) is a symmetric operator when $\mathbf{J}$ and $\mathbf{M}$ are of finite support and the fields they generate satisfy the radiation condition. Alternatively,

Hence, it follows from (2.65) and (2.66) that

$$\langle \mathbf{E}_2, \mathbf{J}_1 \rangle - \langle \mathbf{H}_2, \mathbf{M}_1 \rangle = \langle \mathbf{E}_1, \mathbf{J}_2 \rangle - \langle \mathbf{H}_1, \mathbf{M}_2 \rangle. \qquad (2.71)$$

The above is known as the reciprocity theorem. As can be seen, it is a consequence of the symmetry (in a mathematical sense) of the electromagnetic equations, under the assumption that $\overline{\mu}$ and $\overline{\epsilon}$ are symmetric tensors (not Hermitian!).

We can physically interpret $\langle \mathbf{E}_i, \mathbf{J}_j \rangle$ as the electric field due to sources $\mathbf{J}_i$ and $\mathbf{M}_i$ "measured" by source $\mathbf{J}_j$, and $-\langle \mathbf{H}_i, \mathbf{M}_j \rangle$ as the magnetic field due to sources $\mathbf{J}_i$ and $\mathbf{M}_i$ "measured" by source $\mathbf{M}_j$. This "measurement" is symmetric according to the reciprocity theorem if the medium is also symmetric or reciprocal.

The proof can be generalized easily to bianisotropic media of finite extent. For this case, the constitutive relation is

$$\begin{bmatrix} \mathbf{D} \\ \mathbf{B} \end{bmatrix} = \begin{bmatrix} \overline{\epsilon} & \overline{\xi} \\ \overline{\zeta} & \overline{\mu} \end{bmatrix} \cdot \begin{bmatrix} \mathbf{E} \\ \mathbf{H} \end{bmatrix}. \qquad (2.72)$$

The corresponding modification to the $\mathbb{L}$ operator is

$$\mathbb{L} = \begin{bmatrix} 0 & \nabla \times \overline{\mathbf{I}} \\ \nabla \times \overline{\mathbf{I}} & 0 \end{bmatrix} + i\omega \begin{bmatrix} \overline{\epsilon} & \overline{\xi} \\ -\overline{\zeta} & -\overline{\mu} \end{bmatrix}. \qquad (2.73)$$

To retain the symmetry of the $\mathbb{L}$, in addition to requiring symmetry in $\overline{\epsilon}$ and $\overline{\mu}$, we require $\overline{\xi} = -\overline{\zeta}^t$. These are also the requirements for reciprocity in bianisotropic media [23].

---

[2] We are associating $[\mathbf{E}_1, \mathbf{H}_1]^t$ with g and $[\mathbf{E}_2, \mathbf{H}_2]$ with f in (2.19) to arrive at this.

2.10.1   Lorentz Reciprocity Theorem

The reciprocity theorem is sometimes written with the surface integrals from (2.69) augmenting the terms in (2.71). In this case, it is commonly known as the Lorentz reciprocity theorem:

$$\langle \mathbf{E}_2, \mathbf{J}_1 \rangle - \langle \mathbf{H}_2, \mathbf{M}_1 \rangle = \langle \mathbf{E}_1, \mathbf{J}_2 \rangle - \langle \mathbf{H}_1, \mathbf{M}_2 \rangle$$
$$+ \oint_S dS\hat{n} \cdot (\mathbf{E}_2 \times \mathbf{H}_1) - \oint_S dS\hat{n} \cdot (\mathbf{E}_1 \times \mathbf{H}_2). \tag{2.74}$$

The implication of the above is that the $\mathbb{L}$ operator is not a symmetric operator when the integration of the inner product is taken over a finite volume, but one can augment the $\mathbb{L}$ operator to make it symmetric when the inner product is defined over a finite volume.

The nonsymmetric part for finite volumes arises from the differential operator, which can be symmetrized. To this end, we define the curl differential operator to be

$$\mathbb{C} = \begin{bmatrix} 0 & \nabla \times \bar{\mathbf{I}} \\ \nabla \times \bar{\mathbf{I}} & 0 \end{bmatrix}. \tag{2.75}$$

The above operator is symmetric if the inner product is defined over an infinite volume, and if the field satisfies the radiation condition at infinity. To make the above operator symmetric even when the inner product is defined over a finite volume, we augment $\mathbb{C}$ with an additional term, defining a new curl differential operator to be

$$\hat{\mathbb{C}} = \mathbb{C} + \mathbb{S}, \tag{2.76}$$

where $\mathbb{S}$ is an operator with the property that a surface integral will be induced when it is sandwiched in the inner product definition. Namely,

$$\left\langle [\mathbf{E}_2, \mathbf{H}_2], \mathbb{S} \begin{bmatrix} \mathbf{E}_1 \\ \mathbf{H}_1 \end{bmatrix} \right\rangle = -\frac{1}{2} \oint_S dS\hat{n} \cdot (\mathbf{E}_2 \times \mathbf{H}_1 - \mathbf{E}_1 \times \mathbf{H}_2). \tag{2.77}$$

The $\mathbb{S}$ operator has the sifting property of the Dirac delta function, but in a three-dimensional space. Hence, it converts a volume integral on the left-hand side to a surface integral. With this definition,

$$\left\langle [\mathbf{E}_2, \mathbf{H}_2], \hat{\mathbb{C}} \begin{bmatrix} \mathbf{E}_1 \\ \mathbf{H}_1 \end{bmatrix} \right\rangle = \langle \mathbf{E}_2, \nabla \times \mathbf{H}_1 \rangle + \langle \mathbf{H}_2, \nabla \times \mathbf{E}_1 \rangle$$
$$-\frac{1}{2} \oint_S dS\hat{n} \cdot (\mathbf{E}_2 \times \mathbf{H}_1 - \mathbf{E}_1 \times \mathbf{H}_2), \tag{2.78}$$

$$\left\langle [\mathbf{E}_1, \mathbf{H}_1], \hat{\mathbb{C}} \begin{bmatrix} \mathbf{E}_2 \\ \mathbf{H}_2 \end{bmatrix} \right\rangle = \langle \mathbf{E}_1, \nabla \times \mathbf{H}_2 \rangle + \langle \mathbf{H}_1, \nabla \times \mathbf{E}_2 \rangle$$
$$-\frac{1}{2} \oint_S dS\hat{n} \cdot (\mathbf{E}_1 \times \mathbf{H}_2 - \mathbf{E}_2 \times \mathbf{H}_1). \tag{2.79}$$

Using (2.69), it can be shown that (2.78) and (2.79) are the same. Hence, $\hat{\mathbb{C}}$ is a symmetric operator even when the integral is taken over a finite volume.

The term $\mathbb{S}[\mathbf{E}, \mathbf{H}]^t$ can be added to both sides of (2.61) to symmetrize the operator on the left-hand side, and the Lorentz reciprocity theorem indicated in (2.74) can be derived.

The $\mathbb{S}$ operator can be thought of as an operator that will produce an equivalence surface current on S when it operates on a field. When these sources are tested with another field, a surface integral is produced. Away from the surface S, this operator has no effect. Physically, one can think that the $\mathbb{S}$ operator places an equivalence current on the surface S so as to generate a field external to S that cancels the original external field there. By the extinction theorem [10], this equivalence surface current does not generate a field inside S. Consequently, one needs only integrate over the finite volume V.

2.11   Energy Conservation Theorem—A New Look

The electromagnetic energy conservation theorem in the time domain is expressed using the Poynting theorem, which was derived by Poynting [24] in 1884 and Heaviside in 1885 [25]. The energy conservation theorem in the Fourier space, making use of complex fields, was probably first derived by Stratton [13].

We can look at the energy conservation theorem from a different perspective using our understanding of vector spaces, operators, and their eigenvalues. We know that Hermitian operators have real eigenvalues, and they are intimately related to the lossless nature of the systems they describe, as is the case with a quantum system. We rewrite electromagnetic equations into a form that produces a Hermitian system when the medium is lossless. To this end, we rewrite (2.60) as

$$\begin{bmatrix} 0 & -i\nabla \times \overline{\mathbf{I}} \\ i\nabla \times \overline{\mathbf{I}} & 0 \end{bmatrix} \cdot \begin{bmatrix} \mathbf{E} \\ \mathbf{H} \end{bmatrix} + \omega \begin{bmatrix} \overline{\boldsymbol{\epsilon}} & 0 \\ 0 & \overline{\boldsymbol{\mu}} \end{bmatrix} \cdot \begin{bmatrix} \mathbf{E} \\ \mathbf{H} \end{bmatrix} = -i \begin{bmatrix} \mathbf{J} \\ \mathbf{M} \end{bmatrix}. \tag{2.80}$$

The above can be compactly rewritten as

$$\mathbb{L} \begin{bmatrix} \mathbf{E} \\ \mathbf{H} \end{bmatrix} = -i \begin{bmatrix} \mathbf{J} \\ \mathbf{M} \end{bmatrix}, \tag{2.81}$$

where

$$\mathbb{L} = \begin{bmatrix} 0 & -i\nabla \times \overline{\mathbf{I}} \\ i\nabla \times \overline{\mathbf{I}} & 0 \end{bmatrix} + \omega \begin{bmatrix} \overline{\boldsymbol{\epsilon}} & 0 \\ 0 & \overline{\boldsymbol{\mu}} \end{bmatrix}. \tag{2.82}$$

When $\overline{\boldsymbol{\epsilon}}$ and $\overline{\boldsymbol{\mu}}$ are Hermitian, it can be easily shown, using the definition for self-adjointness, that $\mathbb{L}$ is Hermitian under appropriate boundary conditions. The appropriate boundary condition is obtained if the above equation is restricted to a resonant modes enclosed by a surface S and the perfect-electric or perfect-magnetic boundary condition is imposed on the wall of the resonant modes. In other words, the resonant modes must have no wall loss.

If the resonant modes is source free, (2.80) reduces to

$$\begin{bmatrix} 0 & -i\nabla \times \overline{\mathbf{I}} \\ i\nabla \times \overline{\mathbf{I}} & 0 \end{bmatrix} \cdot \begin{bmatrix} \mathbf{E} \\ \mathbf{H} \end{bmatrix} = -\omega \begin{bmatrix} \overline{\boldsymbol{\epsilon}} & 0 \\ 0 & \overline{\boldsymbol{\mu}} \end{bmatrix} \cdot \begin{bmatrix} \mathbf{E} \\ \mathbf{H} \end{bmatrix}. \tag{2.83}$$

The above can be thought of as a generalized eigenvalue problem where $-\omega$ is the eigenvalue. If $\overline{\boldsymbol{\epsilon}}$ and $\overline{\boldsymbol{\mu}}$ are Hermitian, and because the differential operator above is also Hermitian for a resonant modes without wall loss, the above equation can only have real eigenvalues. Real eigenvalues correspond to modes whose fields do not decay with time. This implies that the resonant modes of the resonant modes do in fact correspond to those of a lossless cavity. We shall see below that the Hermitian nature of $\overline{\epsilon}$ and $\overline{\mu}$ is in fact necessary for a lossless medium.

Multiplying (2.80) by $[\mathbf{E}^*, \mathbf{H}^*]$, and integrating over a volume V, we have

$$\left\langle [\mathbf{E}^*, \mathbf{H}^*], \mathbb{L} \begin{bmatrix} \mathbf{E} \\ \mathbf{H} \end{bmatrix} \right\rangle = \left\langle [\mathbf{E}^*, \mathbf{H}^*], -i \begin{bmatrix} \mathbf{J} \\ \mathbf{M} \end{bmatrix} \right\rangle = -i \left(\langle \mathbf{E}^*, \mathbf{J} \rangle + \langle \mathbf{H}^*, \mathbf{M} \rangle \right). \qquad (2.84)$$

The above inner product is defined by integrating over a volume V. Writing out the left-hand sides of (2.84) explicitly, we have

$$\text{LHS}(2.84) = -i \langle \mathbf{E}^*, \nabla \times \mathbf{H} \rangle + i \langle \mathbf{H}^*, \nabla \times \mathbf{E} \rangle + \omega \langle \mathbf{E}^*, \overline{\epsilon} \cdot \mathbf{E} \rangle + \omega \langle \mathbf{H}^*, \overline{\mu} \cdot \mathbf{H} \rangle. \qquad (2.85)$$

Using integration by parts, it can be shown that

$$\begin{aligned} -i \langle \mathbf{E}^*, \nabla \times \mathbf{H} \rangle &= -i \langle \nabla \times \mathbf{E}^*, \mathbf{H} \rangle + i \oint_S dS \hat{n} \cdot (\mathbf{E}^* \times \mathbf{H}) \\ &= -\omega \langle \overline{\mu}^* \cdot \mathbf{H}^*, \mathbf{H} \rangle + i \langle \mathbf{M}^*, \mathbf{H} \rangle + i \oint_S dS \hat{n} \cdot (\mathbf{E}^* \times \mathbf{H}), \end{aligned} \qquad (2.86)$$

where the normal $\hat{n}$ points outward from the surface S. Similarly, it can be shown that

$$i \langle \mathbf{H}^*, \nabla \times \mathbf{E} \rangle = -\omega \langle \overline{\epsilon}^* \cdot \mathbf{E}^*, \mathbf{E} \rangle + i \langle \mathbf{J}^*, \mathbf{E} \rangle + i \oint_S dS \hat{n} \cdot (\mathbf{E} \times \mathbf{H}^*). \qquad (2.87)$$

Consequently, using (2.86) and (2.87) in (2.85), we have

$$\begin{aligned} \text{LHS}(2.84) &= 2i \Re e \left[\oint_S dS \hat{n} \cdot (\mathbf{E}^* \times \mathbf{H}) \right] \\ &\quad + 2i\omega \Im m \{ \langle \mathbf{E}^*, \overline{\epsilon} \cdot \mathbf{E} \rangle + \langle \mathbf{H}^*, \overline{\mu} \cdot \mathbf{H} \rangle \} \\ &\quad + i \langle \mathbf{M}^*, \mathbf{H} \rangle + i \langle \mathbf{J}^*, \mathbf{E} \rangle. \end{aligned} \qquad (2.88)$$

Using Eq. (2.88), Eq. (2.84) can be rewritten as

$$\begin{aligned} \Re e \left[\oint_S dS \hat{n} \cdot (\mathbf{E} \times \mathbf{H}^*) \right] &= -\omega \Im m \{ \langle \mathbf{E}^*, \overline{\epsilon} \cdot \mathbf{E} \rangle + \langle \mathbf{H}^*, \overline{\mu} \cdot \mathbf{H} \rangle \} \\ &\quad - \Re e \{ \langle \mathbf{E}, \mathbf{J}^* \rangle + \langle \mathbf{H}, \mathbf{M}^* \rangle \}. \end{aligned} \qquad (2.89)$$

The above is reminiscent of the complex Poynting theorem, except that it makes a statement only on the real part of the complex power $\mathbf{E} \times \mathbf{H}^*$. It makes no statement on the imaginary part of the complex power, which is expressed in complex Poynting theorem. This is because the differential operator in (2.80) is Hermitian, and it will produce a real quantity under the inner product we have used here.

In the above, the real part of a complex power corresponds to time-average power. If the region enclosed by S is source-free, so that $\mathbf{J}$ and $\mathbf{M}$ are zero, and if $\overline{\epsilon}$ and $\overline{\mu}$ are Hermitian, so that $\langle \mathbf{E}^*, \overline{\boldsymbol{\epsilon}} \cdot \mathbf{E} \rangle + \langle \mathbf{H}^*, \overline{\boldsymbol{\mu}} \cdot \mathbf{H} \rangle$ is pure real, then the right-hand side of (2.89) is zero, and then there is no time-average power flowing into S, implying that the region is lossless. Hence, the Hermitian nature of $\overline{\boldsymbol{\epsilon}}$ and $\overline{\boldsymbol{\mu}}$ is necessary for a lossless medium.

For a bianisotropic medium, Eq. (2.82) becomes

$$\mathbb{L} = \begin{bmatrix} 0 & -i\nabla \times \overline{\mathbf{I}} \\ i\nabla \times \overline{\mathbf{I}} & 0 \end{bmatrix} + \omega \begin{bmatrix} \overline{\boldsymbol{\epsilon}} & \overline{\boldsymbol{\xi}} \\ \overline{\boldsymbol{\zeta}} & \overline{\boldsymbol{\mu}} \end{bmatrix}. \tag{2.90}$$

The conditions for a lossless medium are that $\overline{\boldsymbol{\epsilon}}$ and $\overline{\boldsymbol{\mu}}$ are Hermitian and $\overline{\boldsymbol{\xi}} = \overline{\boldsymbol{\zeta}}^\dagger$.

To derive a statement on the imaginary part of the complex power, we replace the differential operator in (2.80) with a skew Hermitian one (we define this to be $\mathbb{A}^\dagger = -\mathbb{A}$). In this case, the electromagnetic equations are written as

$$\begin{bmatrix} 0 & i\nabla \times \overline{\mathbf{I}} \\ i\nabla \times \overline{\mathbf{I}} & 0 \end{bmatrix} \cdot \begin{bmatrix} \mathbf{E} \\ \mathbf{H} \end{bmatrix} - \omega \begin{bmatrix} \overline{\boldsymbol{\epsilon}} & 0 \\ 0 & -\overline{\boldsymbol{\mu}} \end{bmatrix} \cdot \begin{bmatrix} \mathbf{E} \\ \mathbf{H} \end{bmatrix} = i \begin{bmatrix} \mathbf{J} \\ -\mathbf{M} \end{bmatrix}. \tag{2.91}$$

Upon multiplying the above by $[\mathbf{E}^*, \mathbf{H}^*]$, and integrating over a volume V, we have a new equation that becomes

$$i\langle \mathbf{E}^*, \nabla \times \mathbf{H} \rangle + i\langle \mathbf{H}^*, \nabla \times \mathbf{E} \rangle - \omega \langle \mathbf{E}^*, \overline{\boldsymbol{\epsilon}} \cdot \mathbf{E} \rangle + \omega \langle \mathbf{H}^*, \overline{\boldsymbol{\mu}} \cdot \mathbf{H} \rangle = i\langle \mathbf{E}^*, \mathbf{J} \rangle - i\langle \mathbf{H}^*, \mathbf{M} \rangle. \tag{2.92}$$

Upon simplifying the above using (2.86) and (2.87), we have

$$\begin{aligned} \Im m \left[\oint_S dS \hat{n} \cdot (\mathbf{E} \times \mathbf{H}^*) \right] &= \omega \Re e \{ \langle \mathbf{H}^*, \overline{\boldsymbol{\mu}} \cdot \mathbf{H} \rangle - \langle \mathbf{E}^*, \overline{\boldsymbol{\epsilon}} \cdot \mathbf{E} \rangle \} \\ &\quad + \Im m \{ \langle \mathbf{H}, \mathbf{M}^* \rangle - \langle \mathbf{E}, \mathbf{J}^* \rangle \}. \end{aligned} \tag{2.93}$$

The above is the energy conservation theorem regarding the imaginary part of the complex power, which is the reactive power. The reactive power is proportional to the difference between the stored energy in the magnetic field and the electric field, which is indicated by the first term on the right-hand side. The second term on the right-hand side corresponds to the reactive power absorbed or exuded by the sources $\mathbf{M}$ and $\mathbf{J}$. The above derivation can also be generalized to bianisotropic media.

In a time-harmonic sinusoidal system, reactive power is the power that one pumps into a system in one cycle, and then withdraws from it in the next cycle. This happens, for instance, when a capacitor is hooked to a sinusoidal generator source, where in one cycle the capacitor is charged (and hence taking power from the generator), while in another cycle the capacitor is discharged (hence returning power back to the generator).

It is interesting to note that the energy conservation theorem can be derived using our understanding of Hermitian operators. Hermitian operators have the physical meaning of "energy" conserving operators. To derive energy conservation statements, we are also motivated to construct an operator equation whose operator becomes Hermitian in the lossless case.

It is to be noted that in wave physics, a lossless medium does not necessarily imply a Hermitian operator. It is only in the case of a closed cavity with lossless media as well as no wall loss that the resulting system is Hermitian. Hence, in (2.83), an appropriate boundary condition corresponding to no wall loss needs to be imposed before the operator is Hermitian. For instance, if the radiation condition is imposed on the wave at infinity, the resultant operator is non-Hermitian. Hence, when the free-space Green's function, $\exp(ik|\mathbf{r} - \mathbf{r}'|)/|\mathbf{r} - \mathbf{r}'|$), forms the kernel of an integral operator, the operator is symmetric, but not Hermitian. A radiating system, even though it is embedded in a lossless medium of infinite extent, is not lossless, because it leaks energy to infinity.

2.12   Conclusions

Although most works do not relate the reciprocity theorem and energy conservation theorem to the properties of linear operators in linear vector spaces, we show that, in fact, they are intimately related. It will be better to express these theorems from a modern perspective using linear vector space concepts, which are used frequently in computational sciences. The reciprocity theorem is related to the symmetry of the electromagnetic operators, and the lossless nature of a medium is related to the Hermitian nature of the associated electromagnetic operators. Moreover, one can define a skew-Hermitian electromagnetic operator from which the energy conservation of the reactive component of complex power can be derived.

Bibliography

[1] J. A. Kong, *Theory of Electromagnetic Waves,* New York: Wiley-Interscience, 1975.

[2] W. C. Chew, "A new look at reciprocity and energy conservation theorems in electromagnetics," *Research Report No.: CCEML 11-06,* Oct. 19, 2006. also in *IEEE Trans. Antennas Propag.,* vol. 56, no. 4, pp. 970–975, April 2008.

[3] M. C. Reed and B. Simon, *Methods of Modern Mathematical Physics, Vol. III: Scattering Theory,* New York: Academic Press, 1979. Section XI.10.

[4] L. B. Felsen and N. Marcuvitz, *Radiation and Scattering of Waves,* Englewood Cliffs, NJ: Prentice Hall, 1973. Sections 1.4c and 1.5b.

[5] G. Hanson and A. Yakovlev, *Operator Theory for Electromagnetics: An Introduction,* New York: Springer-Verlag, 2002.

[6] R. E. Collin, *Field Theory of Guided Waves,* Second edition, New York: Wiley-IEEE Press, 1990.

[7] L. Cairo and T. Kahan, *Variational Techniques in Electromagnetism,* New York: Gordon and Breach, 1965.

[8] R. Harrington, *Field Computation by Moment Method,* Malabar, FL: Krieger, 1983. (First printing 1968.)

[9] C. H. Chen and C. D. Lien, "The variational formulation for non-self-adjoint electromagnetic problems." *IEEE Trans. Microwave Theory Tech.,* vol. 28, pp. 878–886, 1980.

[10] W. C. Chew *Waves and Fields in Inhomogeneous Media,* New York: Van Nostrand Reinhold, 1990. Reprinted by Piscataway, NJ: IEEE Press, 1995.

[11] V. H. Rumsey, "Reaction concept in electromagnetic theory," *Phys. Rev.,* vol. 94, pp. 1483–1491 (1954), Errata, vol. 95, pp. 1705, 1954.

[12] M. Artin, *Algebra,* Englewood Cliffs, NJ: Prentice Hall, 1991.

[13] J. A. Stratton *Electromagnetic Theory,* New York: McGraw-Hill Book Company Inc., 1941.

[14] I. Stakgold, *Boundary Value Problems of Mathematical Physics*, vol. I, London: Macmillan Co., 1967.

[15] E. Merzbacher, *Quantum Mechanics*, New York: John Wiley & Sons, 1970.

[16] J. J. Sakurai, *Modern Quantum Mechanics*, New York: Wiley-Interscience, 1985.

[17] T. K. Moon and W. C. Stirling, *Mathematical Methods and Algorithms for Signal Processing*, Englewood Cliffs, NJ: Prentice Hall, 2000.

[18] C. T. Tai, *Dyadic Green's Functions in Electromagnetic Theory*, second edition, Piscataway, NJ: IEEE Press, 1993.

[19] H. A. Lorentz, "The theorem of Poynting concerning the energy in the electromagnetic field and two general propositions concerning the propagation of light," *Amsterdammer Akad. Weten.*, vol. 4, pp. 176, 1896.

[20] N. M. Ferrers (Editor), *Mathematical Papers of George Green*, New York: Chelsea Publ., 1997.

[21] C. Altman and K. Suchy, *Reciprocity, Spatial Mapping, and Time Reversal in Electromagnetics*, Boston: Kluwer Academic Pub., 1991.

[22] I. V. Lindell, A. H. Sihvola, and K. Suchy, "Six-vector formalism in electromagnetics of bi-anisotropic media," *J. Electromagn. Waves and Appl.*, vol. 9, no. 7/8, pp. 887–903, 1995.

[23] J. A. Kong, "Theorems of bianisotropic media," *Proc. IEEE*, vol. 60, pp. 1036–1046, 1972.

[24] J. H. Poynting, "On the transfer of energy in the electromagnetic field," *Philos. Trans. R. Soc. London*, vol. 175, Part II, pp. 343–361, 1884.

[25] O. Heaviside, "Electromagnetic induction and its propagation," *The Electrician*, 1885.

CHAPTER 3

Introduction to Integral Equations

3.1 Introduction

This chapter serves to give a pedestrian introduction of integral equations to a newcomer in computational electromagnetics (CEM) who will learn most of surface integral equations from this chapter. The more detailed subjects will be dealt with in later chapters. The derivation of various theorems, principles, and operators will be done in the frequency domain.

In the previous chapters, we consider frequency domain solutions to be the solutions of the electromagnetic equations in the Fourier space. One can also consider frequency domain solutions when the fields or signals are all purely time harmonic. All time-harmonic vector fields can be regarded as

$$\mathbf{F}(\mathbf{r}, t) = \Re e\left[\tilde{\mathbf{F}}(\mathbf{r}, \omega)e^{-i\omega t}\right] = \tilde{\mathbf{F}}_r(\mathbf{r}, \omega)\cos(\omega t) + \tilde{\mathbf{F}}_i(\mathbf{r}, \omega)\sin(\omega t) \tag{3.1}$$

where $\tilde{\mathbf{F}}(\mathbf{r}, \omega) = \tilde{\mathbf{F}}_r(\mathbf{r}, \omega) + i\tilde{\mathbf{F}}_i(\mathbf{r}, \omega)$ is a complex field which is independent of time. The complex function $\tilde{\mathbf{F}}(\mathbf{r}, \omega)$ is also known as the phasor representation of the time harmonic field $\mathbf{F}(\mathbf{r}, t)$. It is related to the Fourier transform of the time harmonic field. By substituting the above into the electromagnetic equations, they can be simplified just as Fourier transform is used to simplify these equations as was done in the previous chapters.

3.2 The Dyadic Green's Function

Before we can derive integral equations for vector electromagnetic field, it is most convenient to express these integral equations in terms of the dyadic Green's function [1–4]. The dyadic Green's function is the point source response to the vector wave equation,

$$\nabla \times \nabla \times \mathbf{E}(\mathbf{r}) - k^2\, \mathbf{E}(\mathbf{r}) = i\omega\mu\mathbf{J}(\mathbf{r}). \tag{3.2}$$

derivable from the electromagnetic equations for a homogeneous medium where $k^2 = \omega^2 \mu \epsilon$ is a constant.

By letting

$$\mathbf{E}(\mathbf{r}) = i\omega\mu \int_V d\mathbf{r}' \, \mathbf{J}(\mathbf{r}') \cdot \overline{\mathbf{G}}(\mathbf{r}', \mathbf{r}), \qquad (3.3)$$

where $\overline{\mathbf{G}}(\mathbf{r}', \mathbf{r})$ is the dyadic Green's function, and using the above in (3.2), assuming that the current $\mathbf{J}(\mathbf{r}')$ is arbitrary, one can show that the homogeneous medium dyadic Green's function is the solution to the equation

$$\nabla \times \nabla \times \overline{\mathbf{G}}(\mathbf{r}, \mathbf{r}') - k^2 \, \overline{\mathbf{G}}(\mathbf{r}, \mathbf{r}') = \overline{\mathbf{I}} \, \delta(\mathbf{r} - \mathbf{r}'). \qquad (3.4)$$

Its solution is given by

$$\overline{\mathbf{G}}(\mathbf{r}, \mathbf{r}') = \left(\overline{\mathbf{I}} + \frac{\nabla\nabla}{k^2} \right) \frac{e^{ik|\mathbf{r}-\mathbf{r}'|}}{4\pi |\mathbf{r} - \mathbf{r}'|}. \qquad (3.5)$$

Physically, a dyadic Green's function, when operating on an electric current, yields the electric field produced by the current. By the principle of linear superposition, it can be used to calculate the field due to a distributed current source via (3.3)   As a consequence of the reciprocity theorem, one can show that [7]

$$\overline{\mathbf{G}}^t(\mathbf{r}', \mathbf{r}) = \overline{\mathbf{G}}(\mathbf{r}, \mathbf{r}'). \qquad (3.6)$$

Then, taking its transpose, (3.3) becomes

$$\mathbf{E}(\mathbf{r}) = i\omega\mu \int_V d\mathbf{r}' \, \overline{\mathbf{G}}(\mathbf{r}, \mathbf{r}') \cdot \mathbf{J}(\mathbf{r}'). \qquad (3.7)$$

In the above, $\mathbf{r}$ is referred to as the field point (or observation point), while $\mathbf{r}'$ is the source point.

Because this dyadic Green's function generates electric field from electric current, it is also known as the electric dyadic Green's function. Because of the double derivative on the scalar Green's function, the dyadic Green's function is very singular or hypersingular, and hence, its numerical evaluation via integration often poses a problem when the field point is close to the source point. However, in CEM, the practice has been to avoid the evaluation of the dyadic Green's function in its above form. The singularity is reduced by various mathematical tricks such as integration by parts (see Chapter 2).

3.3   Equivalence Principle and Extinction Theorem

The building blocks of surface integral equations, which appear in different forms, are the equivalence principle and the extinction theorem. The equivalence principle is similar to Huygens' principle, but its mathematical form is not attributable Huygens. The mathematical form for scalar wave was derived by Kirchhoff and Green [5]. The modern vector electromagnetic form was contributed by Stratton and Chu [6], and also Franz [8]. To derive

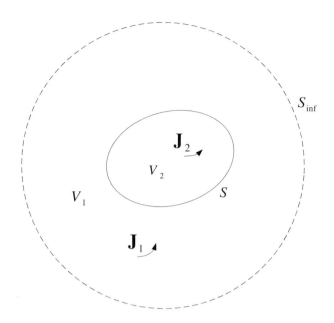

Figure 3.1: Derivation of equivalence principle and extinction theorem.

the equivalence principle and extinction theorem, we start with the vector wave equation (3.2) where k^2 is assumed to be constant everywhere. Figure 3.1 illustrates the setup for deriving the equivalence principle and the extinction theorem. Here, (3.2), $\mathbf{J}(\mathbf{r})$ represents the currents $\mathbf{J}_1(\mathbf{r})$ and $\mathbf{J}_2(\mathbf{r})$, where $\mathbf{J}_1(\mathbf{r})$ is in V_1 and $\mathbf{J}_2(\mathbf{r})$ is in V_2. Here, V_1 is bounded by surface S and S_{inf} while V_2 is bounded by S.

Premultiplying (3.4) by $\mathbf{E}(\mathbf{r})$ and postmultiplying (3.2) by $\overline{\mathbf{G}}(\mathbf{r}, \mathbf{r}')$, and then subtracting the resultant equations, we have

$$\mathbf{E}(\mathbf{r}) \cdot \nabla \times \nabla \times \overline{\mathbf{G}}(\mathbf{r}, \mathbf{r}') - \nabla \times \nabla \times \mathbf{E}(\mathbf{r}) \cdot \overline{\mathbf{G}}(\mathbf{r}, \mathbf{r}') = \delta(\mathbf{r} - \mathbf{r}')\mathbf{E}(\mathbf{r})$$
$$- i\omega\mu\mathbf{J}(\mathbf{r}) \cdot \overline{\mathbf{G}}(\mathbf{r}, \mathbf{r}') \tag{3.8}$$

and integrating the difference over a volume V_1, we have

$$\mathbf{E}(\mathbf{r}') - \mathbf{E}_1(\mathbf{r}')$$
$$= \int_{V_1} dV \left[\mathbf{E}(\mathbf{r}) \cdot \nabla \times \nabla \times \overline{\mathbf{G}}(\mathbf{r}, \mathbf{r}') - \nabla \times \nabla \times \mathbf{E}(\mathbf{r}) \cdot \overline{\mathbf{G}}(\mathbf{r}, \mathbf{r}') \right].$$
$$\tag{3.9}$$

where

$$\mathbf{E}(\mathbf{r}') = \int_{V1} d\mathbf{r}\, \delta(\mathbf{r} - \mathbf{r}')\mathbf{E}(\mathbf{r}), \tag{3.10}$$

$$\mathbf{E}_1(\mathbf{r}') = i\omega\mu \int_{V_1} d\mathbf{r}\, \mathbf{J}_1(\mathbf{r}) \cdot \overline{\mathbf{G}}(\mathbf{r}, \mathbf{r}') \tag{3.11}$$

The above is field produced by $\mathbf{J}_1$ in V_1. Because $\mathbf{J}_2$ is not in V_1, it does not contribute to the above volume integral.

Using the identity that

$$-\nabla \cdot \left[\mathbf{E}(\mathbf{r}) \times \nabla \times \overline{\mathbf{G}}(\mathbf{r}, \mathbf{r}') + \nabla \times \mathbf{E}(\mathbf{r}) \times \overline{\mathbf{G}}(\mathbf{r}, \mathbf{r}')\right]$$
$$= \mathbf{E}(\mathbf{r}) \cdot \nabla \times \nabla \times \overline{\mathbf{G}}(\mathbf{r}, \mathbf{r}') - \nabla \times \nabla \times \mathbf{E}(\mathbf{r}) \cdot \overline{\mathbf{G}}(\mathbf{r}, \mathbf{r}'), \tag{3.12}$$

Eq. (3.9), with the help of Gauss' divergence theorem, can be rewritten as

$$\mathbf{E}(\mathbf{r}') - \mathbf{E}_1(\mathbf{r}')$$
$$= \oint_{S+S_{\text{inf}}} dS\, \hat{n} \cdot \left[\mathbf{E}(\mathbf{r}) \times \nabla \times \overline{\mathbf{G}}(\mathbf{r}, \mathbf{r}') + \nabla \times \mathbf{E}(\mathbf{r}) \times \overline{\mathbf{G}}(\mathbf{r}, \mathbf{r}')\right]$$
$$= \oint_{S+S_{\text{inf}}} dS\, \left[\hat{n} \times \mathbf{E}(\mathbf{r}) \cdot \nabla \times \overline{\mathbf{G}}(\mathbf{r}, \mathbf{r}') + i\omega\mu\, \hat{n} \times \mathbf{H}(\mathbf{r}) \cdot \overline{\mathbf{G}}(\mathbf{r}, \mathbf{r}')\right].$$
$$\tag{3.13}$$

where the normal $\hat{n}$ is a normal on the surface S that points outward. The term $\mathbf{E}(\mathbf{r}')$ is zero if $\mathbf{r}' \notin V_1$ because of the sifting property of delta function (one can appreciate this only by going through the detail of the derivation that arrives at the above). Hence, the above equation can be written as

$$\left.\begin{array}{cc} \mathbf{r}' \in V_1, & \mathbf{E}(\mathbf{r}') \\ \mathbf{r}' \notin V_1, & 0 \end{array}\right\} = \mathbf{E}_1(\mathbf{r}') + \oint_{S+S_{\text{inf}}} dS\, \left[\hat{n} \times \mathbf{E}(\mathbf{r}) \cdot \nabla \times \overline{\mathbf{G}}(\mathbf{r}, \mathbf{r}')\right.$$
$$\left. + i\omega\mu\, \hat{n} \times \mathbf{H}(\mathbf{r}) \cdot \overline{\mathbf{G}}(\mathbf{r}, \mathbf{r}')\right]. \tag{3.14}$$

By making use of the far field or radiation condition, the integral over S_{inf} is shown to vanish.

It can be shown that a finite source will generate a field that decays as $1/r$ as $r \to \infty$, where r is the source point to field point separation. Hence, $\mathbf{E}$, $\mathbf{H}$, $\overline{\mathbf{G}}$, and $\nabla \times \overline{\mathbf{G}}$ individually will decay as $1/r$. Therefore, each of the terms on the right-hand side of (3.14) decays as $1/r^2$. But this reason alone is not sufficient to ignore the contribution from the integral over S_{inf}, as the surface area of S_{inf} grows as r^2 when $r \to \infty$. However, it can be shown that the two terms on the right-hand side of (3.14) cancel each other to leading order, causing the total integrand to decay as $1/r^3$. This is the correct reason that the integral over S_{inf} vanishes when $r \to \infty$.

Then by swapping $\mathbf{r}$ and $\mathbf{r}'$, the above finally becomes

$$\left.\begin{array}{cc} \mathbf{r} \in V_1, & \mathbf{E}(\mathbf{r}) \\ \mathbf{r} \in V_2, & 0 \end{array}\right\} = \mathbf{E}_1(\mathbf{r}) + \oint_S dS'\, \left[\hat{n}' \times \mathbf{E}(\mathbf{r}') \cdot \nabla' \times \overline{\mathbf{G}}(\mathbf{r}', \mathbf{r})\right.$$
$$\left. + i\omega\mu\, \hat{n}' \times \mathbf{H}(\mathbf{r}') \cdot \overline{\mathbf{G}}(\mathbf{r}', \mathbf{r})\right]. \tag{3.15}$$

Taking the transpose of the above equation, and making use of the following (which can be proven by using the reciprocity theorem [3, 7])

$$\nabla \times \overline{\mathbf{G}}(\mathbf{r}, \mathbf{r}') = \left[\nabla' \times \overline{\mathbf{G}}(\mathbf{r}', \mathbf{r})\right]^t, \quad \overline{\mathbf{G}}(\mathbf{r}, \mathbf{r}') = \left[\overline{\mathbf{G}}(\mathbf{r}', \mathbf{r})\right]^t \tag{3.16}$$

we have

$$\left.\begin{array}{cc} \mathbf{r} \in V_1, & \mathbf{E}(\mathbf{r}) \\ \mathbf{r} \in V_2, & 0 \end{array}\right\} = \mathbf{E}_1(\mathbf{r}) + \oint_S dS' \left[\nabla \times \overline{\mathbf{G}}(\mathbf{r}, \mathbf{r}') \cdot \hat{n}' \times \mathbf{E}(\mathbf{r}')\right.$$
$$\left. + i\omega\mu \,\overline{\mathbf{G}}(\mathbf{r}, \mathbf{r}') \cdot \hat{n}' \times \mathbf{H}(\mathbf{r}')\right]. \tag{3.17}$$

In the above, we can let $\mathbf{M}_s(\mathbf{r}') = -\hat{n}' \times \mathbf{E}(\mathbf{r}')$ and $\mathbf{J}_s(\mathbf{r}') = \hat{n}' \times \mathbf{H}(\mathbf{r}')$, which can be thought of as equivalence surface magnetic and electric currents respectively impressed on the surface S, and the above then becomes

$$\left.\begin{array}{cc} \mathbf{r} \in V_1, & \mathbf{E}(\mathbf{r}) \\ \mathbf{r} \in V_2, & 0 \end{array}\right\} = \mathbf{E}_1(\mathbf{r}) - \oint_S dS' \left[\nabla \times \overline{\mathbf{G}}(\mathbf{r}, \mathbf{r}') \cdot \mathbf{M}_s(\mathbf{r}')\right.$$
$$\left. - i\omega\mu \,\overline{\mathbf{G}}(\mathbf{r}, \mathbf{r}') \cdot \mathbf{J}_s(\mathbf{r}')\right]. \tag{3.18}$$

The physical meaning of the above is as follows: The total field $\mathbf{E}$ in V_1 is generated by the source $\mathbf{J}_1$ inside V_1 as well as source outside V_1 such as $\mathbf{J}_2$ in our example, and it can be decomposed into the field $\mathbf{E}_1$ on the right-hand side which is generated by $\mathbf{J}_1$, plus field generated by equivalence surface currents $\mathbf{M}_s$ and $\mathbf{J}_s$ impressed on the surface S. The equivalence surface currents generate the same field produced by $\mathbf{J}_2$ or any source in V_2 enclosed by S.

Moreover, the total field is zero if the field point is placed in V_2: this is reflected by the lower part of the left-hand side of (3.18). This implies that the equivalence surface currents produce a field that exactly cancels $\mathbf{E}_1$ in V_2. This is known as the *extinction theorem.*

It may be instructive to go through different cases to see how the extinction theorem manifests itself.

Case (a): $\mathbf{J}_1 = 0$

In this case, $\mathbf{E}_1 = 0$ and $\mathbf{E}$ is entirely due to $\mathbf{J}_2$, and hence, $\mathbf{E} = \mathbf{E}_2$. Consequently, (3.18) becomes

$$\left.\begin{array}{cc} \mathbf{r} \in V_1, & \mathbf{E}_2(\mathbf{r}) \\ \mathbf{r} \in V_2, & 0 \end{array}\right\} = - \oint_S dS' \left[\nabla \times \overline{\mathbf{G}}(\mathbf{r}, \mathbf{r}') \cdot \mathbf{M}_{2s}(\mathbf{r}')\right.$$
$$\left. - i\omega\mu \,\overline{\mathbf{G}}(\mathbf{r}, \mathbf{r}') \cdot \mathbf{J}_{2s}(\mathbf{r}')\right]. \tag{3.19}$$

From the above, one notices that the equivalence surface currents produce the correct field outside the volume V_2, but produces zero field inside V_2. In other words, if we think of the equivalence surface currents as propagating information away from the source $\mathbf{J}_2$, it correctly does so for field points outside V_2, but extincts the field for points inside V_2 that contains the source $\mathbf{J}_2$.

Case (b): $\mathbf{J}_2 = 0$

In this case, $\mathbf{E} = \mathbf{E}_1$, and (3.18) after some rearrangement becomes

$$\left.\begin{array}{ll} \mathbf{r} \in V_1, & 0 \\ \mathbf{r} \in V_2, & \mathbf{E}_1(\mathbf{r}) \end{array}\right\} = \oint_S dS' \left[\nabla \times \overline{\mathbf{G}}(\mathbf{r}, \mathbf{r}') \cdot \mathbf{M}_{1s}(\mathbf{r}') \right.$$
$$\left. - i\omega\mu\, \overline{\mathbf{G}}(\mathbf{r}, \mathbf{r}') \cdot \mathbf{J}_{1s}(\mathbf{r}') \right]. \qquad (3.20)$$

From the above, we see that the equivalence currents generate no field inside V_1, the region that contains the source $\mathbf{J}_1$, whereas it generates the correct field (save for a minus sign) in region V_2 that does not contain the source. The minus sign can be absorbed by redefining the surface normals. From the above, one notes that the physical interpretation of (3.20) is actually no different from that of (3.19).

As a final note, in deriving Eq. (3.9), if we integrate over V_2 instead, a similar equation for the above for V_2 can be derived. By invoking duality, similar equations for the magnetic field can be derived. The above is the mathematical statement of equivalence principles in electromagnetics as espoused by Harrington [9].

3.4   Electric Field Integral Equation—A Simple Physical Description

To simply see how an integral equation can arise physically, let us consider a scattering problem involving a metallic scatterer [perfect electric conductor (PEC)]. Consider an incident electric field from an electromagnetic wave that impinges on the scatterer. The first phenomenon that happens is that electric current will be induced on the scatterer. The current produces an electric field that exactly cancels the incident electric field inside the scatterer, as no electric field should exist inside.

In addition to canceling the incident electric field, the current also generates an electromagnetic field outside the scatterer yielding a scattered field. The currents have to cooperate with each other to produce a field that cancels the incident field inside the scatterer. To cooperate, the currents communicate with each other via the Green's function. Their cooperation with each other can be mathematically described via the integral equation

$$-\mathbf{E}_{inc}(\mathbf{r}) = i\omega\mu \int_S \overline{\mathbf{G}}(\mathbf{r}, \mathbf{r}') \cdot \mathbf{J}_s(\mathbf{r}') d\mathbf{r}', \quad \mathbf{r} \in S^- \qquad (3.21)$$

where $\overline{\mathbf{G}}(\mathbf{r}, \mathbf{r}')$ is the dyadic Green's function defined in (3.5) that produces an electric field due to a point electric current source, $\mathbf{J}_s(\mathbf{r}')$ is the electric surface current on the surface of the scatterer, $\mathbf{E}_{inc}(\mathbf{r})$ is the incident electric field, and S^- represents the surface of the scatterer that is just inside the surface S. In the above, the unknown is the surface current $\mathbf{J}_s(\mathbf{r}')$ on the surface of the scatterer.

As $\hat{n} \times \mathbf{E}$ is zero on the surface of the PEC, $\mathbf{M}_s = 0$. By assuming that $\mathbf{E}_1 = \mathbf{E}_{inc}$, we arrive at the above integral equation. The above equation has three vector components, but $\mathbf{J}_s$, a surface current, has only two vector components. Hence, a vector equation with two components is sufficient to solve for $\mathbf{J}_s$. Consequently, we can reduce (3.21) to have two vector components by taking its tangential components, namely,

$$-\hat{n} \times \mathbf{E}_{inc}(\mathbf{r}) = i\omega\mu\hat{n} \times \int_S \overline{\mathbf{G}}(\mathbf{r},\mathbf{r}') \cdot \mathbf{J}_s(\mathbf{r}')d\mathbf{r}', \quad \mathbf{r} \in S \qquad (3.22)$$

As tangential $\mathbf{E}$ is continuous across a current sheet, the above integral equation need only be applied on S. This integral equation is also known as the electric field integral equation (EFIE) [10–14].

3.4.1 EFIE—A Formal Derivation

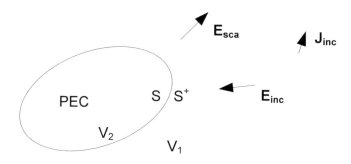

Figure 3.2: Scattering by a PEC (perfect electric conductor) object.

The above integral equation can also be derived more rigorously by invoking the extinction theorem in the lower part of Eq. (3.15). Figure 3.2 shows the scattering by a PEC. We assume that $\mathbf{J}_{inc}$ is a current that generates an incident field $\mathbf{E}_{inc}$ that impinges on a PEC object. Electric current will be induced on the PEC object, which in turn radiates to generate the scattered field. We can imagine a surface S^+ that is just large enough to contain S, and apply the equivalence principle and extinction theorem to it.

When equivalence currents are placed on S^+, they produce the total field outside S^+ and zero field inside S^-. Hence, the PEC object can be removed without disturbing the field outside. The equivalence currents are given by

$$\mathbf{J}_{eq} = \hat{n} \times \mathbf{H}, \quad \mathbf{M}_{eq} = \mathbf{E} \times \hat{n} \quad \text{on} \quad S^+ \qquad (3.23)$$

where $\mathbf{E}$ and $\mathbf{H}$ represent total field. But as $\mathbf{E} \times \hat{n} = 0$ on S^+, only $\mathbf{J}_{eq}$ is needed to represent the field outside. Hence, the equivalence principle and extinction theorem becomes

$$\left.\begin{array}{ll}\mathbf{r} \in V_1, & \mathbf{E}(\mathbf{r}) \\ \mathbf{r} \in V_2, & 0\end{array}\right\} = \mathbf{E}_{inc}(\mathbf{r}) + i\omega\mu\int_{S^+} dS'\overline{\mathbf{G}}(\mathbf{r},\mathbf{r}') \cdot \mathbf{J}_{eq}(\mathbf{r}') \qquad (3.24)$$

When we apply the extinction theorem part of the above equation to S, we obtain the integral equation of scattering, viz.,

$$\mathbf{r} \in S, \quad 0 = \mathbf{E}_{inc}(\mathbf{r}) + i\omega\mu\int_{S^+} dS'\overline{\mathbf{G}}(\mathbf{r},\mathbf{r}') \cdot \mathbf{J}_{eq}(\mathbf{r}') \qquad (3.25)$$

The magnetic field version of (3.24) can be derived by taking the curl of (3.24), namely,

$$\left.\begin{array}{ll} \mathbf{r} \in V_1, & \mathbf{H}(\mathbf{r}) \\ \mathbf{r} \in V_2, & 0 \end{array}\right\} = \mathbf{H}_{inc}(\mathbf{r}) + \nabla \times \int_{S+} dS' \overline{\mathbf{G}}(\mathbf{r}, \mathbf{r}') \cdot \mathbf{J}_{eq}(\mathbf{r}') \tag{3.26}$$

The corresponding integral equation of scattering is

$$\mathbf{r} \in S, \quad 0 = \mathbf{H}_{inc}(\mathbf{r}) + \nabla \times \int_{s+} dS' \overline{\mathbf{G}}(\mathbf{r}, \mathbf{r}') \cdot \mathbf{J}_{eq}(\mathbf{r}') \tag{3.27}$$

In the above, $\mathbf{J}_{eq}$ is practically the same as $\mathbf{J}_s$ because S^+ and S are infinitesimally close together, and the tangential $\mathbf{H}$ on both these surfaces are practically equal to each other.

For a resonant modes whose surface is S, the requirement that $\hat{n} \times \mathbf{E} = 0$ on the cavity wall will ensure null solution inside the resonant modes except at the cavity resonance. Hence, away from cavity resonances, it is sufficient to include the tangential component of (3.25) and (3.27) to get

$$\mathbf{r} \in S, \quad 0 = \hat{n} \times \mathbf{E}_{inc}(\mathbf{r}) + i\omega\mu\hat{n} \times \int_S dS' \overline{\mathbf{G}}(\mathbf{r}, \mathbf{r}') \cdot \mathbf{J}_{eq}(\mathbf{r}') \tag{3.28}$$

$$\mathbf{r} \in S, \quad 0 = \hat{n} \times \mathbf{H}_{inc}(\mathbf{r}) + \hat{n} \times \nabla \times \int_{S+} dS' \overline{\mathbf{G}}(\mathbf{r}, \mathbf{r}') \cdot \mathbf{J}_{eq}(\mathbf{r}') \tag{3.29}$$

Eq. (3.28) is known as the electric field integral equation (EFIE) for a PEC, whereas Eq. (3.29) is known as the magnetic field integral equation (MFIE) for a PEC. They break down at the resonant frequency of the resonant modes enclosed by the PEC surface S, because at the resonance frequency, the extinction of the tangential components of the $\mathbf{E}$ and $\mathbf{H}$ fields does not guarantee the extinction of the field inside the scatterer. This point will be further elaborated in Section 3.8.

A distinction between EFIE and MFIE is that the current on the surface of the PEC gives rise to a discontinuous $\mathbf{H}$ field across the surface, but a continuous $\mathbf{E}$ field across the same surface. Hence, the EFIE (3.28) holds true irrespective of whether we impose $\mathbf{r}$ to be on S, S^+, or S^-. However, for the MFIE, (3.29) holds true because of the extinction theorem, and it is important that we impose $\mathbf{r}$ on S such that S is contained in S^+.

In the literature, it is a common practice to define the $\mathcal{L}$ operator such that

$$\mathbf{E}(\mathbf{r}) = \mathcal{L}(\mathbf{r}, \mathbf{r}') \cdot \mathbf{J}(\mathbf{r}') = i\omega\mu \int \overline{\mathbf{G}}(\mathbf{r}, \mathbf{r}') \cdot \mathbf{J}(\mathbf{r}') d\mathbf{r}' \tag{3.30}$$

and $\mathcal{K}$ operator such that

$$\mathbf{H}(\mathbf{r}) = \mathcal{K}(\mathbf{r}, \mathbf{r}') \cdot \mathbf{J}(\mathbf{r}') = \nabla \times \int \overline{\mathbf{G}}(\mathbf{r}, \mathbf{r}') \cdot \mathbf{J}(\mathbf{r}') d\mathbf{r}' \tag{3.31}$$

$$= \int \nabla g(\mathbf{r}, \mathbf{r}') \times \mathbf{J}(\mathbf{r}') d\mathbf{r}'$$

In the above, one uses $\nabla \times \nabla g(\mathbf{r}, \mathbf{r}') = 0$ to simplify the $\mathcal{K}$ operator, and integration is implied over repeated variable $\mathbf{r}'$.

3.5 Understanding the Method of Moments—A Simple Example

Method of moments (MOM) [10] is a way of solving an integral equation by converting it into a matrix equation. In other words, it is a way of finding the matrix representation of an integral operator (see Chapter 2), which can be quite complex. In the previous chapter, we have discussed this in an abstract manner, but in this section, we will give a concrete example as an illustration. We will illustrate MOM with a simple PEC scatterer.

MOM finds the matrix representation of an integral operator by first expanding and approximating the unknown current in terms of a finite set of expansion functions:

$$\mathbf{J}(\mathbf{r}) = \sum_{n=1}^{N} I_n \mathbf{J}_n(\mathbf{r}) \tag{3.32}$$

In the above, $\mathbf{J}_n(\mathbf{r})$ is a known as expansion function or a basis function, while I_n's are unknowns yet to be sought. So the search for the unknown current $\mathbf{J}_s$ has been transferred to the search for the finite set of unknown numbers denoted by I_n's.

Substituting the above into the integral equation, we have

$$-\hat{n} \times \mathbf{E}_{inc}(\mathbf{r}) = i\omega\mu \sum_{n=1}^{N} I_n \int_s \hat{n} \times \overline{\mathbf{G}}(\mathbf{r}, \mathbf{r}') \cdot \mathbf{J}_n(\mathbf{r}') d\mathbf{r}', \quad \mathbf{r} \in S \tag{3.33}$$

The above is not quite a matrix equation yet. It is an equation that depends on $\mathbf{r}$. To remove the $\mathbf{r}$ dependence, we dot multiply the above by $-\hat{n} \times \mathbf{J}_m(\mathbf{r})$, $m = 1, \cdots, N$, and integrate the resultant dot product over the surface of the scatterer. By so doing, we have, after using that $\hat{n} \times \mathbf{J}_m(\mathbf{r}) \cdot \hat{n} \times \mathbf{E}_{inc}(\mathbf{r}) = -\mathbf{J}_m(\mathbf{r}) \cdot \mathbf{E}_{inc}(\mathbf{r})$

$$-\int_s \mathbf{J}_m(\mathbf{r}) \cdot \mathbf{E}_{inc}(\mathbf{r}) d\mathbf{r} = i\omega\mu \sum_{n=1}^{N} I_n \int_s d\mathbf{r} \mathbf{J}_m(\mathbf{r}) \cdot \int_s \overline{\mathbf{G}}(\mathbf{r}, \mathbf{r}') \cdot \mathbf{J}_n(\mathbf{r}') d\mathbf{r}',$$
$$m = 1, ..., N \tag{3.34}$$

Now the left-hand side is only dependent on m and the double integral on the right-hand side is only dependent on m, n. The above procedure is known as testing the equation with testing functions. Thus the above now is a set of linear algebraic equation that can be written more concisely as

$$V_m = \sum_{n=1}^{N} Z_{mn} I_n, \quad m = 1, ..., N \tag{3.35}$$

or more compactly in vector notation

$$\mathbf{V} = \overline{\mathbf{Z}} \cdot \mathbf{I} \tag{3.36}$$

where $\mathbf{V}$ is a column vector that contains V_n as its elements, $\overline{\mathbf{Z}}$ is a matrix that contains Z_{mn} as its elements, and $\mathbf{I}$ is a column vector that contains I_n. More explicitly,

$$Z_{mn} = i\omega\mu \langle \mathbf{J}_m, \overline{\mathbf{G}}, \mathbf{J}_n \rangle \tag{3.37}$$

$$V_m = - \langle \mathbf{J}_m, \mathbf{E}_{inc} \rangle \tag{3.38}$$

where we have used the short hand

$$\langle \mathbf{J}_m, \overline{\mathbf{G}}, \mathbf{J}_n \rangle = \int_s d\mathbf{r} \mathbf{J}_m(\mathbf{r}) \cdot \int_s \overline{\mathbf{G}}(\mathbf{r}, \mathbf{r}') \cdot \mathbf{J}_n(\mathbf{r}') d\mathbf{r}' \tag{3.39}$$

$$\langle \mathbf{J}_m, \mathbf{E}_{inc} \rangle = \int_s \mathbf{J}_m(\mathbf{r}) \cdot \mathbf{E}_{inc}(\mathbf{r}) d\mathbf{r} \tag{3.40}$$

The above inner product is the reaction inner product discussed in Chapter 2.

The above method for converting an integral equation into a matrix equation is alternatively known as the Galerkin's method. In the above derivation, the testing functions need not be from the same basis set as the expansion functions. However, if the operator is symmetric, which is true of the above, choosing the testing functions to be the same as the expansion functions, assuming the reaction inner product, gives rise to a symmetric matrix system, reducing the storage requirements of the system.

3.6   Choice of Expansion Function

In general, the testing function in MOM can be different from the expansion function. Because in the above, we have chosen them to be the same, we will talk about the choice of expansion function here. One popular choice of expansion function is the Rao-Wilton-Glisson (RWG) function [15], which is the workhorse of most MOM computations. To use this expansion function, the surface of a scatterer has to be discretized into small elements each of which is a triangle.

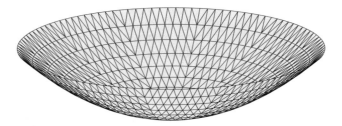

Figure 3.3: A geometrical discretization of a parabola with triangular patches.

Figure 3.3 shows a geometrical discretization (also called tessellation, meshing, gridding) of an object with triangular patches. Triangular patches are known as simplices in a two dimensional manifold, and all two-dimensional manifolds can be tessellated with triangles [16]. Hence, they are very versatile in describing surfaces of geometrical objects. For metallic scatterers, only surface currents are induced on the scatterer, and hence, we need to only mesh the surface of the scatterer.

After the object is meshed, the RWG function can be described. The RWG function straddles two adjacent triangles, and hence, its description requires two contiguous triangles as shown in the Figure 3.4. The expression for the expansion function describing the current on the triangle is

$$\mathbf{\Lambda}_n^s(\mathbf{r}) = \begin{cases} \frac{\ell_n}{2A_n^{\pm}} \boldsymbol{\rho}_n^{\pm}(\mathbf{r}), & \mathbf{r} \in T_n^{\pm} \\ 0, & \text{otherwise} \end{cases} \tag{3.41}$$

where $\pm$ is used to denote the respective triangles, ℓ_n is the edge length of the contiguous edge between the two triangles, $A_n^{\pm}$ is the area of the respective triangles, and $\pm\boldsymbol{\rho}_n^{\pm}(\mathbf{r})$ is the vector from the point $\mathbf{r}$ to the apex of the respective triangles, and $T_n^{\pm}$ is the support of the respective triangles.

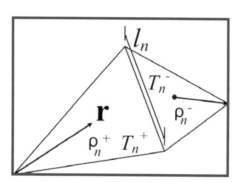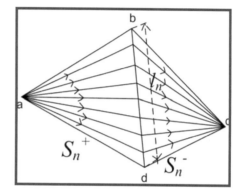

Figure 3.4: The geometry description and the current flowing on a Rao-Wilton-Glisson (RWG) function.

As the current function $\mathbf{J}_n(\mathbf{r})$ is a vector function, the RWG function is also a vector function. A current that flows from one triangle to another triangle via the contiguous edge represents the RWG function. Hence, there is one RWG function for every common edge of two contiguous triangles. Consequently, only inner edges of a surface require an RWG function.

One thing is clear—for the RWG function to reside on the surface of the object, the object must be meshed with good triangles. The triangle mesh shown in Figure 3.5 is considered a bad mesh, because more than two triangles are sharing an edge, and hence, an RWG function cannot be described on such a mesh.

The RWG function is vastly popular for a simple reason: an arbitrary surface can be approximated fairly well with a triangulated surface. This is because a triangle is a simplex for a 2D manifold (while a line segment is a simplex for a line, and a tetrahedron is a simplex for a volume). If a surface is not approximated well, it can be further improved by mesh refinement. This is not the case for a surface that is described by a quad patch, where frustration can occur. Another reason for its popularity is its simplicity: it is of the lowest order, is divergence conforming function[1] and is defined on a pair of triangles. Its divergence conforming property makes its representation of the current physical. One may argue that a

---

[1]A divergence conforming function is one whose divergence is finite—more of it in the next Chapter.

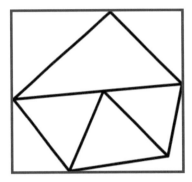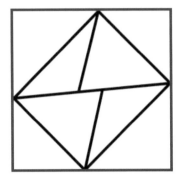

Figure 3.5: Defective meshes happen when an RWG function cannot be defined on such meshes.

rooftop function defined over two quad patches has similar properties, but it does not have the versatility of triangular patches.

If one views the center of a triangular patch as a node in a graph, then the RWG functions allow the flow of current from one node to another, and each node is connected to three other nodes in the graph (see Figure 3.6). This graph forms a circuit network on which current can flow. Hence, RWG functions replace a current on a surface approximately by a circuit network where each node is connected to every other nodes (except for the boundary nodes). Hence, a smooth-flowing current is replaced by a somewhat tortuous current flow which is not in a smooth line. Because of the nature of electromagnetic physics, when the scale of this "circuitous" flow is happening at a lengthscale much smaller than wavelength, this flow can still capture the physics of the current flow. The graininess of this current flow is due to the low-order nature of the RWG basis function. One minimizes the effect of this graininess by making the patch size smaller, or alternatively, one can go to higher-order basis, albeit with more work.

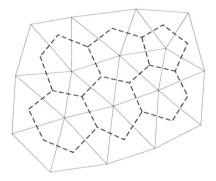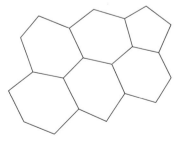

Figure 3.6: The current in a mesh flows from triangle center to triangle center (left). The centers of the triangles can be connected to form a network on which the current flows (right).

3.7   Closed Surface versus Open Surface

When an object is described by a surface, there are two classes of surfaces: closed surfaces, or open surfaces. A closed surface is a surface that encloses a nonzero volume (see Figure 3.7). A spherical surface is an example of a closed surface, so is a doughnut surface, or a doughnut with two holes. All topological equivalence of a sphere is also a closed surface.

An open surface is like a thin disk: it is a surface that encloses no volume. Hence, sharp or zero thickness edges occur in open surfaces (see Figure 3.3). An annular disk, or a disk with two holes is still an open surface. All topological equivalence of these thin surfaces are also open surfaces.

A thin sheet object may be thought of as a limiting case of an open surface object whose volume has collapsed to zero. In this case, the thin sheet object has two original surfaces that have collapsed to one, and these two surfaces coincide in space.

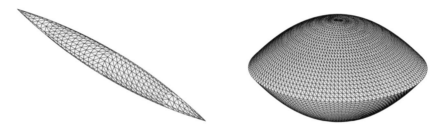

Figure 3.7: Example of closed surfaces: (left) An ogive. (right) A flying saucer.

3.7.1   EFIE for Open Surfaces

The EFIE introduced in Eq. (3.22) is valid for both open and closed surface. But when it is applied to an open surface, one has to treat both sides of an open surface as one surface, and the surface current is the sum of the surface currents on both sides of the surface. This is because tangential $\mathbf{E}$ field is continuous across a current-carrying open surface, and hence, Eq. (3.22) is valid on both sides of an open surface.

If one collapses a closed surface into a open surface, for example, by reducing a flying saucer into one with zero thickness, in the case of EFIE, the equation before the collapse has been imposed on both the upper and lower surfaces of the flying saucer, but after the collapse, it is only imposed on the collapsed surface which is one surface. Hence, the number of independent equations (one can imagine the EFIE has been reduced to a set of linear algebraic equations by using MOM) has seemingly decreased as now we have more unknowns than equations. However, one can sum the currents on the top and bottom surfaces into one current, making the number of equations equal to the number of unknowns.

3.7.2 MFIE and More

The MFIE is obtained by the invocation of the extinction theorem. One can rewrite the MFIE equation (3.29) with the simplification in (3.31)

$$-\hat{n} \times \mathbf{H}_{inc}(\mathbf{r}) = \hat{n} \times \int_S \nabla g(\mathbf{r}, \mathbf{r}') \times \mathbf{J}_s(\mathbf{r}') d\mathbf{r}', \quad \mathbf{r} \in S^- \qquad (3.42)$$

since S and $S^{\pm}$ are infinitesimally close to each other. MFIE can only be written for closed surfaces because it is based on the extinction theorem which is valid only for closed surfaces. Hence, the field points in the above equation cannot be taken outside the closed surface S. When a closed surface collapses to a thin open surface as in the case of a flying saucer, the field points become redundant, and the number of equations is smaller than the number of unknowns. Alternatively, one can argue that the extinction theorem cannot be invoked for an open surface. Hence, the resultant equation becomes ill-conditioned, and not solvable.

The gradient on the Green's function makes it more singular, and the principal value integral method can be used to evaluate this singular integral [17]. After applying such technique, the form of MFIE becomes

$$\hat{n} \times \mathbf{H}_{inc}(\mathbf{r}) = \frac{1}{2}\mathbf{J}_s(\mathbf{r}) - \hat{n} \times P.V. \int_s \nabla g(\mathbf{r}, \mathbf{r}') \times \mathbf{J}_s(\mathbf{r}') d\mathbf{r}', \quad \mathbf{r} \in S$$

$$(3.43)$$

where the first term on the right-hand side is the residue contribution of the singular integral, and the second term, a principal value integral. The second term converges to the same value irrespective of if the field point is on S, S^+, or S^-. This equation can be solved by using MOM as for EFIE. In the above, since the unknown current appears both inside and outside the integral, it is known as the second-kind integral equation. (In contrast, EFIE is known as the first-kind integral equation.)

3.8 Internal Resonance and Combined Field Integral Equation

For closed surfaces, EFIE is only imposed on the surface of the scatterer. Hence, even if we are solving for the scattering solution from a solid PEC object, EFIE does not distinguish that from the solution of the thin-shell PEC object. Though simple that it may seem, this model has a problem. A thin-shell PEC object is also a resonance cavity. Since MOM approximates an actual physical object with facets and approximates its solution, electromagnetic energy will leak into the thin-shell PEC object, exciting the internal resonance modes of the cavity. These excited internal resonances contribute to the extraneous surface current on the surface S which is quite different from the actual physical surface current that one wants to solve for, that is, that of a solid PEC scatterer. Moreover, when the operating frequency is near the resonance of the interior cavity, the interior resonance mode will also leak energy to the exterior, giving rise to erroneous scattered field.

One may argue that the internal resonance surface current of a resonant modes is a non-radiating current, and hence, the scattered field should not be overly affected. But in an approximate model of facets and approximate expansion functions, this is not a perfect non-radiating current. It is a weakly radiating current.

MFIE also suffers from the internal resonance problem as the EFIE, but the internal resonance is of a different nature. MFIE is derived from the invocation of the extinction theorem, such that the total field is extinct at S^-. The nature of internal resonance for MFIE is quite different from that of EFIE, and we will discuss this in greater detail in the next chapter.

The internal resonance problem is related to the existence of a null-space solution for the EFIE. Even if we can solve the EFIE exactly, the current from the internal resonance mode belongs to the null space of the EFIE integral operator. For a PEC cavity with lossless medium, this internal resonance frequency is pure real, and can coincide (collide) with the operating frequency of one's calculation. Hence, the current is nonunique when the operating frequency is exactly at the internal resonance frequency. At frequencies close to the internal resonance frequency, the equation is ill-conditioned because the pertinent matrix representation of the EFIE operator is also ill-conditioned: small eigenvalue exists because the resonance condition is almost satisfied. The internal resonance mode of the EFIE operator is characterized by the boundary condition $\hat{n} \times \mathbf{E} = 0$ on S.

The ill-conditioned matrix system due to internal resonances is deleterious to iterative solvers as the iteration count needed to solve such problems becomes exceedingly large (see last section of this chapter). For noniterative solvers, roundoff error can exist in the solution because of ill conditioning. The internal resonance gives rise to spurious currents flowing on the surface of the scatterer, and hence, the current obtained can be very wrong when this occurs. Luckily, there is a way to remove this problem. The remedy is to formulate the scattering problem so that the solution external to the scatterer is the same, but the interior solution corresponds to a lossy wall cavity resonance [18]. Hence, the interior resonance solution can only occur as a damped sinusoidal signal. In other words, the internal resonances of the cavity correspond to complex resonances. The operating frequency of the calculation, which is usually real, does not coincide (or collide) with the complex resonance frequencies (which correspond to poles in the complex plane below the real axis). Consequently, when the scatterer is excited by a sinusoidal time-harmonic signal, the interior resonance solution does not grow to a large value to corrupt the scattered field in the exterior.

Alternatively, the remedy is to use a combined field integral equation (CFIE) [11], which is a linear superposition of EFIE and MFIE. Usually, the integral equations are superposed as

$$\text{CFIE} = \alpha \text{EFIE} + \eta(1 - \alpha)\text{MFIE} \tag{3.44}$$

with α chosen to be about 0.5 and $\eta = \sqrt{\mu/\epsilon}$ is the intrinsic impedance of space. It is important that the above equations be combined in a manner so that an effective lossy boundary condition ensues. As a consequence, the null-space solution of CFIE can occur only at complex resonance frequencies. (We omit the proof here as it is similar to that for scalar wave equation [19]. It will be discussed in more detail in the next chapter.) Hence, the equation is free from internal resonance problem. A wrongly combined CFIE can correspond to reactive boundary condition on the wall rather than dissipative wall losses. Reactive boundary condition is lossless. In this case, there will still be real resonance modes rather than complex resonance modes, plaguing calculations with real frequencies.

3.9   Other Boundary Conditions—Impedance Boundary Condition, Thin Dielectric Sheet, and R-Card

So far, we have described the scattering solution to a scatterer where the surface scatterer is modeled by a PEC. This is an idealization, and most likely, some surfaces are not PEC surfaces. In this case, we can model a lossy surface or a reactive surface with an impedance boundary condition (IBC) [20]. More of this will be discussed in Chapter 4.

For an IBC, instead of imposing the boundary condition that the tangential electric field is zero on the surface of the scatterer, one imposes the condition that the tangential electric field is [21]

$$\mathbf{E}_{tan} = \eta_s \mathbf{J}_s \tag{3.45}$$

where η_s is the surface impedance in ohms. When the surface impedance is zero, we recover the PEC case. For a thinly coated PEC surface, we can use the following equation to estimate the surface impedance:

$$\eta_s = -i\sqrt{\frac{\mu}{\epsilon}} \tan\left(\sqrt{\frac{\epsilon\mu}{\epsilon_0\mu_0}} k_0 d\right) \tag{3.46}$$

where d is the thickness, ϵ and μ are the complex permittivity and permeability of the material coating, ϵ_0 and μ_0 are free-space permittivity and permeability, and k_0 is the free-space wavenumber. The above is obtained using a simple transmission-line model or a layered-medium model, and should be quite accurate when the surface is almost flat, and the thickness is thin.

For lossy, bulky, large dielectric objects one may use the formula

$$\eta_s = \sqrt{\frac{\mu}{\epsilon}} \tag{3.47}$$

which is only an approximate concept. A thin dielectric sheet (TDS) can be modeled as a zero thickness surface with a special impedance condition [12, 13]. Such an impedance condition is

$$Z = \frac{i}{\omega d \Delta\epsilon} \tag{3.48}$$

where ω is the operating angular frequency, $\Delta\epsilon = \epsilon - \epsilon_0$ and d is its thickness. Notice that the above impedance is purely reactive when ϵ is pure real. When the permittivity is made to be lossy, the TDS can be used to model a resistive sheet or an R-card. The difference between an IBC and a TDS is that IBC is best used for modeling coating of an impenetrable scatterer. On the other hand, a TDS can be used to model a semitransparent dielectric sheet that allows the transmission of electromagnetic field through it. For instance, one may want to model the windscreen of a car or a cockpit with a TDS. A more elaborate version of the TDS has recently been developed [22].

3.10   Matrix Solvers–A Pedestrian Introduction

A pedestrian introduction to matrix solver is given here so that one can distinguish the difference between solving an equation iteratively and noniteratively. This section does not do justice to the vast amount of literature that have been written on matrix solvers. It is hoped that the reader will gain at least a pedestrian knowledge of matrix solvers here.

Once the matrix equation is obtained, a whole sleuth of methods can solve the matrix equation. The more recent advances have been on fast solvers in solving the matrix system iteratively. But before describing the iterative solvers for the matrix equation, we will describe some traditional solvers for the matrix system.

The simplest method of solving a matrix system is to use Gaussian elimination or LUD (lower-upper triangular decomposition) [23]. These solvers are also known as direct inversion solvers or noniterative solvers. One reason for their convenience is that there have been many canned routines for such methods. However, their complexities are bad: If the matrix system has N unknowns, the CPU time grows as N^3. Hence, it uses a humongous amount of CPU resource when N is large. One advantage of such a method is that when the LU factorization is performed, the CPU cost of solving for a new right-hand side is proportional to N^2. However, the matrix system has to be available, and hence, all the matrix elements have to be stored in the computer memory, requiring storage proportional to N^2.

3.10.1   Iterative Solvers and Krylov Subspace Methods

Another way of solving a matrix system is to use an iterative method [24]. This method allows one to solve a matrix system by performing a small number of matrix-vector multiplies at each iteration. As one can produce the matrix-vector product without generating or storing the matrix, the method can be made matrix free, greatly reducing the storage requirements. In this case, only the unknowns need to be stored, and the memory requirement is proportional to N instead of N^2. For large N, this can be a tremendous saving in memory. For sparse matrices, and fast algorithms for dense matrices, such matrix-vector product can be effected in $O(N)$ or $O(N \log N)$ operations, instead of $O\left(N^2\right)$ operations. So if the number of iterations can be kept small in these methods, one can greatly improve the solution time of the equation when N is large.

There are umpteen ways to solve a matrix equation iteratively. But we will illustrate the concept with a very simple iterative solver. Consider a matrix system

$$\overline{\mathbf{A}} \cdot \mathbf{x} = \mathbf{b} \tag{3.49}$$

where $\overline{\mathbf{A}} = \overline{\mathbf{D}} + \overline{\mathbf{T}}$, the matrix $\overline{\mathbf{D}}$ is the diagonal part of $\overline{\mathbf{A}}$ while $\overline{\mathbf{T}}$ is its off-diagonal part. The above equation can be rewritten as

$$\overline{\mathbf{A}} = \overline{\mathbf{D}} + \overline{\mathbf{T}}$$
$$\overline{\mathbf{D}} \cdot \mathbf{x} = \mathbf{b} - \overline{\mathbf{T}} \cdot \mathbf{x}$$
$$\mathbf{x} = \overline{\mathbf{D}}^{-1} \cdot \left(\mathbf{b} - \overline{\mathbf{T}} \cdot \mathbf{x}\right) \tag{3.50}$$

Since $\overline{\mathbf{D}}$ is diagonal, finding its inverse is trivial. We can use the last equation to seek the solution iteratively by writing it as

$$\mathbf{x}_n = \overline{\mathbf{D}}^{-1} \cdot \left(\mathbf{b} - \overline{\mathbf{T}} \cdot \mathbf{x}_{n-1}\right) \tag{3.51}$$

In the equation above, we can start with an initial guess for $\mathbf{x}$, substitute it into the right-hand side and then solve for its new estimate on the left-hand side. The new value can then be substituted into the right-hand side to obtain a newer estimate of $\mathbf{x}$. This iteration goes on until the value of $\mathbf{x}$ ceases to change. This method is called a fixed-point iteration also known as Jacobi iteration.

In the above equation, the major cost of each iteration is a matrix-vector product $\overline{\mathbf{T}} \cdot \mathbf{x}$. For a dense matrix with N unknowns, the cost of the matrix-vector product involves N^2 operations. So if the iterative process converges in N_i iterations, then the cost of solving the above equation involves $N^2 N_i$ operations. If $N_i \ll N$, this method can be much faster than the direct solver, which involves N^3 operations.

The above is only a simple illustration of an iteration solver, which is based on fixed point iteration. There are a whole host of different iterative solvers making an alphabet soup such as CG (conjugate gradient), BCG or BiCG (biconjugate gradient), BCGSTAB (BiCG stabilized), GMRES (generalized minimal residual), QMR (quasi-minimal residual), TFQMR (transpose free QMR) etc. Unfortunately, there is no one iterative solver that is the panacea for all problems. One pretty much has to try different ones to see which one works the best for our problems.

Many iterative solvers are based on the Krylov subspace method. One can understand a lot about the properties of these iterative solvers through the Krylov subspace concept without going through their full machineries [25–28]. In the Krylov subspace method, one estimates the best solution to a matrix equation (3.49) by performing a number of matrix-vector multiplies. One defines the initial estimate of the solution as $\mathbf{x}_0$, and the initial residual error as $\mathbf{r}_0 = \mathbf{b} - \overline{\mathbf{A}} \cdot \mathbf{x}_0$. After K matrix-vector multiplies, K independent vectors, each of length N, are generated that span a space called the Krylov subspace $\mathcal{K}^K\left(\overline{\mathbf{A}}, \mathbf{r}_0\right)$, or

$$\mathcal{K}^K\left(\overline{\mathbf{A}}, \mathbf{r}_0\right) = \text{span}\left\{\mathbf{r}_0, \overline{\mathbf{A}} \cdot \mathbf{r}_0, \overline{\mathbf{A}}^2 \cdot \mathbf{r}_0, \ldots, \overline{\mathbf{A}}^{K-1} \cdot \mathbf{r}_0\right\} \tag{3.52}$$

An iterative solver finds the optimal solution to (3.49) by making use of these K independent vectors to minimize the residual error. It finds the optimal solution at the K-th iteration by choosing $\mathbf{x}_K = \mathbf{x}_0 + \mathbf{z}_K$ such that $\mathbf{z}_K \in \mathcal{K}^K\left(\overline{\mathbf{A}}, \mathbf{r}_0\right)$. In this manner, the residual at the K-th iteration is

$$\begin{aligned}
\mathbf{r}_K &= \mathbf{b} - \overline{\mathbf{A}} \cdot \mathbf{x}_K \\
&= \mathbf{r}_0 - \overline{\mathbf{A}} \cdot \mathbf{z}_K \in \text{span}\left\{\mathbf{r}_0, \overline{\mathbf{A}} \cdot \mathbf{r}_0, \overline{\mathbf{A}}^2 \cdot \mathbf{r}_0, \ldots, \overline{\mathbf{A}}^K \cdot \mathbf{r}_0\right\} \\
&\in \mathcal{K}^{K+1}\left(\overline{\mathbf{A}}, \mathbf{r}_0\right)
\end{aligned} \tag{3.53}$$

In other words, the best estimate solution has the residual error in the K-th iteration as

$$\mathbf{r}_K = \mathbf{r}_0 + a_1 \overline{\mathbf{A}} \cdot \mathbf{r}_0 + a_2 \overline{\mathbf{A}}^2 \cdot \mathbf{r}_0 + \cdots + a_2 \overline{\mathbf{A}}^K \cdot \mathbf{r}_0$$

$$= \sum_{k=0}^{K} a_k \overline{\mathbf{A}}^k \cdot \mathbf{r}_0 = P_K^0 \left(\overline{\mathbf{A}} \right) \cdot \mathbf{r}_0 \qquad (3.54)$$

where $a_0 = 1$, and the residual error for exact solution is zero, whereas that of the optimal one is as small as possible. In the above, $P_K^0(x)$ is a K-th order polynomial with $P_K^0(0) = 1$, and hence, $P_K^0 \left(\overline{\mathbf{A}} \right)$ is a polynomial of the matrix $\overline{\mathbf{A}}$.

Different iterative solvers use different strategies in finding the a_k's. For instance, in the CG method (applicable to a positive definite Hermitian system), at every new iteration, when a new independent vector is added to the Krylov subspace by performing a matrix-vector product, it constructs a new residual error vector that is orthogonal to all the previous ones. Because in an N dimensional space, there could be only N independent orthogonal vectors of length N, the exact solution is found after N iterations.

In practice, however, the arithmetics in a computer is not exact, and numerical roundoff error often precludes the convergence of the CG method in N steps. Numerical roundoff causes the loss of orthogonality in these vectors. As a result, different methods are constructed to remedy this convergence issue. Moreover, methods are developed for non-Hermitian systems.

To understand the convergence issue, one can expand $\mathbf{r}_0$ in terms of the eigenvectors of $\overline{\mathbf{A}}$, namely,

$$\mathbf{r}_0 = \sum_{n=1}^{N} \mathbf{v}_n \left(\mathbf{w}_n^t \cdot \mathbf{r}_0 \right) \qquad (3.55)$$

assuming that the eigenvalues and eigenvectors are distinct and that a reciprocal set of vectors $\mathbf{w}_n$ exists such that $\mathbf{w}_n \cdot \mathbf{v}_n = \delta_{ij}$. Using the above in the k-th term of (3.54), one gets

$$\overline{\mathbf{A}}^k \cdot \mathbf{r}_0 = \sum_{n=1}^{N} \lambda_n^k \mathbf{v}_n \left(\mathbf{w}_n^t \cdot \mathbf{r}_0 \right) \qquad (3.56)$$

From the above, one sees that when $k \to \infty$, the eigenvectors with small eigenvalues are deemphasized in the k-th term, whereas the large eigenvalue terms are emphasized swamping the small eigenvalue terms. Hence, the new independent vector that can be generated by this process is limited by numerical roundoff. In the extreme case when one eigenvector has a very large eigenvalue, while all the other eigenvalues are very small or close to zero, each matrix-vector product does not add a new independent eigenvector to the Krylov subspace. Therefore, when the dynamic range of eigenvalues is very large, the iterative method is not effective in generating new independent vectors after a while because of numerical roundoff.

The condition number of a matrix is the ratio of the largest singular value of the matrix to the smallest singular value (For positive definite Hermitian matrix, the eigenvalues are also the singular values). For matrices with large condition numbers, the eigenvector constituents in the system become imbalanced after a number of iterations. Therefore, the number of iterations needed to solve the matrix equation is related to the condition number of the matrix system.

Iterative method based on Krylov subspace method seeks the minimum of $\|\mathbf{r}_K\|$ subject to $\mathbf{z}_K \in \mathcal{K}^K\left(\overline{\mathbf{A}}, \mathbf{r}_0\right)$, or

$$\min_{\mathbf{z}_K \in \mathcal{K}^K\left(\overline{\mathbf{A}}, \mathbf{r}_0\right)} \|\mathbf{r}_K\| = \min_{P_K^0(x)} \left\|P_K^0\left(\overline{\mathbf{A}}\right) \cdot \mathbf{r}_0\right\| \leq \min_{P_K^0(x)} \left\|P_K^0\left(\overline{\mathbf{A}}\right)\right\| \|\mathbf{r}_0\| \tag{3.57}$$

The vector norm $\|\mathbf{r}\| = |\mathbf{r}| = \sqrt{\mathbf{r}^\dagger \cdot \mathbf{r}}$ or the magnitude of the vector, while the matrix norm

$$\left\|\overline{\mathbf{A}}\right\| = \max_{\mathbf{x} \neq 0}\left\{\left\|\overline{\mathbf{A}} \cdot \mathbf{x}\right\| / \|\mathbf{x}\|\right\}$$

If the matrix $\overline{\mathbf{A}}$ is diagonalizable such that $\overline{\mathbf{A}} = \overline{\mathbf{U}} \cdot \overline{\mathbf{\Lambda}} \cdot \overline{\mathbf{U}}^{-1}$, where $\overline{\mathbf{\Lambda}}$ is a diagonal matrix that contains the eigenvalues of $\overline{\mathbf{A}}$ on its diagonal, then $P_K^0\left(\overline{\mathbf{A}}\right) = \overline{\mathbf{U}} \cdot P_K^0\left(\overline{\mathbf{\Lambda}}\right) \cdot \overline{\mathbf{U}}^{-1}$. As a result,

$$\left\|P_K^0\left(\overline{\mathbf{A}}\right)\right\| \leq \left\|\overline{\mathbf{U}}\right\| \left\|P_K^0\left(\overline{\mathbf{\Lambda}}\right)\right\| \left\|\overline{\mathbf{U}}^{-1}\right\| \tag{3.58}$$

where the inequality $\left\|\overline{\mathbf{A}} \cdot \overline{\mathbf{B}}\right\| \leq \left\|\overline{\mathbf{A}}\right\| \left\|\overline{\mathbf{B}}\right\|$ has been used. Furthermore

$$\left\|P_K^0\left(\overline{\mathbf{\Lambda}}\right)\right\| = \max_{\lambda_i \in \Lambda\left(\overline{\mathbf{A}}\right)} \left|P_K^0(\lambda_i)\right| \tag{3.59}$$

where $\Lambda\left(\overline{\mathbf{A}}\right)$ means the set of all eigenvalues of $\overline{\mathbf{A}}$. Finally, one gets

$$\min_{\mathbf{z}_K \in \mathcal{K}^K\left(\overline{\mathbf{A}}, \mathbf{r}_0\right)} \|\mathbf{r}_K\| \leq \left\|\overline{\mathbf{U}}\right\| \left\|\overline{\mathbf{U}}^{-1}\right\| \|\mathbf{r}_0\| \min_{P_K^0(x)} \max_{\lambda_i \in \Lambda\left(\overline{\mathbf{A}}\right)} \left|P_K^0(\lambda_i)\right| \tag{3.60}$$

Therefore, to minimize the residual error, one picks the polynomial $P_K^0(x)$ such that it has min-max value at the point when x is evaluated at the eigenvalues of $\overline{\mathbf{A}}$. It is seen that when the eigenvalues are close to the origin, because $P_K^0(0) = 1$, it is hard to find such a min-max polynomial. Also, if the eigenvalues are scattered widely over the complex plane, finding such a polynomial is difficult. However, if the eigenvalues are clustered around points away from the origin, finding such a polynomial is easier.

The factor $\left\|\overline{\mathbf{U}}\right\| \left\|\overline{\mathbf{U}}^{-1}\right\|$ is the condition number of the matrix $\overline{\mathbf{U}}$. For ill-conditioned $\overline{\mathbf{U}}$, this number can be fairly large, and (3.60) is not a very tight bound. For a normal matrix[2] for which Hermitian matrix is a special case, the matrix $\overline{\mathbf{A}}$ can be diagonalized such that $\overline{\mathbf{U}}$ is unitary and $\overline{\mathbf{U}}^{-1} = \overline{\mathbf{U}}^\dagger$. The norm of a unitary matrix is 1, and the above bound is tight. For a diagonalizable complex symmetric system, we can show that $\overline{\mathbf{U}}^{-1} = \overline{\mathbf{U}}^t$ implying that $\overline{\mathbf{U}}$ is complex orthogonal, we expect the bound to be tight since complex orthogonal matrices generally have good condition numbers.

3.10.2   Effect of the Right-Hand Side—A Heuristic Understanding

The convergence of an iterative solver often depends on the right-hand side of the equation. In scattering, this corresponds to the nature of the exciting source that generates the incident

---

[2]A normal matrix is one for which $\overline{\mathbf{A}}^\dagger \cdot \overline{\mathbf{A}} = \overline{\mathbf{A}} \cdot \overline{\mathbf{A}}^\dagger$.

field. We can gain a heuristic understanding of the effect of the right-hand side by performing an eigenvalue analysis of the pertinent equation (3.49). We let

$$\mathbf{b} = \sum_{n=1}^{N} \mathbf{v}_n (\mathbf{w}_n^t \cdot \mathbf{b}) \tag{3.61}$$

and let

$$\mathbf{x} = \sum_{n=1}^{N} c_n \mathbf{v}_n \tag{3.62}$$

Using the above in (3.49), and using similar technique to (3.55), we deduce that $c_n = \mathbf{w}_n^t \cdot \mathbf{b}/\lambda_n$ or

$$\mathbf{x} = \sum_{n=1}^{N} \mathbf{v}_n \mathbf{w}_n^t \cdot \mathbf{b}/\lambda_n \tag{3.63}$$

From the above, we observe that the importance of the eigenvector in a solution depends very much on the right-hand side and the eigenvalue. The right-hand side will excite different eigenvectors of the system depending on its nature or the inner product $\mathbf{w}_n^t \cdot \mathbf{b}$. When eigenvectors with small eigenvalues are excited, poor convergence of the iterative solution will ensue because of the swamping of small eigenvalues by large eigenvalues in the Krylov subspace (see discussion after Eq. (3.55)).

A way to reduce the condition number of a matrix is to precondition it by multiplying the matrix equation by its approximate inverse. It is hoped that the resultant matrix equation, with a smaller condition number, requires a smaller number of iterations to solve. The construction of a preconditioner remains an art rather than a science. The matrix is often constructed with the aid of heuristics or physical insight.

The disadvantage of an iterative solver is that the number of iterations needed is unpredictable, and very much problem dependent. Also, whenever the right-hand side of the equation changes, the process has to begin all over again.

3.11   Conclusions

In this chapter, we introduce the concept of integral equations and the method used to solve them. A very popular method is MOM which has been used with great success in solving a number of practical problems. Coupled with fast solvers and parallel computers, it can solve up to tens of millions of unknowns making electrically large problems solvable by numerical methods, that were previously unsolvable [19, 29–31].

Figure 3.8 shows the current on an 83 Ford Camaro when it is illuminated by an electromagnetic field generated by a Hertzian dipole at 1 GHz. It is solved with the Fast Illinois Solver Code (FISC) [32], a MOM code where the multilevel fast multipole algorithm (MLFMA) is applied to accelerate the solution. The body of the car is modeled as a PEC, the windscreens are modeled as TDS, and the tires are modeled with IBC.

Figure 3.9 shows the current distribution on a 02 Cadillac Deville when an XM antenna is mounted on its roof [33]. Because of the disparate mesh sizes involved, the matrix system is extremely ill-conditioned. A specially designed preconditioner, the self-box inclusion (SBI)

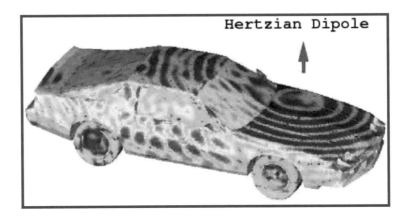

Figure 3.8: The current on a 83 Ford Camaro solved with MOM and the MLFMA accelerated fast solver called the Fast Illinois Solver Code (FISC) [32].

preconditioner [34], has to be used in order to arrive at a converged solution. The windows of the car are modeled with TDS.

The XM antenna consists of a probe-fed square patch with edges cleaved so that a circularly polarized wave can be generated by the antenna. The antenna has a dielectric substrate which is modeled as a volume integral equation (VIE). It has a radome that is modeled as a TDS.

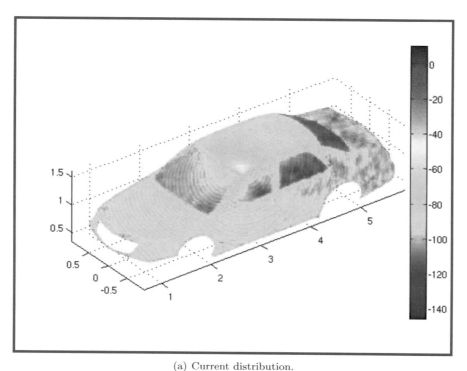

(a) Current distribution.

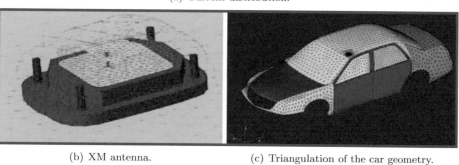

(b) XM antenna.   (c) Triangulation of the car geometry.

Figure 3.9: (a) The current distribution on an 02 Cadillac Deville when (b) an XM antenna is mounted on its roof, solved with MOM and the MLFMA accelerated fast solver called the FastAnt [33]. (c) shows the triangulated geometry description of the car.

Bibliography

[1] P. M. Morse and H. Feshbach, *Methods of Theoretical Physics*, New York: McGraw-Hill, 1953.

[2] C. T. Tai, *Dyadic Green's Functions in Electromagnetic Theory*, Piscataway, NJ: IEEE Press, 2nd ed., 1994.

[3] J. A. Kong, *Theory of Electromagnetic Waves,* New York: Wiley-Interscience, 1975.

[4] A. D. Yaghjian, "Electric dyadic Green's functions in the source region," *Proc. IEEE*, vol. 68, pp. 248–263, 1980.

[5] G. Green, *An Essay on the Application of Mathematical Analysis to the Theories of Electricity and Magnetism*, T. Wheelhouse, Nottingham, 1828. Also, see L. Challis and F. Sheard, "The green of Green functions," *Phys. Today*, vol. 56, no. 2, pp. 41-46, Dec. 2003.

[6] J. A. Stratton and L. J. Chu, "Diffraction theory of electromagnetic waves," *Phys. Rev.*, vol. 56, pp. 99-107, Jul. 1939.

[7] W. C. Chew *Waves and Fields in Inhomogeneous Media*, New York: Van Nostrand Reinhold, 1990. Reprinted by Piscataway, NJ: IEEE Press, 1995.

[8] W. Franz, "Zur formulierung des huygenschen prinzips," *Z. Naturforsch.*, vol. 3a, pp. 500-506, 1948.

[9] R. F. Harrington, *Time-Harmonic Electromagnetic Fields*, New York: McGraw-Hill, 1961.

[10] R. F. Harrington, *Field Computation by Moment Methods*, Malabar, FL: Krieger, 1968.

[11] R. F. Harrington and J. R. Mautz, "E-field, H-field, and combined field solutions for conducting bodies of revolution," *A.E.Ü.*, vol. 32, no. 4, pp. 159–164, Apr. 1978.

[12] R. F. Harrington and J. R. Mautz, "An impedance sheet approximation for thin dielectric shells," *IEEE Trans. Antennas Propag.*, vol. AP-23, pp. 531534, Jul. 1975.

[13] E. Bleszynski, M. Bleszynski, and T. Jaroszewicz, "Surface integral equations for electromagnetic scattering from impenetrable and penetrable sheets," *IEEE Antennas Propag. Mag.*, vol. 35, no. 6, pp. 14–26, Dec. 1993.

[14] R. Mittra, editor, *Computer Techniques for Electromagnetics*, New York: Hemisphere Publisher, 1978.

[15] S. M. Rao, G. R. Wilton, and A. W. Glisson, "Electromagnetic scattering by surfaces of arbitrary shape," *IEEE Trans. Antennas Propag.*, vol. 30, no. 3, pp. 409–418, May 1982.

[16] P. P. Silvester and R. L. Ferrari, *Finite Elements for Electrical Engineers*, Cambridge, MA: Cambridge University Press, 1986.

[17] D. R. Wilton, S. M. Rao, A. W. Glisson, D. Schaubert, O. Al-Bundak, C. Butler, "Potential Integrals for Uniform and Linear Source Distributions on Polygonal and Polyhedral Domains," *IEEE Trans. Antennas Propag.*, AP-32, No. 3, pp. 276–281, Mar. 1984.

[18] D. R. Wilton, "Review of current status and trends in the use of integral equations in computational electromagnetics," *Electromagn.*, vol. 12, pp. 287-341, 1992.

[19] W. C. Chew, J. M. Jin, E. Michielssen, and J. M. Song, eds., *Fast and Efficient Algorithms in Computational Electromagnetics*, Boston, MA: Artech House, 2001.

[20] K. M. Mitzner, "Effective boundary conditions for reflection and transmission by an absorbing shell of arbitrary shape," *IEEE Trans. Antennas Propag.*, vol. AP-16, pp. 706–712, 1968.

[21] J. A. Stratton, *Electromagnetic Theory*, New York: McGraw-Hill, 1941.

[22] I. T. Chiang and W. C. Chew, "Thin dielectric sheet simulation by surface integral equation using modified RWG and pulse bases," *IEEE Trans. Antennas Propag.*, vol. 54, no. 7, pp. 1927-1934, Jul. 2006.

[23] G. H. Golub and C. F. Van Loan, *Matrix Computations*, Baltimore, MD: Johns Hopkins University Press, 1983.

[24] R. Barrett, M. Berry, T. F. Chan, J. Demmel, J. Donato, J. Dongarra, V. Eijkout, R. Pozo, C. Romine, and H. van der Vorst, *Templates for the Solution of Linear Systems: Building Blocks for Iterative Methods*, Second Edition, Philadelphia, PA: SIAM, 1994.

[25] Y. Saad and M. H. Schultz, "GMRES: a generalized minimal residual algorithm for solving nonsymmetric linear systems," *SIAM J. Sci. Stat. Comput.*, vol. 7, pp. 856–869, 1986.

[26] T. A. Driscoll, K. C. Toh, L. N. TREFETHEN, "From potential theory to matrix iterations in six steps," *SIAM Rev.*, vol. 40, no. 3, pp. 547–578, 1998.

[27] J. Liesen and P. Tichy, "Convergence analysis of Krylov subspace methods," *GAMM-Mitteilungen*, vol. 27, no. 2, pp. 153–173, 2004.

[28] C. Siefert and E. de Sturler, "Preconditioners for generalized saddle-point problems," *SIAM Journal on Numerical Analysis*, vol. 44, no. 3, pp. 1275-1296, 2006.

[29] S. Velamparambil, W. C. Chew, and J. M. Song, "10 million unknowns, is it that large," *IEEE Antennas Propag. Mag.*, vol. 45, no. 2, pp. 43–58, Apr. 2003.

[30] M. L. Hastriter, "A study of MLFMA for large scale scattering problems," *Ph.D. Thesis*, Dept. ECE, U. Illinois, Urbana-Champaign, Jun. 2003.

[31] L. Gurel and O. Ergul, "Fast and accurate solutions of extremely large integral-equation problems discretized with tens of millions of unknowns," *Electron. Lett.*, accepted for publication.

[32] J. M. Song, C. C. Lu, W. C. Chew, and S. W. Lee, "Fast Illinois solver code (FISC)," *IEEE Antennas Propag. Mag.*, vol. 40, no. 3, pp. 27–33, 1998.

[33] H. Y. Chao, K. Pirapaharan, V. Bodrov, T. J. Cui, H. P. Hsu, G. Huff, X. J. Zhang, J. S. Zhao, J. Bernhard, and W. C. Chew, "Simulation of vehicle antennas by the multilevel fast multipole algorithm," *Antennas Appl. Symp.*, Allerton Park, Illinois, Sept. 18–20, 2002.

[34] H. Y. Chao, "A multilevel fast multipole algorithm for analyzing radiation and scattering from wire antennas in a complex environment," *Ph.D. Thesis*, Dept. ECE, U. Illinois, Urbana-Champaign, Jan. 2003.

CHAPTER 4

Integral Equations
for Penetrable Objects

4.1 Introduction

The previous chapter illustrates the derivation of integral equations for a perfect electric conductor (PEC) object. PEC objects are some of the simplest scattering objects we can think of. They are also termed impenetrable objects as the electromagnetic energy does not penetrate into them. Another case of an impenetrable object is the perfect magnetic conductor (PMC) object. It is the dual of the PEC. In these objects, either the tangential electric field or tangential magnetic field is zero on the entire surface, so that there could be no net power flow (real or complex) into them.

However, many objects are penetrable such as dielectric objects, magnetic objects, or lossy objects. It is possible to have net power flow into or through these objects. Hence, the tangential electric field or the tangential magnetic field cannot be zero on the entire surface of the object.

The subject of scattering by penetrable scatterers has been fervently studied by many workers throughout the years. Before numerical methods were in vogue, approximate analytic methods were popular such as the high frequency methods [1]. At one time, the extended boundary condition (EBC) method [2] was popular, because of its simplicity. Other methods that have been used are generalized multipole method [3], and more recently, differential equation solvers such as finite element (FEM) and finite difference methods (FDM) [4–8].

There are two main types of integral equations for penetrable scatterers: surface integral equation (SIE) when the inhomogeneity is piecewise constant, and volume integral equation (VIE) when the inhomogeneity is highly heterogeneous. Early works were done in two dimensions [9–11] in the 1960s. Three dimensional integral equation work based on the method of moments (MOM) has only become popular in the 1970s to the 1990s (see an account by Poggio and Miller [12] and Miller *et al.* [13]). The first comprehensive paper on SIE method for 3D electromagnetic scatterer was reported by Midgyesi-Mitschang *et al.* [14]. The readers are referred to these works plus the references there in for a historical account.

We will first derive the SIEs for penetrable bodies. We shall assume that the permittivity ϵ and the permeability μ are different from their free-space values, but are constant or homogeneous inside these bodies. The case of lossy bodies is easily included by making the ϵ or (and) the μ complex. Then later, we will derive the volume integral equations (VIEs) for inhomogeneous penetrable scatterers.

4.2   Scattering by a Penetrable Object Using SIE

We assume that $\mathbf{J}_{inc}$ is a current that generates an incident field $\mathbf{E}_{inc}$ that impinges on a penetrable object (see Figure 4.1). Polarization currents will be induced on the penetrable object, which in turn radiates to generate the scattered field. We can imagine a surface S^+ that is just large enough to enclose S, the surface of the penetrable object and apply the equivalence principle and extinction theorem to it [2, 15]. The equivalence principle and extinction theorem have been stated in words by Harrington [15]. But we will show their mathematical forms here (see derivation in the previous chapter).

When equivalence surface currents are placed on S^+, they produce the total field outside S^+ and zero field inside S^- (which is a surface just small enough to be enclosed in S). Hence, the scattering object can be removed without disturbing the field outside. The equivalence surface currents are given by

$$\mathbf{J}_s = \hat{n} \times \mathbf{H}, \quad \mathbf{M}_s = \mathbf{E} \times \hat{n} \quad \text{on} \quad S^+ \tag{4.1}$$

where $\mathbf{E}$ and $\mathbf{H}$ represent total fields and $\hat{n}$ is a outward normal on S^+. Consequently, the equivalence principle and the extinction theorem become [16]

$$\left.\begin{matrix} \mathbf{r} \in V_1, & \mathbf{E}(\mathbf{r}) \\ \mathbf{r} \in V_2, & 0 \end{matrix}\right\} = \mathbf{E}_{inc}(\mathbf{r}) + i\omega\mu \int_{S^+} dS' \overline{\mathbf{G}}(\mathbf{r},\mathbf{r}') \cdot \mathbf{J}_s(\mathbf{r}') - \nabla \times \int_{S^+} dS' \overline{\mathbf{G}}(\mathbf{r},\mathbf{r}') \cdot \mathbf{M}_s(\mathbf{r}') \tag{4.2}$$

When we apply the extinction theorem part of the above equation to S^-, we obtain the integral equation of scattering, viz.,

$$\mathbf{r} \in S^-, \quad 0 = \mathbf{E}_{inc}(\mathbf{r}) + i\omega\mu \int_{S^+} dS' \overline{\mathbf{G}}(\mathbf{r},\mathbf{r}') \cdot \mathbf{J}_s(\mathbf{r}') - \nabla \times \int_{S^+} dS' \overline{\mathbf{G}}(\mathbf{r},\mathbf{r}') \cdot \mathbf{M}_s(\mathbf{r}') \tag{4.3}$$

The above is also known as the electric field integral equation (EFIE) for a penetrable object.[1] Understanding the equivalence principle and the extinction theorem above is key to the understanding of the internal resonance problem as shall be shown in the next section.

Using operator notations, we can rewrite (4.3) more compactly as

$$-\mathbf{E}_{inc}(\mathbf{r}) = \mathcal{L}_{1E}(\mathbf{r},\mathbf{r}') \cdot \mathbf{J}_s(\mathbf{r}') + \mathcal{K}_{1E}(\mathbf{r},\mathbf{r}') \cdot \mathbf{M}_s(\mathbf{r}'), \quad \mathbf{r} \in S^-, \quad \mathbf{r}' \in S^+ \tag{4.4}$$

where integrations are implied over repeated variables (extension of the index notation to continuous variables), and $\mathcal{L}_{1E}$ and $\mathcal{K}_{1E}$ assume the property of medium 1. More explicitly, they are

$$\mathcal{L}_{1E}(\mathbf{r},\mathbf{r}') = i\omega\mu_1 \overline{\mathbf{G}}_1(\mathbf{r},\mathbf{r}') = i\omega\mu_1 \left(\overline{\mathbf{I}} + \frac{\nabla\nabla}{k_1^2} \right) g_1(\mathbf{r},\mathbf{r}') \tag{4.5}$$

---

[1] When only the tangential components of the above equation is taken, the equation is still known as EFIE in the literature.

$$\mathcal{K}_{1E}(\mathbf{r}, \mathbf{r}') = -\nabla \times \overline{\mathbf{G}}_1(\mathbf{r}, \mathbf{r}') = -\nabla g_1(\mathbf{r}, \mathbf{r}') \times \overline{\mathbf{I}} \qquad (4.6)$$

where

$$g_1(\mathbf{r}, \mathbf{r}') = \frac{e^{ik_1|\mathbf{r}-\mathbf{r}'|}}{4\pi|\mathbf{r} - \mathbf{r}'|} \qquad (4.7)$$

The subscripts E of the above operators imply that they are generating $\mathbf{E}$ field from either the electric current $\mathbf{J}_s$ or from the magnetic current $\mathbf{M}_s$ respectively.

The above constitutes one equation with two unknowns, and the solution for $\mathbf{J}_s$ and $\mathbf{M}_s$ cannot be obtained by solving only one equation alone—more equations are needed. To derive more equations, we apply the extinction theorem to region 2 in an inside-out manner. As a result, we have

$$0 = \mathcal{L}_{2E}(\mathbf{r}, \mathbf{r}') \cdot \mathbf{J}_s(\mathbf{r}') + \mathcal{K}_{2E}(\mathbf{r}, \mathbf{r}') \cdot \mathbf{M}_s(\mathbf{r}'), \quad \mathbf{r} \in S^+, \quad \mathbf{r}' \in S^-. \qquad (4.8)$$

In the above, we are assuming that $\mathbf{J}_s$ and $\mathbf{M}_s$ are residing on S^-, namely, $\mathbf{r}' \in S^-$, but they are the same as those for (4.4) since $\mathbf{J}_s = \hat{n} \times \mathbf{H}$, and $\mathbf{M}_s = \mathbf{E} \times \hat{n}$, and tangential $\mathbf{E}$ and $\mathbf{H}$ are continuous across the interface S from S^- to S^+. We can impose (4.4) for $\mathbf{r} \in S^-$, and (4.8) for $\mathbf{r} \in S^+$ and obtain two integral equations for two unknowns $\mathbf{J}_s$ and $\mathbf{M}_s$. Since $\mathbf{J}_s$ and $\mathbf{M}_s$ are surface currents on a 2D manifold (or surface), only the tangential components of the Eqs. (4.4) and (4.8) are imposed to solve for the current unknowns.

Even though we have enough equations to solve for the number of unknown functions, the solution to the penetrable scatterer problem is fraught with nonuniqueness problem at the internal resonance of the resonant modes formed by the surface S filled with exterior material. This is because we impose the extinction theorem on a surface S, for the electric field only or for the magnetic field only, which does not guarantee the extinction of the field in the entire volume surrounded by S. We shall show this point later.

To derive equations that are free of internal resonances, we need the aid of the magnetic field equivalence of the above equations. The magnetic field version of (4.2) can be derived by taking the curl of (4.2) or by invoking the duality principle.

$$\begin{array}{l} \mathbf{r} \in V_1, \quad \mathbf{H}(\mathbf{r}) \\ \mathbf{r} \in V_2, \quad 0 \end{array} \Bigg\} = \mathbf{H}_{inc}(\mathbf{r}) + i\omega\epsilon \int_{S^+} dS' \overline{\mathbf{G}}(\mathbf{r}, \mathbf{r}') \cdot \mathbf{M}_s(\mathbf{r}') + \nabla \times \int_{S^+} dS' \overline{\mathbf{G}}(\mathbf{r}, \mathbf{r}') \cdot \mathbf{J}_s(\mathbf{r}') \quad (4.9)$$

The corresponding integral equation of scattering is

$$\mathbf{r} \in S^-, \quad 0 = \mathbf{H}_{inc}(\mathbf{r}) + i\omega\epsilon \int_{S^+} dS' \overline{\mathbf{G}}(\mathbf{r}, \mathbf{r}') \cdot \mathbf{M}_s(\mathbf{r}') + \nabla \times \int_{S^+} dS' \overline{\mathbf{G}}(\mathbf{r}, \mathbf{r}') \cdot \mathbf{J}_s(\mathbf{r}') \quad (4.10)$$

The above is the magnetic field integral equation (MFIE) for a penetrable object.

In a similar manner, MFIE can be compactly written using operator notation

$$-\mathbf{H}_{inc}(\mathbf{r}) = \mathcal{L}_{1H}(\mathbf{r}, \mathbf{r}') \cdot \mathbf{M}_s(\mathbf{r}') + \mathcal{K}_{1H}(\mathbf{r}, \mathbf{r}') \cdot \mathbf{J}_s(\mathbf{r}'), \quad \mathbf{r} \in S^-, \quad \mathbf{r}' \in S^+ \qquad (4.11)$$

where integral over repeated variables is implied, and more explicitly, the operators are

$$\mathcal{L}_{1H}(\mathbf{r}, \mathbf{r}') = i\omega\epsilon_1 \overline{\mathbf{G}}_1(\mathbf{r}, \mathbf{r}') = i\omega\epsilon_1 \left(\overline{\mathbf{I}} + \frac{\nabla\nabla}{k_1^2} \right) g_1(\mathbf{r}, \mathbf{r}') = \frac{1}{\eta_1^2} \mathcal{L}_{1E}(\mathbf{r}, \mathbf{r}') \qquad (4.12)$$

Figure 4.1: Scattering by a penetrable object inside which the electric field and the magnetic field are nonzero.

$$\mathcal{K}_{1H}(\mathbf{r}, \mathbf{r}') = \nabla \times \overline{\mathbf{G}}_1(\mathbf{r}, \mathbf{r}') = \nabla g_1(\mathbf{r}, \mathbf{r}') \times \overline{\mathbf{I}} = -\mathcal{K}_{1E}(\mathbf{r}, \mathbf{r}') \tag{4.13}$$

Similar to the EFIE case, we apply extinction theorem to region 2 in an inside-out manner to obtain

$$0 = \mathcal{L}_{2H}(\mathbf{r}, \mathbf{r}') \cdot \mathbf{M}_s(\mathbf{r}') + \mathcal{K}_{2H}(\mathbf{r}, \mathbf{r}') \cdot \mathbf{J}_s(\mathbf{r}'), \quad \mathbf{r} \in S^+, \quad \mathbf{r}' \in S^- \tag{4.14}$$

It is to be noted that the MFIE is not independent from the EFIE since they are derivable from one another by the action of the curl operator. However, we now have four equations and two unknowns to solve for.

To prevent the internal resonance problem, a myriad of remedies have been proposed. One is to use the combined field integral equation (CFIE) as has been used for the PEC scatterer case. This will reduce the four equations we have derived to two. Alternatively, a weighted sum of the external and internal equations is obtained both for the EFIE and the MFIE to reduce the number of equations to two. This is known as either the PMCHWT [14, 17–19] method or the Müller method [20] depending on how the weights are chosen. It can be shown that these weighted-sum methods are free of internal resonance problems.

In the weighted-sum method, we add (4.4) and (4.8) with a weighting factor to obtain

$$-\alpha_1 \mathbf{E}_{inc}(\mathbf{r}) = [\alpha_1 \mathcal{L}_{1E} + \alpha_2 \mathcal{L}_{2E}] \cdot \mathbf{J}_s(\mathbf{r}') + [\alpha_1 \mathcal{K}_{1E} + \alpha_2 \mathcal{K}_{2E}] \cdot \mathbf{M}_s(\mathbf{r}'), \quad \mathbf{r} \in S^- \tag{4.15}$$

Again, we can add (4.11) and (4.14) with a weighting factor to get

$$-\eta_1 \beta_1 \mathbf{H}^{inc}(\mathbf{r}) = \eta_1 [\beta_1 \mathcal{K}_{1H} + \beta_2 \mathcal{K}_{2H}] \cdot \mathbf{J}_s(\mathbf{r}') + \eta_1 [\beta_1 \mathcal{L}_{1H} + \beta_2 \mathcal{L}_{2H}] \cdot \mathbf{M}_s(\mathbf{r}'), \quad \mathbf{r} \in S^+ \tag{4.16}$$

where η_1 is used to ensure that (4.15) and (4.16) are of the same dimension. Again, only the tangential components of the above equations need to be imposed to generate enough equations to solve for the surface currents $\mathbf{J}_s$ and $\mathbf{M}_s$, since they live on a 2D manifold.

When $\alpha_1 = \alpha_2 = \beta_1 = \beta_2 = 1$, the above formulation is known as the PMCHWT (Poggio, Miller, Chang, Harrington, Wu, Tsai) formulation. When $\alpha_1 = \beta_1 = 1$, $\alpha_2 = -\frac{\epsilon_2}{\epsilon_1}$, and

$\beta_2 = \frac{\mu_2}{\mu_1}$, it is known as the Müller formulation. When the tangential component of the above equations are extracted, the corresponding $\mathcal{K}$ operators have the principal value terms (see previous chapter). Because of the difference in the direction of the normals, the principal value terms cancel in the the PMCHWT formulation. They do not cancel in the Müller formulation. The Müller formulation is good for low contrasts, whereas the PMCHWT formulation is good for high contrasts.

4.3 Gedanken Experiments for Internal Resonance Problems

The internal resonance problem of integral equations is a problem that has plagued solutions of scalar integral equations and vector integral equations [18–29]. Although the existence of these internal resonances can be easily proved for some cases, their existence for other cases is less well understood.

For example in electromagnetics, the existence of internal resonance for the EFIE for perfect electric conducting (PEC) scatterer can be easily shown. But the existence of this problem for MFIE for PEC scatterer is less obvious. Their existence for penetrable scatterers is even less obvious. The internal resonance problem in EFIE for PEC scatterers is related to the internal cavity resonance and nonradiating current of a resonance cavity. The physical character of the internal resonance can be easily understood. However, the physical character of the internal resonance problem for MFIE for PEC scatterers is less well understood. For penetrable scatterers, this is even less well understood.

In this section, we will prove, via Gedanken experiments (thought experiment), the existence of the internal resonance problems for MFIE for PEC, and for penetrable scatterers when EFIE or MFIE alone is applied [30]. These proofs reveal with physical clarity the reasons of the internal resonance problem, as well as the physical character of the internal resonance problem.

The internal resonance problem exists both for impenetrable as well as penetrable scatterers when certain integral equations are used to solve for their scattering solutions. For simplicity, we shall discuss the case of impenetrable scatterers first.

4.3.1 Impenetrable Objects

Some integral equations for impenetrable objects exhibit internal resonance problems. When the solution of the integral equation is sought at the frequency that corresponds to the internal resonance of the resonant modes formed by the object, the solution to the integral equation is nonunique. Such is the case with the EFIE and the MFIE. However, this internal resonance problem can be remedied.

There are two major kinds of impenetrable objects in electromagnetics, the PEC and the PMC. Since the PMC is the dual of the PEC, we will consider the PEC here only. Other impenetrable scatterers are those with reactive boundary conditions such that all energy is reflected from the scatterer.

EFIE-PEC Case

The internal resonance problem for an EFIE when solving a PEC scattering problem will be reviewed here. The EFIE for a PEC scattering problem is given as

$$\hat{n} \times \hat{n} \times \mathcal{L}_E(\mathbf{r}, \mathbf{r}') \cdot \mathbf{J}_{eq}(\mathbf{r}') = -\hat{n} \times \hat{n} \times \mathbf{E}_{inc}(\mathbf{r}), \quad \mathbf{r} \in S \qquad (4.17)$$

where

$$\mathcal{L}_E(\mathbf{r}, \mathbf{r}') \cdot \mathbf{J}_{eq}(\mathbf{r}') = i\omega\mu \int_S dS' \overline{\mathbf{G}}(\mathbf{r}, \mathbf{r}') \cdot \mathbf{J}_{eq}(\mathbf{r}') \qquad (4.18)$$

In the above, the $-\hat{n} \times \hat{n}\times$ extracts the tangential components of the field. It can be replaced by the dyad $\overline{\mathbf{I}}_t = \overline{\mathbf{I}} - \hat{n}\hat{n}$ where $\overline{\mathbf{I}}$ is a unit dyad, and the above can be rewritten as

$$\overline{\mathbf{I}}_t \cdot \mathcal{L}_E(\mathbf{r}, \mathbf{r}') \cdot \mathbf{J}_{eq}(\mathbf{r}') = -\overline{\mathbf{I}}_t \cdot \mathbf{E}_{inc}(\mathbf{r}), \quad \mathbf{r} \in S \qquad (4.19)$$

When the EFIE is solved by imposing tangential $\hat{n} \times \mathbf{E} = 0$ on the surface of the scatterer, the field is not extinct inside the scatterer if the operating frequency of the time-harmonic incident wave is at one of the resonance frequencies of the PEC cavity formed by the surface S. We can also imagine that the scattering solution from a thin PEC shell is being solved, for which nontrivial interior solution to this PEC shell exists at the resonance frequencies of this resonant modes. Consequently, a solution to the EFIE exists even when no exciting incident field is present. In other words, the EFIE in the following has a nontrivial solution at internal resonance, indicating the presence of the null-space solution to (4.19):

$$\overline{\mathbf{I}}_t \cdot \mathcal{L}_E(\mathbf{r}, \mathbf{r}') \cdot \mathbf{J}_{eq}(\mathbf{r}') = 0, \quad \mathbf{r} \in S \qquad (4.20)$$

The existence of a null-space solution makes the solution to the equation nonunique. Hence, when the matrix representation of the EFIE operator is sought, the matrix equation is ill conditioned.

To elaborate further, Eq. (4.20) is the equation for the internal resonance current of a PEC cavity at its resonance frequencies. This current produces nontrivial electromagnetic field inside the cavity, but it produces no field outside the cavity. This can also be understood from the equivalence principle viewpoint. Hence, the resonance current is nonradiating.

Because the null-space current is a nonradiating current, we may be led to think that this resonance current will not contribute to the scattered field, and hence, may still obtain the correct scattered field even though the current is wrong and nonunique. However, the scattered field is still fraught with small error. The reason is that the numerical solution is an approximate solution, and this approximate null-space current is not a perfect nonradiating source. At the resonance frequency of the resonant modes, this nonradiating current will still leak some energy to the outside of the resonant modes contributing to error in the scattered field [31].

Because of the nonuniqueness of the integral equation at resonance, the matrix system representing the integral equation becomes ill conditioned. Hence, when iterative solvers are used to solve for the solution, the number of iterations required is large, or the iterative solution is nonconvergent. However, a noniterative solver such as lower-upper triangular decomposition (LUD) can be used to solve the matrix system near resonance. The solution usually exhibits larger error in a narrow band near the resonance frequencies of the resonant modes.

Scattering from a 2D PEC circular cylinder is used as an example here [30], with pulse expansion functions and point-matching. The diagonal elements of the MOM matrix are calculated by small argument approximation of the Hankel function and the off-diagonal elements are evaluated using the one-point quadrature rule [17, p. 41].

The first resonance of TM (transverse magnetic) modes in a circular cylinder is $ka = 2.40482555769577$ for TM_{01} mode, where k is the wave number and a is the radius of the circle. Figure 4.2(a) shows the current distribution calculated by solving EFIE using MOM and by Mie series at $ka = 2.40482555769577$. Large difference (relative RMS error: 0.436) between the MOM results and Mie series is observed. The difference is the resonant electric current induced on the PEC circular cylinder surface due to the incident plane wave. Because the resonant electric current does not radiate, very accurate scattered fields as shown in Figure 4.2(b) are obtained with a relative error as small as 0.001 in comparison with the Mie series.

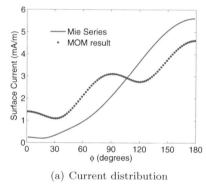
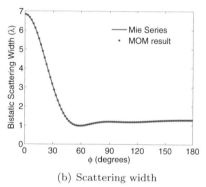

(a) Current distribution   (b) Scattering width

Figure 4.2: (a) Current distribution and (b) scattering width at $ka = 2.40482555769577$ for a circular cylinder due to a TM plane wave incidence.

MFIE-PEC Case

The MFIE is defined by

$$\hat{n} \times \mathcal{K}_H(\mathbf{r}, \mathbf{r}') \cdot \mathbf{J}_{eq}(\mathbf{r}') = -\hat{n} \times \mathbf{H}_{inc}(\mathbf{r}), \quad \mathbf{r} \in S^- \tag{4.21}$$

where

$$\mathcal{K}_H(\mathbf{r}, \mathbf{r}') \cdot \mathbf{J}_{eq}(\mathbf{r}') = \nabla \times \int_S dS' \overline{\mathbf{G}}(\mathbf{r}, \mathbf{r}') \cdot \mathbf{J}_{eq}(\mathbf{r}') \tag{4.22}$$

The internal resonance problem of an MFIE for a PEC scattering problem is less well understood. At the internal resonance of the MFIE, the scattered field solution has larger error compared to that of EFIE. We shall provide a lucid physical explanation for this phenomenon.

To understand this, we perform a Gedanken experiment (see Figure 4.3) to construct a null-space solution to the MFIE. We assume that the incident field on the scatterer, instead

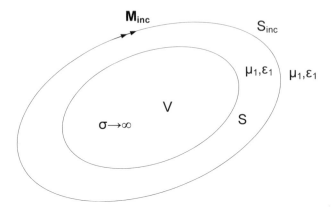

Figure 4.3: Presence of null-space solution for some integral equations for PEC scatterers can be proved by performing a Gedanken experiment.

of coming from far away, is generated by equivalence currents impressed on a surface S_{inc}, where S_{inc} is larger than the surface of the PEC scatterer so that the PEC scatterer is completely enclosed inside S_{inc}. In general, by the equivalence principle, two impressed currents $\mathbf{J}_{inc} = \hat{n} \times \mathbf{H}_{inc}$ and $\mathbf{M}_{inc} = \mathbf{E}_{inc} \times \hat{n}$ are required on the surface S_{inc}.

First we pick the operating frequency of the Gedanken experiment to be at the resonance frequency of the PMC formed by S_{inc}. In this manner, a nontrivial resonance solution exists inside this PMC. Next we pick the equivalence currents to be the resonance current on the surface of this PMC. Then only magnetic equivalence current, $\mathbf{M}_{inc} = \mathbf{E}_{inc} \times \hat{n}$, is needed, because the electric equivalence current, $\mathbf{J}_{inc} = \hat{n} \times \mathbf{H}_{inc} = 0$ on S_{inc}. Also, just inside S_{inc}, $\hat{n} \times \mathbf{H}_{inc} = 0$ on S_{inc} since this is the PMC solution. However, the incident field (both electric and magnetic) inside S_{inc} is not zero. Therefore, when a scatterer is placed inside S_{inc}, scattered field is generated and exists outside the scatterer.

Now, with this experimental picture in mind, we can gradually shrink the size of S_{inc} such that it becomes S^+, where S^+ is a surface just infinitesimally larger than S to enclose it. As the size of S_{inc} is gradually shrunk, we also alter the operating frequency of the impressed current so that it becomes the resonance frequency of the new PMC after shrinkage. Because of this, $\hat{n} \times \mathbf{H}_{inc} = 0$ always on the surface of the cavity formed by S^+, but $\hat{n} \times \mathbf{E}_{inc} \neq 0$, and only the equivalence magnetic current is needed. Also, there is always an incident field inside S_{inc} even when it becomes S^+. So when we put a PEC inside S^+, electric current will be induced on the PEC, and a scattered field is generated outside S^+.

By the above Gedanken experiment, we have constructed a nontrivial solution to the MFIE. The driving term of the equation, $\hat{n} \times \mathbf{H}_{inc}$ is zero, because it corresponds to the tangential magnetic field on the surface of the PMC cavity formed by S^+ at resonance. Tangential magnetic field is continuous from S^+ to S^-; therefore this field is also zero on S^-. Hence, the right-hand side of Eq. (4.21) is zero. Therefore, this nontrivial solution is also the null-space solution to the MFIE, which is the solution to the following equation:

$$\hat{n} \times \mathcal{K}_H(\mathbf{r}, \mathbf{r}') \cdot \mathbf{J}_{eq}(\mathbf{r}') = 0, \quad \mathbf{r} \in S \qquad (4.23)$$

Two facts become clear in the above Gedanken experiment:

- The MFIE has a null-space solution when the operating frequency is at the resonance frequency of the PMC formed by S. This frequency is exactly the same as the resonance frequency of a similar PEC cavity by duality principle. Therefore, MFIE has the same breakdown frequencies as EFIE.

- Unlike the null-space solution or current of the EFIE, this null-space current is quite different physically. It is not the resonance current of the PEC cavity formed by S. But it is an induced current on the surface S due to an incident field with $\hat{n} \times \mathbf{H}_{inc} = 0$ and $\hat{n} \times \mathbf{E}_{inc} \neq 0$ on S^{+}. Hence, this current is a good radiator, generating a strong scattered field outside the scatterer. This is actually observed by MFIE numerical calculation near the internal resonance frequency of the resonant modes.

The MFIE for a TE (transverse electric) wave scattering from a 2D PEC circular cylinder is studied in the same way [30]. The diagonal elements are 0.5 and the off-diagonal elements are evaluated using one-point rule. Figure 4.4 shows the current distribution and scattering width at $ka = 2.40482555769577$. The difference is the electric current induced on the surface due to the incident wave, which radiates and leads to large error in the scattered fields.

By the duality principle, the above can also be applied to prove the existence of internal resonance problem when EFIE or MFIE is used to solve for the scattering solution of a PMC scatterer.

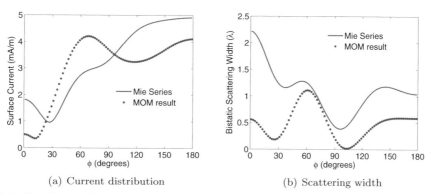

(a) Current distribution   (b) Scattering width

Figure 4.4: Similar to Figure 4.2, solution of the magnetic field integral equation for a TE plane wave.

4.3.2 Penetrable Objects

When a penetrable object such as a dielectric scatterer is solved by MFIE or EFIE alone, internal resonance problem occurs. On first thought, this is not obvious as no PEC cavity seems to be formed here. Moreover, the resonance frequency of a dielectric scatterer must be complex because of radiation damping. Hence, this resonance problem is coming from somewhere else. We again shall provide a lucid physical explanation as to why this is happening.

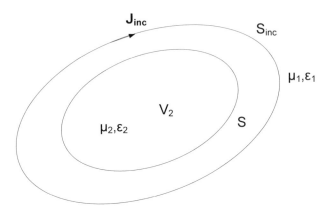

Figure 4.5: The presence of null-space solution for some integral equations for penetrable scatterers can be proved by performing a Gedanken experiment.

Since the surface currents in (4.4) and (4.8) live in a 2D manifold, to maintain the same number of equations as unknowns in the discretized equation, only the tangential components of (4.4) and (4.8) are imposed when solving them. These are known as the EFIE form of the equation for penetrable scatterers. We can show that $\bar{\mathbf{I}}_t \cdot (4.4)$ and $\bar{\mathbf{I}}_t \cdot (4.8)$ have internal resonance problem by a similar Gedanken experiment.

To prove the existence of the null-space solution for the EFIE equation when applied to the penetrable scatterer, we consider an incident field generated by an equivalence current $\mathbf{J}_{inc}$ impressed on a surface S_{inc} (Figure 4.5). Again, we pick the frequency such that it is the resonance frequency of the PEC cavity formed by S_{inc} in the absence of the scatterer. Then we make $\mathbf{J}_{inc}$ the resonance current of the PEC cavity formed by S_{inc}. Only electric current will be needed by the equivalence principle, since $\hat{n} \times \mathbf{E}_{inc} = 0$ on the surface of S_{inc}.

But $\mathbf{H}_{inc} \neq 0$, and $\mathbf{E}_{inc} \neq 0$ inside S_{inc}. There will be a scattered field $\mathbf{H}_{sca}$ and $\mathbf{E}_{sca}$ in this problem when a penetrable scatterer such as a dielectric scatterer is placed inside S_{inc}.

Now, we adjust and shrink S_{inc} so that it just becomes S^+, which is infinitesimally larger than S so as to enclose it, and also we alter the frequency continuously so that the operating frequency is always equal to the resonance frequency of the resonant modes formed by S_{inc}. Then, there exists a trivial $\hat{n} \times \mathbf{E}_{inc}$ in (4.4) such that the scattering solution is nontrivial. Namely $\mathbf{J}_s$ and $\mathbf{M}_s$ are nontrivial, even when $\hat{n} \times \mathbf{E}_{inc} = 0$. This proves the existence of null-space solutions to the tangential components of (4.4) and (4.8).

For the case when MFIE alone is used to solve for the scattering solution of a penetrable scatterer, we can repeat the above Gedanken experiment with a PMC resonance cavity that generates an equivalence magnetic current. We use this resonance current as the equivalence current to generate an incident field and a scattered field that have associated currents on the surface of the scatterer. These currents are in the null space of the tangential components of the equations (4.11) and (4.14).

A few things become clear in this Gedanken experiment.

- The scattering solutions of a penetrable scatterer have internal resonance problem when only EFIE or MFIE alone is used.

- The solution is nonunique at the resonance frequencies of the resonant modes formed by the surface S. These are exactly the same frequencies that one encounters in solving the PEC or PMC scatterer case when EFIE or MFIE is used to solve the problem.

- The incident field acts like the field from a resonant electric current of a PEC cavity (for EFIE alone) or a resonant magnetic current of a PMC cavity (for MFIE alone) on the penetrable scatterer. They will appear as zero on the left-hand sides of (4.4) and (4.11) when tangential components of the equations are taken. However, these incident fields are capable of inducing a scattered field from a scatterer. Therefore, there will be nonzero equivalent surface currents $\mathbf{J}_s$ and $\mathbf{M}_s$ that satisfy these equations, which form the null-space solutions of the equations. These null-space currents are good radiators that produce erroneous scattered fields at these frequencies, hence contributing to the scattered field errors.

4.3.3   A Remedy

With the above insight, the source of the internal resonance problem becomes clear: The extinction of tangential electric field or tangential magnetic field on a surface S does not guarantee the extinction of the total field inside the surface S. This is because internal resonance solutions exist inside the cavity formed by S at its resonance frequencies. These resonance fields, depending on whether the resonant modes is a PMC or a PEC, have either $\hat{n} \times \mathbf{H} = 0$ or $\hat{n} \times \mathbf{E} = 0$ on S.

The remedy for this is to arrive at equations such that when the appropriate field components are extinct on the surface S, it guarantees the extinction of the total field inside S. The CFIE is such an equation. It is a linear superposition of the EFIE and MFIE.

$$\alpha \bar{\mathbf{I}}_t \cdot \mathcal{L}_E(\mathbf{r}, \mathbf{r}') \cdot \mathbf{J}_{eq}(\mathbf{r}') + (1-\alpha)\eta \hat{n} \times \mathcal{K}_H(\mathbf{r}, \mathbf{r}') \cdot \mathbf{J}_{eq}(\mathbf{r}')$$
$$= -\alpha \bar{\mathbf{I}}_t \cdot \mathbf{E}_{inc}(\mathbf{r}) - (1-\alpha)\eta \hat{n} \times \mathbf{H}_{inc}(\mathbf{r}), \quad \mathbf{r} \in S^- \qquad (4.24)$$

The null-space solution of the above has to satisfy the equation

$$\alpha \bar{\mathbf{I}}_t \cdot \mathcal{L}_E(\mathbf{r}, \mathbf{r}') \cdot \mathbf{J}_{eq}(\mathbf{r}') + (1-\alpha)\eta \hat{n} \times \mathcal{K}_H(\mathbf{r}, \mathbf{r}') \cdot \mathbf{J}_{eq}(\mathbf{r}') = 0, \quad \mathbf{r} \in S^- \qquad (4.25)$$

The above is the equation for a resonant modes solution with boundary condition that

$$\alpha \mathbf{E}_t(\mathbf{r}) + (1-\alpha)\eta \hat{n} \times \mathbf{H}(\mathbf{r}) = 0, \quad \mathbf{r} \in S^- \qquad (4.26)$$

The above is equivalent to that the tangential electric field and the tangential magnetic field satisfy an impedance boundary condition (IBC) together on the surface S. If the IBC is made lossy, then the resonance solutions inside the surface S have complex resonance frequencies [32]. These complex resonances correspond to damped sinusoidal signals inside the resonant modes formed by the surface S. Hence, when the solution is sought at a real time harmonic frequency, this complex resonance solution is not possible and the integral equation does not have a null space.

4.3.4   Connection to Cavity Resonance

The proof of the uniqueness of CFIE has been given [18]. However, no proof of its connection to the cavity resonance problem has been reported, even though assertion has been made in the literature [17]. The partial proof is shown by considering a cavity resonance problem where the resonant modes is filled with material of region 1. We can formulate the integral equation for cavity resonance by using (4.8) but with material of region 1.

$$0 = \mathcal{L}_{1E}(\mathbf{r}, \mathbf{r}') \cdot \mathbf{J}_s(\mathbf{r}') + \mathcal{K}_{1E}(\mathbf{r}, \mathbf{r}') \cdot \mathbf{M}_s(\mathbf{r}'), \quad \mathbf{r} \in S^+ \qquad (4.27)$$

Now the boundary condition on a lossy cavity wall is

$$\mathbf{E}_s = Z_s \mathbf{J}_s, \quad \text{or} \quad \mathbf{M}_s = -Z_s \hat{n} \times \mathbf{J}_s \qquad (4.28)$$

Using the above in (4.27), then

$$0 = \mathcal{L}_{1E}(\mathbf{r}, \mathbf{r}') \cdot \mathbf{J}_s(\mathbf{r}') - Z_s \mathcal{K}_{1E}(\mathbf{r}, \mathbf{r}') \cdot \hat{n}' \times \mathbf{J}_s(\mathbf{r}'), \quad \mathbf{r} \in S^+ \qquad (4.29)$$

The above is the integral equation for the cavity resonance problem. Since it is a source-free problem, it does not have a nontrivial solution except when it is solved at the resonance frequencies of the cavity. In other words, the integral operator in (4.29) does not have a null space except at the resonance frequencies of the cavity. When the resonant modes is lossy, its resonance frequency is complex. Hence, at real ω, the integral operator in (4.29) does not have a null space.

We can show that (4.29) and (4.25) are related. To this end, we test the above with a tangential current $\mathbf{J}_T$ which is tangential to S and is supported at $\mathbf{r} \in S^+$.

$$0 = \langle \mathbf{J}_T(\mathbf{r}), \mathcal{L}_{1E}(\mathbf{r}, \mathbf{r}'), \mathbf{J}_s(\mathbf{r}') \rangle - Z_s \langle \mathbf{J}_T(\mathbf{r}), \mathcal{K}_{1E}(\mathbf{r}, \mathbf{r}') \cdot \hat{n}' \times, \mathbf{J}_s(\mathbf{r}') \rangle,$$
$$\mathbf{r}' \in S, \quad \mathbf{r} \in S^+ \qquad (4.30)$$

From the above, the transpose of the integral operator shall be derived. Because $\mathcal{L}_{1E}(\mathbf{r}, \mathbf{r}')$ is a symmetric operator, its transpose is just itself. Looking more carefully at the second part of the operator,

$$\langle \mathbf{J}_T(\mathbf{r}), \mathcal{K}_{1E}(\mathbf{r}, \mathbf{r}') \cdot \hat{n}' \times, \mathbf{J}_s(\mathbf{r}') \rangle$$
$$= \int_{S^+} dS \int_S dS' \mathbf{J}_T(\mathbf{r}) \cdot \nabla g(\mathbf{r}, \mathbf{r}') \times (\hat{n}' \times \mathbf{J}_s(\mathbf{r}'))$$
$$= \int_{S^+} dS \int_S dS' \mathbf{J}_T(\mathbf{r}) \times \nabla g(\mathbf{r}, \mathbf{r}') \cdot (\hat{n}' \times \mathbf{J}_s(\mathbf{r}'))$$
$$= \int_S dS' \int_{S^+} dS (\hat{n}' \times \mathbf{J}_s(\mathbf{r}')) \cdot \nabla' g(\mathbf{r}, \mathbf{r}') \times \mathbf{J}_T(\mathbf{r})$$
$$= -\int_S dS' \int_{S^+} dS \mathbf{J}_s(\mathbf{r}') \cdot \hat{n}' \times (\nabla' g(\mathbf{r}, \mathbf{r}') \times \mathbf{J}_T(\mathbf{r}))$$
$$= -\langle \mathbf{J}_s(\mathbf{r}'), \hat{n}' \times \mathcal{K}_{1H}(\mathbf{r}', \mathbf{r}), \mathbf{J}_T(\mathbf{r}) \rangle \qquad (4.31)$$

Notice that in the above $g(\mathbf{r}, \mathbf{r}') = g(\mathbf{r}', \mathbf{r})$. Moreover, $\mathbf{r}$ and $\mathbf{r}'$ are dummy variables. Hence, it can be seen that the transpose of the operator $\mathcal{K}_{1E}(\mathbf{r}, \mathbf{r}') \cdot \hat{n}' \times$ is $-\hat{n}' \times \mathcal{K}_{1H}(\mathbf{r}', \mathbf{r})$ [16].

Hence, if $Z_s = (1 - \alpha)\eta/\alpha$ is substituted in (4.29) and the resultant equation multiplied by α, then the integral operators in (4.29) and (4.25) are related to each other by a transpose relation.

When the matrix representations of these operators are sought with the proper expansion functions for the range and the domain spaces, then the matrices are also the transpose of each other. It is well known in matrix theory that if a matrix has a right null space, it will also have a left null space, and vice versa. However, this concept is not easily extendible to linear operators in infinite dimensional space. But we shall assume this to be true for this class of problems.

Therefore, we shall assume that if the transpose of an operator has no null space, the operator itself has no null space. Hence, if (4.29) has no null space when ω is real by physical argument, then (4.25) has no null space. However, if Z_s is chosen wrongly to give rise to a lossless, reactive IBC, internal cavity resonances at real ω still exist, and hence (4.29) will still have a null space, and so would the transpose operator. Hence, the proper choice of the combination parameter in CFIE is important so that the right physics occurs in the solution.

It is to be noted that the combined source integral equation (CSIE) produces an operator that is the transpose of the CFIE operator [33]. Hence, CSIE is free of the internal resonance problem because it is directly connected to cavity resonance.

4.3.5 Other remedies

The above use of CFIE is one remedy to the internal resonance problem. Other remedies are:

1. Extinction of both the tangential and the normal components of the electric or the magnetic field to guarantee the removal of the resonance solution.

2. Extinction of the tangential components of both the electric and the magnetic field to guarantee the removal of the resonance solution.

3. Extinction of the tangential component of the electric or the magnetic field at two surfaces close to each other to remove the internal resonance solution.

4. Insertion of shorts in the resonance cavity to push the resonance frequency of the resonant modes higher (see [29] and references therein).

5. Insertion of a lossy object inside the resonance cavity to cause the resonance frequency to become complex [34].

However, the above remedies require more equations than unknowns for an impenetrable scatterer. Hence the CFIE for impenetrable scatterers offers a solution where the equation number is not increased compared to the number of unknowns.

For a penetrable scatterer, remedy 2 above can be used effectively. The tangential components of (4.4), (4.8), (4.11), and (4.14) constitute four equations and two surface unknowns $\mathbf{J}_s$ and $\mathbf{M}_s$. The integral equations can be added to the exterior equations to reduce the number of equations to two, but at the same time, both the tangential components of the electric and magnetic field are extinct at the surface S^-. This will guarantee the extinction of the total field inside the resonant modes even at internal resonances, and hence, remove the null-space solutions that we have constructed by our Gedanken experiment. Numerical calculations with the weighted-sum methods of PMCHWT (Poggio, Miller, Chang, Harrington, Wu, Tsai) and Müller do not indicate internal resonance problem for this reason.

4.4 Volume Integral Equations

In this section, the VIE is derived similar to the derivation given in [16]. The vector wave equation for a general inhomogeneous, anisotropic medium is given by

$$\nabla \times \overline{\boldsymbol{\mu}}_r^{-1} \cdot \nabla \times \mathbf{E}(\mathbf{r}) - \omega^2 \overline{\boldsymbol{\epsilon}}_r(\mathbf{r}) \cdot \mu_0 \epsilon_0 \mathbf{E}(\mathbf{r}) = i\omega\mu_0 \mathbf{J}(\mathbf{r}) \tag{4.32}$$

where $\overline{\boldsymbol{\mu}}_r(\mathbf{r}) = \frac{\overline{\boldsymbol{\mu}}(\mathbf{r})}{\mu_0}$, and $\overline{\boldsymbol{\epsilon}}_r(\mathbf{r}) = \frac{\overline{\boldsymbol{\epsilon}}(\mathbf{r})}{\epsilon_0}$. After subtracting $\nabla \times \nabla \times \mathbf{E}(\mathbf{r}) - \omega^2 \mu_0 \epsilon_0 \mathbf{E}(\mathbf{r})$ from the sides of the equation, the above can be rewritten as

$$\nabla \times \nabla \times \mathbf{E}(\mathbf{r}) - \omega^2 \mu_0 \epsilon_0 \mathbf{E}(\mathbf{r}) = i\omega\mu_0 \mathbf{J}(\mathbf{r}) + \nabla \times \left[\overline{\mathbf{I}} - \overline{\boldsymbol{\mu}}_r^{-1}(\mathbf{r}) \right] \cdot \nabla \times \mathbf{E}(\mathbf{r})$$
$$- \omega^2 \mu_0 \epsilon_0 \left[\overline{\mathbf{I}} - \overline{\boldsymbol{\epsilon}}_r(\mathbf{r}) \right] \cdot \mathbf{E}(\mathbf{r}) \tag{4.33}$$

The dyadic Green's function corresponding to the differential operator on the left hand side can be derived, viz.,

$$\nabla \times \nabla \times \overline{\mathbf{G}}(\mathbf{r}, \mathbf{r}') - k_0^2 \overline{\mathbf{G}}(\mathbf{r}, \mathbf{r}') = \overline{\mathbf{I}}\delta(\mathbf{r} - \mathbf{r}') \tag{4.34}$$

where $k_0^2 = \omega^2 \mu_0 \epsilon_0$, and

$$\overline{\mathbf{G}}(\mathbf{r}, \mathbf{r}') = \left(\overline{\mathbf{I}} + \frac{\nabla\nabla}{k_0^2} \right) g(\mathbf{r}, \mathbf{r}') \tag{4.35}$$

$\overline{\mathbf{G}}(\mathbf{r}, \mathbf{r}') \cdot \mathbf{a}$ can be thought of as the field due to a point source $\mathbf{a}\delta(\mathbf{r} - \mathbf{r}')$ located at $\mathbf{r} = \mathbf{r}'$. By the principle of linear superposition, we treat the right hand side of (4.33) as equivalence volume sources, and write down the solution $\mathbf{E}(\mathbf{r})$ as

$$\mathbf{E}(\mathbf{r}) = i\omega\mu_0 \int_{V+} \overline{\mathbf{G}}(\mathbf{r}, \mathbf{r}') \cdot \mathbf{J}(\mathbf{r}) d\mathbf{r}'$$
$$+ \int_{V+} \overline{\mathbf{G}}(\mathbf{r}, \mathbf{r}') \cdot \nabla' \times \left[\overline{\mathbf{I}} - \overline{\boldsymbol{\mu}}_r^{-1}(\mathbf{r}') \right] \cdot \nabla' \times \mathbf{E}(\mathbf{r}') d\mathbf{r}'$$
$$- k_0^2 \int_{V+} \overline{\mathbf{G}}(\mathbf{r}, \mathbf{r}') \cdot \left[\overline{\mathbf{I}} - \overline{\boldsymbol{\epsilon}}_r(\mathbf{r}') \right] \cdot \mathbf{E}(\mathbf{r}') d\mathbf{r}' \tag{4.36}$$

where V^+ is a volume that is slightly larger than the support of the scatterer V defined by the volume where $\overline{\boldsymbol{\mu}}_r$ or $\overline{\boldsymbol{\epsilon}}_r$ departs from $\overline{\mathbf{I}}$. The first term can be considered the field generated by the current source $\mathbf{J}$ in the absence of the scatterer; hence, we call it $\mathbf{E}^{inc}(\mathbf{r})$, a known field. Consequently, (4.36) becomes an integral equation for $\mathbf{E}(\mathbf{r})$, or

$$\mathbf{E}(\mathbf{r}) = \mathbf{E}^{inc}(\mathbf{r}) + \int_{V+} \overline{\mathbf{G}}(\mathbf{r}, \mathbf{r}') \cdot \nabla' \times \left[\overline{\mathbf{I}} - \overline{\boldsymbol{\mu}}_r^{-1}(\mathbf{r}') \right] \cdot \nabla' \times \mathbf{E}(\mathbf{r}') d\mathbf{r}'$$
$$- k_0^2 \int_{V+} \overline{\mathbf{G}}(\mathbf{r}, \mathbf{r}') \cdot \left[\overline{\mathbf{I}} - \overline{\boldsymbol{\epsilon}}_r(\mathbf{r}') \right] \cdot \mathbf{E}(\mathbf{r}') d\mathbf{r}' \tag{4.37}$$

The second term on the right-hand side represents the scattered field due to induced magnetic polarization current from the inhomogeneous permeability, whereas the third term represents that due to the induced electric polarization current from the inhomogeneous permittivity. A similar derivation has been presented in [16] for isotropic media. This derivation clearly separates the scattered field from the inhomogeneous permeability from that of the inhomogeneous permittivity.

4.4.1 Alternative Forms of VIE

The above is not suitable for computational electromagnetics. However, the equation can be made computationally friendly through mathematical manipulations. The dyadic Green's function in the second term on the right-hand side of (4.37) has a term that involves the double ∇ operator. To reduce numerical quadrature error, it is a general wisdom to move the del operator from more singular terms to less singular terms using integration by parts [35]. But it can be shown that

$$\nabla\nabla \int_{V+} g(\mathbf{r},\mathbf{r}') \cdot \nabla' \times \mathbf{F}(\mathbf{r}')d\mathbf{r}' = -\nabla \int_{V+} \nabla'g(\mathbf{r},\mathbf{r}') \cdot \nabla' \times \mathbf{F}(\mathbf{r}')d\mathbf{r}'$$

$$= -\nabla \int_{V+} \nabla' \times \nabla'g(\mathbf{r},\mathbf{r}') \cdot \mathbf{F}(\mathbf{r}')d\mathbf{r}' = 0 \qquad (4.38)$$

Consequently, (4.37) becomes

$$\begin{aligned}
\mathbf{E}(\mathbf{r}) &= \mathbf{E}^{inc}(\mathbf{r}) + \int_{V+} g(\mathbf{r},\mathbf{r}')\nabla' \times \left[\overline{\mathbf{I}} - \overline{\boldsymbol{\mu}}_r^{-1}(\mathbf{r}')\right] \cdot \nabla' \times \mathbf{E}(\mathbf{r}')d\mathbf{r}' \\
&\quad -k_0^2 \int_{V+} \overline{\mathbf{G}}(\mathbf{r},\mathbf{r}') \cdot \left[\overline{\mathbf{I}} - \overline{\boldsymbol{\epsilon}}_r(\mathbf{r}')\right] \cdot \mathbf{E}(\mathbf{r}')d\mathbf{r}' \\
&= \mathbf{E}^{inc}(\mathbf{r}) + \nabla \times \int_{V+} g(\mathbf{r},\mathbf{r}') \left[\overline{\mathbf{I}} - \overline{\boldsymbol{\mu}}_r^{-1}(\mathbf{r}')\right] \cdot \nabla' \times \mathbf{E}(\mathbf{r}')d\mathbf{r}' \\
&\quad -k_0^2 \int_{V+} \overline{\mathbf{G}}(\mathbf{r},\mathbf{r}') \cdot \left[\overline{\mathbf{I}} - \overline{\boldsymbol{\epsilon}}_r(\mathbf{r}')\right] \cdot \mathbf{E}(\mathbf{r}')d\mathbf{r}' \qquad (4.39)
\end{aligned}$$

where integration by parts and that $\nabla g(\mathbf{r},\mathbf{r}') = -\nabla'g(\mathbf{r},\mathbf{r}')$ have been used. The meaning of $\overline{\mathbf{G}}(\mathbf{r},\mathbf{r}')$ in (4.39) still has to be properly defined because the dyadic Green's function is plagued with singularities whose evaluation has to be taken with great care. However, we can mitigate the effect of the singularity if we define the action of the dyadic Green's operator on $\left[\overline{\mathbf{I}} - \overline{\boldsymbol{\epsilon}}_r(\mathbf{r}')\right] \cdot \mathbf{E}(\mathbf{r}')$ to mean

$$\begin{aligned}
&\int_V d\mathbf{r}'\overline{\mathbf{G}}(\mathbf{r},\mathbf{r}') \cdot \left[\overline{\mathbf{I}} - \overline{\boldsymbol{\epsilon}}_r(\mathbf{r}')\right] \cdot \mathbf{E}(\mathbf{r}') \\
&= \left(\overline{\mathbf{I}} + \frac{\nabla\nabla}{k_0^2}\right) \cdot \int_V g(\mathbf{r},\mathbf{r}') \left[\overline{\mathbf{I}} - \overline{\boldsymbol{\epsilon}}_r(\mathbf{r}')\right] \cdot \mathbf{E}(\mathbf{r}')d\mathbf{r}' \\
&= \int_V d\mathbf{r}'g(\mathbf{r},\mathbf{r}') \left[\overline{\mathbf{I}} - \overline{\boldsymbol{\epsilon}}_r(\mathbf{r}')\right] \cdot \mathbf{E}(\mathbf{r}') + \frac{\nabla}{k_0^2} \int_V g(\mathbf{r},\mathbf{r}')\nabla' \cdot \left[\overline{\mathbf{I}} - \overline{\boldsymbol{\epsilon}}_r(\mathbf{r}')\right] \cdot \mathbf{E}(\mathbf{r}')d\mathbf{r}' \qquad (4.40)
\end{aligned}$$

In other words, we never move both the ∇ operators completely inside the integrand to render it ill defined. Also, integration by parts has been used to arrive at the form of the last integral. The integrals in (4.40) are always well defined.

We can cast the VIE yet into another form which is more suitable for curl-conforming basis. To this end, we make use of the identity for the dyadic Green's function using (4.34) as

$$\overline{\mathbf{G}}(\mathbf{r}, \mathbf{r}') = \frac{1}{k_0^2} \left[\nabla \times \nabla \times \overline{\mathbf{G}}(\mathbf{r}, \mathbf{r}') - \overline{\mathbf{I}} \delta(\mathbf{r} - \mathbf{r}') \right]$$

$$= \frac{1}{k_0^2} \left[\nabla \times \nabla \times \overline{\mathbf{I}} g(\mathbf{r}, \mathbf{r}') - \overline{\mathbf{I}} \delta(\mathbf{r} - \mathbf{r}') \right] \tag{4.41}$$

Using (4.41) in (4.39), we have

$$\overline{\epsilon}_r(\mathbf{r}) \cdot \mathbf{E}(\mathbf{r}) = \mathbf{E}^{inc}(\mathbf{r}) + \nabla \times \int_{V+} g(\mathbf{r}, \mathbf{r}') \left[\overline{\mathbf{I}} - \overline{\boldsymbol{\mu}}_r^{-1}(\mathbf{r}') \right] \cdot \nabla' \times \mathbf{E}(\mathbf{r}') d\mathbf{r}'$$

$$- \nabla \times \int_{V+} \nabla g(\mathbf{r}, \mathbf{r}') \times \left[\overline{\mathbf{I}} - \overline{\epsilon}_r(\mathbf{r}') \right] \cdot \mathbf{E}(\mathbf{r}') d\mathbf{r}' \tag{4.42}$$

where we have moved the delta function contribution from (4.41) in (4.42) to the left-hand side. By using integration by parts, one arrives at

$$\overline{\epsilon}_r(\mathbf{r}) \cdot \mathbf{E}(\mathbf{r}) = \mathbf{E}^{inc}(\mathbf{r}) + \nabla \times \int_{V+} g(\mathbf{r}, \mathbf{r}') \left[\overline{\mathbf{I}} - \overline{\boldsymbol{\mu}}_r^{-1}(\mathbf{r}') \right] \cdot \nabla' \times \mathbf{E}(\mathbf{r}') d\mathbf{r}'$$

$$- \nabla \times \int_{V+} g(\mathbf{r}, \mathbf{r}') \nabla' \times \left\{ \left[\overline{\mathbf{I}} - \overline{\epsilon}_r(\mathbf{r}') \right] \cdot \mathbf{E}(\mathbf{r}') \right\} d\mathbf{r}' \tag{4.43}$$

The above formulations, with the curl operators outside the integrals and some curl operators inside the integrals, are more suitable for curl-conforming expansion and testing functions. When the medium is highly anisotropic, the form in (4.42) is preferable, whereas if the medium is isotropic and piecewise constant, the form in (4.43) is preferable.

We can also derive the magnetic field equivalence of the above equations by duality principle. From (4.39), we have

$$\mathbf{H}(\mathbf{r}) \;\; = \;\; \mathbf{H}^{inc}(\mathbf{r}) + \nabla \times \int_{V+} g(\mathbf{r}, \mathbf{r}') \left[\overline{\mathbf{I}} - \overline{\epsilon}_r^{-1}(\mathbf{r}') \right] \cdot \nabla' \times \mathbf{H}(\mathbf{r}') d\mathbf{r}'$$

$$- k_0^2 \int_{V+} \overline{\mathbf{G}}(\mathbf{r}, \mathbf{r}') \cdot \left[\overline{\mathbf{I}} - \overline{\boldsymbol{\mu}}_r(\mathbf{r}') \right] \cdot \mathbf{H}(\mathbf{r}') d\mathbf{r}' \tag{4.44}$$

From (4.42), we have

$$\overline{\boldsymbol{\mu}}_r(\mathbf{r}) \cdot \mathbf{H}(\mathbf{r}) = \mathbf{H}^{inc}(\mathbf{r}) + \nabla \times \int g(\mathbf{r}, \mathbf{r}') \left[\overline{\mathbf{I}} - \overline{\epsilon}_r^{-1}(\mathbf{r}') \right] \cdot \nabla' \times \mathbf{H}(\mathbf{r}') d\mathbf{r}'$$

$$- \nabla \times \int \nabla g(\mathbf{r}, \mathbf{r}') \times \left[\overline{\mathbf{I}} - \overline{\boldsymbol{\mu}}_r(\mathbf{r}') \right] \cdot \mathbf{H}(\mathbf{r}') d\mathbf{r}' \tag{4.45}$$

The above equations can also be obtained by taking the curl of the EFIEs. They should be used if the magnetic field is the dominant field inside the volumetric scatterer, or when the dominant inhomogeneity is coming from the permeability inside the scatterer.

4.4.2 Matrix Representation of VIE

In this section, we shall derive the matrix representation of one of the integral equations we have derived. The matrix representations of the others should be similar.

When $\overline{\boldsymbol{\mu}}_r = \overline{\mathbf{I}}$, the second term on the right-hand side of (4.39) vanishes, and it is customary to write (4.39) as

$$\mathbf{E}(\mathbf{r}) = \mathbf{E}^{inc}(\mathbf{r}) - k_0^2 \int_{V^+} \overline{\mathbf{G}}(\mathbf{r}, \mathbf{r}') \cdot \overline{\boldsymbol{\xi}}(\mathbf{r}') \cdot \mathbf{D}(\mathbf{r}') d\mathbf{r}' \tag{4.46}$$

where

$$\overline{\boldsymbol{\xi}}(\mathbf{r}') = \frac{1}{\epsilon_0} \left[\overline{\boldsymbol{\epsilon}}_r^{-1}(\mathbf{r}') - \overline{\mathbf{I}} \right] \tag{4.47}$$

The normal component of $\mathbf{D}(\mathbf{r})$ is continuous across the interface of a volume element, and it is expedient to expand $\mathbf{D}(\mathbf{r})$ in terms of divergence conforming expansion functions. Work on this has been reported in the past [36].

However, when $\overline{\boldsymbol{\mu}}_r \neq \overline{\mathbf{I}}$, the existence of $\nabla \times \mathbf{E}(\mathbf{r})$ in (4.39) calls for the use of curl-conforming functions such as the edge elements. Such expansion functions have been used frequently in the finite element literature [6, 7, 37, 38]. When linear edge elements are used, the matrix representation of (4.39) assumes a simple form. Hence, it is prudent to discuss finding the matrix representation of the above when curl-conforming expansion function is used for expanding the electric field $\mathbf{E}$. Assuming that V^+ is infinitesimally larger than V, the support of the scatterer, we can break V^+ into a union of nonoverlapping V_n's. For example, each of the V_n's can represent a tetrahedron if the volume V is tesselated into a union of N tetrahedrons. Consequently, we let

$$\mathbf{E}(\mathbf{r}) = \sum_{n=1}^{N} \sum_{i=1}^{6} I_{ni} \mathbf{E}_{ni}(\mathbf{r}), \quad \mathbf{r} \in V \tag{4.48}$$

where the $\mathbf{E}_{ni}(\mathbf{r})$ is the electric field in the n-th tetrahedron corresponding to the i-th edge, which is represented by the tetrahedral edge element. The expansion function $\mathbf{E}_{ni}(\mathbf{r})$ has several properties that make it useful for representing $\mathbf{E}(\mathbf{r})$.

1. Within each element, the electric field in any direction can be expressed by the six expansion functions defined on the six edges.

2. $\mathbf{E}_{ni}(\mathbf{r})$ has a constant tangential component along edge i, and has no tangential component along the other five edges.

3. The component of $\mathbf{E}_{ni}(\mathbf{r})$ tangential to the i-th edge is constant and continuous, this ensures the continuity of the component of electric field tangent to the edges.

4. I_i represents the amplitude of the tangential component of $\mathbf{E}(\mathbf{r})$ along the i-th edge.

5. The divergence of $\mathbf{E}_{ni}$ is zero inside the tetrahedron for linear edge elements, but it produces delta function singularities on the faces of the tetrahedron.

We assume that the volume of integration is divided into a union of V_n such that $V = \bigcup_n V_n$, where V_n's are nonoverlapping. Within each V_n, $\left[\overline{\mathbf{I}} - \overline{\epsilon}_r(\mathbf{r}')\right] \cdot \mathbf{E}(\mathbf{r}')$ is differentiable, but it is discontinuous from region to region . The action of the divergence operator on $\left[\overline{\mathbf{I}} - \overline{\epsilon}_r(\mathbf{r}')\right] \cdot \mathbf{E}(\mathbf{r}')$ will generate delta function singularities which are equivalent to surface charges. These singularities will introduce a surface integral term. Consequently, using (4.48) in (4.40), we have the scattered field due to the inhomogeneous permittivity part of the scatterer to be

$$
\begin{aligned}
\mathbf{E}^\epsilon(\mathbf{r}) = &-k_0^2 \sum_{n,i} I_{ni} \int_{V_n} d\mathbf{r}' g(\mathbf{r},\mathbf{r}') \left[\overline{\mathbf{I}} - \overline{\epsilon}_r(\mathbf{r}')\right] \cdot \mathbf{E}_{ni}(\mathbf{r}') \\
&- \sum_{n,i} I_{ni} \nabla \int_{V_n^-} g(\mathbf{r},\mathbf{r}') \nabla' \cdot \left[\overline{\mathbf{I}} - \overline{\epsilon}_r(\mathbf{r}')\right] \cdot \mathbf{E}_{ni}(\mathbf{r}') d\mathbf{r}' \\
&- \sum_{n,i} I_{ni} \nabla \int_{S_n} \hat{n}' \cdot g(\mathbf{r},\mathbf{r}') \left[\overline{\mathbf{I}} - \overline{\epsilon}_r(\mathbf{r}')\right] \cdot \mathbf{E}_{ni}(\mathbf{r}') d\mathbf{S}'
\end{aligned}
\tag{4.49}
$$

The surface integral term above can also be derived using careful integration by parts argument, but the derivation is more laborious.

When linear edge elements are used for $\mathbf{E}_{ni}$ such that $\nabla \cdot \mathbf{E}_{ni} = 0$ inside V_n, and we assume that $\overline{\epsilon}_r$ is piecewise constant inside each V_n, the second term in the above vanishes, leaving only the first and the third term. On testing the above with $\mathbf{E}_{mj}(\mathbf{r})$, $m = 1, \ldots, N$, $j = 1, \ldots, 6$, and integrating over V_m^+ which is infinitesimally larger than V_m, we have after using integration by parts that

$$
\begin{aligned}
\langle \mathbf{E}_{mj}, \mathbf{E}^\epsilon \rangle = &-k_0^2 \sum_{n,i} I_{ni} \langle \mathbf{E}_{mj}(\mathbf{r}), g(\mathbf{r},\mathbf{r}') \left[\overline{\mathbf{I}} - \overline{\epsilon}_r(\mathbf{r}')\right], \mathbf{E}_{ni}(\mathbf{r}') \rangle \\
&+ \sum_{n,i} I_{ni} \int_{V_m^+} d\mathbf{r} \nabla \cdot \mathbf{E}_{mj}(\mathbf{r}) \int_{S_n} \hat{n}' \cdot g(\mathbf{r},\mathbf{r}') \left[\overline{\mathbf{I}} - \overline{\epsilon}_r(\mathbf{r}')\right] \cdot \mathbf{E}_{ni}(\mathbf{r}') d\mathbf{r}'
\end{aligned}
\tag{4.50}
$$

Again, if $\mathbf{E}_{mj}$ is a linear edge element, its divergence is zero inside V_m, but has delta function singularities on the faces of V_m, resulting in a surface integral. Consequently, the above becomes

$$
\begin{aligned}
\langle \mathbf{E}_{mj}, \mathbf{E}^\epsilon \rangle = &-k_0^2 \sum_{n,i} I_{ni} \langle \mathbf{E}_{mj}(\mathbf{r}), g(\mathbf{r},\mathbf{r}') \left[\overline{\mathbf{I}} - \overline{\epsilon}_r(\mathbf{r}')\right], \mathbf{E}_{ni}(\mathbf{r}') \rangle \\
&+ \sum_{n,i} I_{ni} \int_{S_m} dS \hat{n} \cdot \mathbf{E}_{mj}(\mathbf{r}) \int_{S_n} \hat{n}' \cdot g(\mathbf{r},\mathbf{r}') \left[\overline{\mathbf{I}} - \overline{\epsilon}_r(\mathbf{r}')\right] \cdot \mathbf{E}_{ni}(\mathbf{r}') d\mathbf{r}'
\end{aligned}
\tag{4.51}
$$

The above surface integral can also be derived using integration by parts if the integral in (4.50) is over V_m instead of over V_m^+.

By applying the same procedure to the part of the scattered field due to the permeability in (4.39), we have

$$
\langle \mathbf{E}_{mj}, \mathbf{E}^\mu \rangle = \sum_{n,i} I_{ni} \langle \nabla \times \mathbf{E}_{mj}(\mathbf{r}), g(\mathbf{r},\mathbf{r}') \left[\overline{\mathbf{I}} - \overline{\mu}_r^{-1}(\mathbf{r}')\right], \nabla' \times \mathbf{E}_{ni}(\mathbf{r}') \rangle
\tag{4.52}
$$

If the edge element tetrahedral expansion functions are used in the above, their curls give rise to a regular field (constant in terms of first order element) inside the tetrahedrons, and surface singularities on the surfaces of the tetrahedrons. However, continuity of the tangential components of the field will cause these surface charges or singularities that come from $\nabla' \times \mathbf{E}_{ni}(\mathbf{r}')$ to cancel each other. Moreover, in testing functions associated with $\nabla \times \mathbf{E}_{mj}(\mathbf{r})$ in the above, the surface singularities will cancel each other except on the surface of the scatterer described by $\overline{\boldsymbol{\mu}}_r^{-1}(\mathbf{r})$.

Consequently, using (4.51) and (4.52) together with the matrix representation of the rest of (4.39), one arrives at the matrix representation of the integral equation (4.39),

$$e_{mj}^{inc} = \sum_{n=1}^{N} \sum_{i=1}^{6} A_{mj,ni} I_{ni}, \quad m = 1, \ldots, N, \quad j = 1, \ldots, 6 \tag{4.53}$$

where

$$e_{mj}^{inc} = \langle \mathbf{E}_{mj}, \mathbf{E}_{inc} \rangle \tag{4.54}$$

$$A_{mj,ni} = \langle \mathbf{E}_{mj}, \mathbf{E}_{ni} \rangle - A_{mj,ni}^{\epsilon} - A_{mj,ni}^{\mu} \tag{4.55}$$

$$A_{mj,ni}^{\epsilon} = -k_0^2 \left\langle \mathbf{E}_{mj}(\mathbf{r}), g(\mathbf{r}, \mathbf{r}') \left[\overline{\mathbf{I}} - \overline{\epsilon}_r(\mathbf{r}') \right], \mathbf{E}_{ni}(\mathbf{r}') \right\rangle$$
$$+ \int_{S_m} dS \hat{n} \cdot \mathbf{E}_{mj}(\mathbf{r}) \int_{S_n} \hat{n}' \cdot g(\mathbf{r}, \mathbf{r}') \left[\overline{\mathbf{I}} - \overline{\epsilon}_r(\mathbf{r}') \right] \cdot \mathbf{E}_{ni}(\mathbf{r}') d\mathbf{r}' \tag{4.56}$$

$$A_{mj,ni}^{\mu} = \left\langle \nabla \times \mathbf{E}_{mj}(\mathbf{r}), g(\mathbf{r}, \mathbf{r}') \left[\overline{\mathbf{I}} - \overline{\boldsymbol{\mu}}_r^{-1}(\mathbf{r}') \right], \nabla' \times \mathbf{E}_{ni}(\mathbf{r}') \right\rangle \tag{4.57}$$

In the above, the tangential continuity of the electric field has not been used. Using the fact that tangential $\mathbf{E}$ is continuous from element to element, the redundancy of I_{ni} can be removed, as the I_{ni}'s from contiguous elements should be the same. This is very much like the assembly process in the FEM with edge elements where redundancies in unknowns are removed. Some numerical results using method presented here are given in [39]. Figure 4.6 shows the comparison of the Mie series solution (exact) with the proposed method. It is seen that as the number of unknowns increases, the agreement with the Mie series solution improves.

4.5   Curl Conforming versus Divergence Conforming Expansion Functions

When we find the matrix representation of an operator such as computing

$$A_{ij} = \langle \mathbf{T}_i, \mathcal{L} \mathbf{F}_j \rangle \tag{4.58}$$

a prerequisite is that such an element exists so that A_{ij} is finite. For instance, if

$$\mathcal{L} = \nabla \nabla \tag{4.59}$$

Then, we can show, after integration by parts that,

$$A_{ij} = -\langle \nabla \cdot \mathbf{T}_i, \nabla \cdot \mathbf{F}_j \rangle \tag{4.60}$$

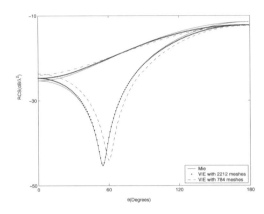

Figure 4.6: Radar cross section for a homogeneous sphere with composite material at moderately low frequency of Mie series (solid line) and the proposed method. The radius of the sphere $a = 0.15\lambda$ with the parameters $\epsilon_r = 1.5$, $\mu_r = 2.2$ with unknowns $N = 784$ and $N = 2,212$ respectively.

If $\mathbf{F}_j(\mathbf{r})$ and $\mathbf{T}_i(\mathbf{r})$ are discontinuous functions, then the $\nabla \cdot \mathbf{F}_j(\mathbf{r})$ and $\nabla \cdot \mathbf{T}_i(\mathbf{r})$ can have delta-function-like singularities and the inner product (4.60) is undefined when these singularities collide in their evaluation. For instance, in one dimension, $\int_{-\infty}^{\infty} dx \delta(x) \delta(x) = \infty$ due to delta function collision. Divergence conforming functions are functions whose divergence is finite and hence A_{ij} exists [40].

Similarly, if

$$\mathcal{L} = \nabla \times \nabla \times \overline{\mathbf{I}} \tag{4.61}$$

then (4.58), after using integration by parts, becomes

$$A_{ij} = \langle \nabla \times \mathbf{T}_i, \nabla \times \mathbf{F}_j \rangle \tag{4.62}$$

If $\mathbf{T}_i(\mathbf{r})$ and $\mathbf{F}_j(\mathbf{r})$ are discontinuous functions such that $\nabla \times \mathbf{T}_i(\mathbf{r})$ and $\nabla \times \mathbf{F}_j(\mathbf{r})$ can have delta-function-like singularities, (4.62) may be infinitely large or undefined. Another possible scenario such that (4.58) is undefined is when

$$\mathcal{L}\mathbf{F}_j = \nabla \int_{V+} g(\mathbf{r}, \mathbf{r}') \nabla' \cdot \mathbf{F}_j(\mathbf{r}') d\mathbf{r}' \tag{4.63}$$

Then the equivalence of (4.60) for this operator is

$$A_{ij} = -\langle \nabla \cdot \mathbf{T}_i, g, \nabla' \cdot \mathbf{F}_j \rangle \tag{4.64}$$

If $g(\mathbf{r}, \mathbf{r}')$ has a singularity, and if both $\nabla \cdot \mathbf{T}_i$ and $\nabla' \cdot \mathbf{F}_j$ have delta-function-like singularities, they will render (4.64) undefined.

4.6 Thin Dielectric Sheet

The thin dielectric sheet (TDS) model can be obtained as a limiting case of a VIE when it is applied to a thin dielectric structure [41–43]. It has also been fervently studied by many workers in recent years. A literature survey is given in [44].

Consider the VIE involving only dielectric inhomogeneity,

$$\mathbf{E}^i(\mathbf{r}) = \frac{-\mathbf{J}_v(\mathbf{r})}{i\omega\epsilon_0[\epsilon_r(\mathbf{r}) - 1]} - i\omega\mu_0 \int_V g(\mathbf{r}, \mathbf{r}')\mathbf{J}_v(\mathbf{r}')dV'$$
$$+ \frac{\nabla}{i\omega\epsilon_0} \int_V g(\mathbf{r}, \mathbf{r}')\nabla' \cdot \mathbf{J}_v(\mathbf{r}')dV' \quad (4.65)$$

where

$$g(\mathbf{r}, \mathbf{r}') = \frac{e^{ik_0|\mathbf{r}-\mathbf{r}'|}}{4\pi|\mathbf{r}-\mathbf{r}'|} \quad (4.66)$$

The VIE can be simplified if we assume that the volume current $\mathbf{J}_v(\mathbf{r})$ inside the TDS is only tangential in nature. This is true when the contrast of the TDS is high, so that the normal component of the field is small in it, and only tangential component dominates. Furthermore, we can assume that the tangential component of $\mathbf{J}_v$ is approximately a constant inside the TDS and that its thickness is much smaller than the wavelength. In this case, we can replace the volume integral with a surface integral and that the volume current can be replaced by a surface current $\mathbf{J}_s(\mathbf{r})$ such that

$$\mathbf{J}_s(\mathbf{r}) = \tau\mathbf{J}_v(\mathbf{r}) \quad (4.67)$$

where τ is the effective thickness of the TDS. Consequently, (4.65) becomes

$$\mathbf{E}^i(\mathbf{r}) \cong Z_s\mathbf{J}_s(\mathbf{r}) - i\omega\mu_0 \int_S g(\mathbf{r}, \mathbf{r}')\mathbf{J}_s(\mathbf{r}')dS'$$
$$+ \frac{\nabla}{i\omega\epsilon_0} \int_S g(\mathbf{r}, \mathbf{r}')\nabla' \cdot \mathbf{J}_s(\mathbf{r}')dS' \quad (4.68)$$

and

$$Z_s(\mathbf{r}) = \frac{i}{\omega\epsilon_0[\epsilon_r(\mathbf{r}) - 1]\tau} \quad (4.69)$$

Eq. (4.68) bears a striking resemblance to EFIE for a surface PEC scatterer. Hence, a computer code for EFIE can be easily modified to accept the TDS model. Moreover, if we let $\epsilon(\mathbf{r}) \sim i\omega\mu\sigma$, when $\sigma \to \infty$, then $Z_s \to 0$, and we must recover the scattering by a PEC object. If the TDS is a lossy dielectric sheet, one can replace the Z_s with a resistive sheet to model the loss inside.

4.6.1   A New TDS Formulation

The aforementioned formulation for the conventional TDS is not valid for low contrast dielectric sheets. Hence a new formulation is required for the low contrast case which has been recently reported [44, 45]. To this end, we write the VIE as

$$\mathbf{E}^i = \frac{\mathbf{D}(\mathbf{r})}{\epsilon_0\epsilon(\mathbf{r})} + \frac{k_0^2}{\epsilon_0} \int_V g(\mathbf{r}, \mathbf{r}')\chi(\mathbf{r}')\mathbf{D}(\mathbf{r}')dV'$$
$$+ \frac{\nabla}{\epsilon_0} \int_V g(\mathbf{r}, \mathbf{r}')\nabla' \cdot [\chi(\mathbf{r}')\mathbf{D}(\mathbf{r}')] dV' \quad (4.70)$$

where

$$\chi(\mathbf{r}) = \epsilon_r^{-1}(\mathbf{r}) - 1 \tag{4.71}$$

Moreover, one can show that

$$\nabla \cdot (\chi \mathbf{D}) = (\nabla \chi) \cdot \mathbf{D} + \chi \nabla \cdot \mathbf{D} = (\nabla \chi) \cdot \mathbf{D} \tag{4.72}$$

since $\nabla \cdot \mathbf{D} = 0$ in a source-free region. Consequently, the last integral in (4.70) becomes

$$I = \int_V g(\mathbf{r}, \mathbf{r}') \nabla' \cdot [\chi(\mathbf{r}') \mathbf{D}(\mathbf{r}')] \, dV' = \int_V g(\mathbf{r}, \mathbf{r}') (\nabla' \chi(\mathbf{r}')) \cdot \mathbf{D}(\mathbf{r}') dV' \tag{4.73}$$

In the above, $\chi(\mathbf{r}') = 0$ outside the TDS volume, and it is a constant inside the TDS volume. It can be shown that

$$\int_V \nabla' \chi(\mathbf{r}') \cdot \mathbf{F}(\mathbf{r}') dV' = \int_S dS' \hat{n}' \cdot [\chi(\mathbf{r}') \mathbf{F}(\mathbf{r}')] \tag{4.74}$$

if $\nabla' \cdot \mathbf{F}(\mathbf{r}')$ is nonsingular. Consequently, we have

$$I = \chi \int_S dS' \hat{n}' \cdot \mathbf{D}(\mathbf{r}') g(\mathbf{r}, \mathbf{r}') \tag{4.75}$$

when χ is assumed constant within TDS. The surface integral can be broken into contribution from the side surface plus the contribution from the top and bottom surfaces. Hence, we can write (4.75) as

$$I = \chi \int_{S_t} dS' \hat{n}_t' \cdot \mathbf{D}_t(\mathbf{r}') g(\mathbf{r}, \mathbf{r}') + \chi \int_{S_n^+} dS' \hat{n}' \cdot \mathbf{D}_n^+(\mathbf{r}') g(\mathbf{r}, \mathbf{r}') dS'$$
$$+ \chi \int_{S_n^-} dS' \hat{n}' \cdot \mathbf{D}_n^-(\mathbf{r}') g(\mathbf{r}, \mathbf{r}') dS' \tag{4.76}$$

where S_t represents the side surface and $\mathbf{D}_t$ represents the component of $\mathbf{D}$ tangential to the TDS, $S_n^{\pm}$ and $\mathbf{D}_n^{\pm}$ represent the top and bottom surfaces and their respective normal components.

We can use the $\nabla \cdot \mathbf{D} = 0$ condition to relate $\mathbf{D}_t$, and $\mathbf{D}_n^{\pm}$, viz.,

$$\nabla \cdot \mathbf{D}(\mathbf{r}) = 0 = \nabla_t \cdot \mathbf{D}_t(\mathbf{r}) + \frac{\partial}{\partial n} \hat{n}_+ \cdot \mathbf{D}_n(\mathbf{r}),$$
$$\approx \nabla_t \cdot \mathbf{D}_t(\mathbf{r}) + \hat{n}_t \cdot \frac{\mathbf{D}_n^+ - \mathbf{D}_n^-}{\tau} \tag{4.77}$$

Using (4.77) in (4.76), we have

$$I \cong \chi \int_{S_t} dS' \hat{n}_t' \mathbf{D}_t(\mathbf{r}') g(\mathbf{r}, \mathbf{r}') + \chi \int_{S_n^+} dS' \hat{n}_+' \cdot \mathbf{D}_n^+(\mathbf{r}') g(\mathbf{r}, \mathbf{r}')$$
$$- \chi \tau \int_{S_n^-} dS' \nabla_t' \cdot \mathbf{D}_t(\mathbf{r}') g(\mathbf{r}, \mathbf{r}') - \chi \int_{S_n^+} dS' \hat{n}_+' \cdot \mathbf{D}_n^+(\mathbf{r}') g(\mathbf{r}, \mathbf{r}') \tag{4.78}$$

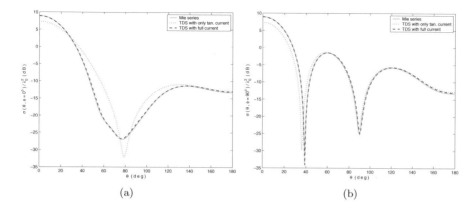

Figure 4.7: Scattering by a dielectric shell with radius 1.0 m, thickness 0.05 m, $\varepsilon_r = 2.6$, and a vertically polarized incident wave from $(\theta, \phi) = (180^0, 0^0)$ at 0.2 GHz. (a) RCS at $\phi = 0^0$. (b) RCS at $\phi = 90^0$.

Note that in the above, we can approximate

$$g(\mathbf{r}, \mathbf{r}')|_{S_n^+} - g(\mathbf{r}, \mathbf{r}')|_{S_n^-} \simeq \tau \frac{\partial g}{\partial n'}(\mathbf{r}, \mathbf{r}')|_{S_n^+} + O(\tau^3) \qquad (4.79)$$

Using (4.79) in (4.78), we have

$$I \cong \chi\tau \int_C dl' \hat{n}_t' \cdot \mathbf{D}_t(\mathbf{r}') g(\mathbf{r}, \mathbf{r}') - \chi\tau \int_{S_n^-} dS' \nabla_t' \cdot \mathbf{D}_t(\mathbf{r}') g(\mathbf{r}, \mathbf{r}')$$
$$+ \chi\tau \int_{S_n^+} dS' \hat{n}_t' \cdot \mathbf{D}_n^+(\mathbf{r}') \frac{\partial g(\mathbf{r}, \mathbf{r}')}{\partial n'} \qquad (4.80)$$

Consequently, the integral equation for TDS can be approximated as

$$\mathbf{E}^i(\mathbf{r}) \simeq \frac{\mathbf{D}(\mathbf{r})}{\epsilon_0 \epsilon_r(\mathbf{r})} - \frac{k_0^2 \tau}{\epsilon_0} \int_S \chi(\mathbf{r}) \mathbf{D}(\mathbf{r}') g(\mathbf{r}, \mathbf{r}') dS'$$
$$+ \frac{\nabla}{\epsilon_0} \left\{ \chi\tau \int_C dl' \hat{n}_t' \cdot \mathbf{D}_t(\mathbf{r}') g(\mathbf{r}, \mathbf{r}') - \chi\tau \int_{S_n^-} dS' \nabla_t' \cdot \mathbf{D}_t(\mathbf{r}') g(\mathbf{r}, \mathbf{r}') \right.$$
$$\left. + \chi\tau \int_{S_n^+} dS' \hat{n}' \cdot \mathbf{D}_n^+(\mathbf{r}') \frac{\partial g(\mathbf{r}, \mathbf{r}')}{\partial n'} \right\} \qquad (4.81)$$

The above integral equation can be solved by expanding $\mathbf{D}_t(\mathbf{r}')$ in terms of divergence conforming expansion function such as RWG function, and $\mathbf{D}_n(\mathbf{r}')$ can be expanded in terms of functions as simple as a pulse function. This integral equation requires less number of unknowns to solve compared to a full-fledge VIE modeling of a TDS (see Figures 4.7, 4.8).

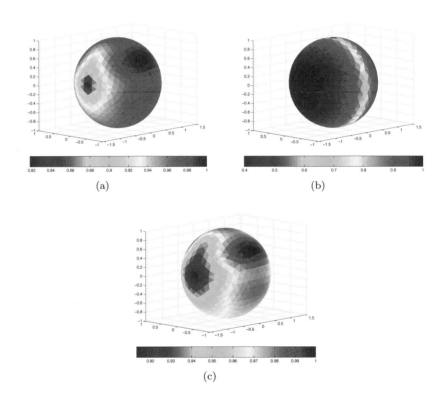

Figure 4.8: Current distribution of the case in the previous figure. (a) Tangential current. (b) Normal current. (c) Full current.

4.7   Impedance Boundary Condition

There have been extensive work done on impedance boundary condition (IBC) [43, 46–50]. The advantage of IBC is that it can greatly simplify the numerical solution of a penetrable object if the penetrable object can be approximated by IBC.

IBCs are used to model a lossy scatterer. In theory, the scattering solution of a lossy scatterer can be formulated in terms of SIEs for penetrable objects. When the conductivities of such objects are high, the integral equations can be approximated using perturbation arguments.

4.7.1   Generalized Impedance Boundary Condition

For a penetrable object, the integral equations for the exterior region can be written using either the electric field or the magnetic field, and similarly for the interior region. Because only current unknowns living on a 2D manifold are solved for, only the tangential components of these equations are needed. For the exterior region, by taking the tangential component of (4.4), we have for EFIE

$$\bar{\mathbf{I}}_t \cdot [\mathcal{L}_{1E}(\mathbf{r}, \mathbf{r}')\mathbf{J}_s(\mathbf{r}') + \mathcal{K}_{1E}(\mathbf{r}, \mathbf{r}')\mathbf{M}_s(\mathbf{r}')] = -\bar{\mathbf{I}}_t \cdot \mathbf{E}^{inc}(\mathbf{r}), \quad \mathbf{r} \in S^- \tag{4.82}$$

where $\bar{\mathbf{I}}_t = \bar{\mathbf{I}} - \hat{n}\hat{n}$ extracts the tangential component of a 3D vector at $\mathbf{r}$. Similarly, by taking the tangential component of (4.11), the MFIE is

$$\bar{\mathbf{I}}_t \cdot [\mathcal{L}_{1H}(\mathbf{r}, \mathbf{r}')\mathbf{M}_s(\mathbf{r}') + \mathcal{K}_{1H}(\mathbf{r}, \mathbf{r}')\mathbf{J}_s(\mathbf{r}')] = -\bar{\mathbf{I}}_t \cdot \mathbf{H}^{inc}(\mathbf{r}), \quad \mathbf{r} \in S^- \tag{4.83}$$

For the interior region, from (4.8), we have for the EFIE

$$\bar{\mathbf{I}}_t \cdot [\mathcal{L}_{2E}(\mathbf{r}, \mathbf{r}')\mathbf{J}_s(\mathbf{r}') + \mathcal{K}_{2E}(\mathbf{r}, \mathbf{r}')\mathbf{M}_s(\mathbf{r}')] = 0, \qquad \mathbf{r} \in S^+ \tag{4.84}$$

From (4.14), we have for the MFIE

$$\bar{\mathbf{I}}_t \cdot [\mathcal{L}_{2H}(\mathbf{r}, \mathbf{r}')\mathbf{M}_s(\mathbf{r}') + \mathcal{K}_{2H}(\mathbf{r}, \mathbf{r}')\mathbf{J}_s(\mathbf{r}')] = 0, \qquad \mathbf{r} \in S^+ \tag{4.85}$$

In principle, one can find $\mathbf{M}_s(\mathbf{r}')$ in terms of $\mathbf{J}_s(\mathbf{r}')$ from (4.84) or (4.85). Using (4.85),

$$\begin{aligned} \mathbf{M}_s(\mathbf{r}') &= -\left[\bar{\mathbf{I}}_t \cdot \mathcal{L}_{2H}(\mathbf{r}', \mathbf{r}'')\right]^{-1} \bar{\mathbf{I}}_t \cdot \mathcal{K}_{2H}(\mathbf{r}'', \mathbf{r}''')\mathbf{J}_s(\mathbf{r}''') \\ &= \mathcal{I}(\mathbf{r}', \mathbf{r}''')\mathbf{J}_s(\mathbf{r}''') \end{aligned} \tag{4.86}$$

The above is the formal solution to the integral equation. The actual numerical solution will be sought by the subspace projection method such as MOM (see previous chapter). The above equation is preferred because in many subspace projection methods, the matrix representation of $\bar{\mathbf{I}}_t \cdot \mathcal{L}_{2H}$ is better conditioned than the matrix representation of $\bar{\mathbf{I}}_t \cdot \mathcal{K}_{2H}$, and hence, its inversion is more easily sought. Also, when medium 2 is highly conductive, $\bar{\mathbf{I}}_t \cdot \mathcal{L}_{2H}$ becomes sparse and quasi diagonal, and its inversion is inexpensive.

Eq. (4.86) represents a nonlocal impedance map. It is a generalized impedance boundary condition (GIBC). It also represents an exact way to solve (4.85). Using the above in (4.82), we can solve for $\mathbf{J}_s$, namely,

$$\mathbf{J}_s = -\left[\bar{\mathbf{I}}_t \cdot \mathcal{L}_{1E} - \bar{\mathbf{I}}_t \cdot \mathcal{K}_{1E}\left(\bar{\mathbf{I}}_t \cdot \mathcal{L}_{2H}\right)^{-1} \bar{\mathbf{I}}_t \cdot \mathcal{K}_{2H}\right]^{-1} \bar{\mathbf{I}}_t \cdot \mathbf{E}^{inc} \tag{4.87}$$

The inversion of the above can also be sought by iterative solvers if needed.

4.7.2   Approximate Impedance Boundary Condition

For highly conductive objects, simplifying approximation to (4.86) can be obtained. For instance, in the case of quasi-planar half-space, in place of (4.86), one may use a simple local impedance map as an approximation, that is,

$$\hat{n} \times \mathbf{M}_s \approx Z_s \mathbf{J}_s \tag{4.88}$$

where $Z_s = \eta_2 \approx \sqrt{\omega \mu_2/(i\sigma_2)}$ is the surface impedance [51] which is the same as the intrinsic impedance of medium 2. In fact, the above can be derived from the GIBC. In this manner, one can solve (4.82) or (4.83) with (4.88) to obtain $\mathbf{J}_s$. Again, (4.82) or (4.83) alone may be fraught with internal resonance problem. In this case, one can combine them and solves the CFIE to avoid internal resonances.

The above is an approximate method to solve the interior equations. We can also simplify the exterior equations. As an example, we rewrite (4.82) as

$$\bar{\mathbf{I}}_t \cdot \mathbf{E}^{inc}(\mathbf{r}) + \bar{\mathbf{I}}_t \cdot \mathcal{L}_{1E}(\mathbf{r}, \mathbf{r}')\mathbf{J}_s(\mathbf{r}') = \bar{\mathbf{I}}_t \cdot \mathcal{K}_{1E}(\mathbf{r}, \mathbf{r}')\mathbf{M}_s(\mathbf{r}'), \quad \mathbf{r} \in S^- \tag{4.89}$$

and we approximate the right-hand side with a local approximation. For example, for a quasi-planar surface, we may approximate the right-hand side as

$$\bar{\mathbf{I}}_t \cdot \mathcal{K}_{1E}(\mathbf{r}, \mathbf{r}')\mathbf{M}_s(\mathbf{r}') \approx -\frac{1}{2}\hat{n} \times \mathbf{M}_s(\mathbf{r}), \quad \mathbf{r} \in S^- \tag{4.90}$$

This second local approximation, together with our first local approximation, (4.88), yields the approximation that

$$\bar{\mathbf{I}}_t \cdot \mathcal{K}_{1E}(\mathbf{r}, \mathbf{r}')\mathbf{M}_s(\mathbf{r}') \approx -\frac{1}{2}Z_s \mathbf{J}_s, \quad \mathbf{r} \in S^- \tag{4.91}$$

When Z_s is very small, and when the above is used in (4.89), the right-hand side of (4.89) is much smaller than the individual terms on the left hand side. Hence, we may also approximate (4.89) as

$$\bar{\mathbf{I}}_t \cdot \mathbf{E}^{inc}(\mathbf{r}) + \bar{\mathbf{I}}_t \cdot \mathcal{L}_{1E}(\mathbf{r}, \mathbf{r}')\mathbf{J}_s(\mathbf{r}') \approx 0, \quad \mathbf{r} \in S^- \tag{4.92}$$

and solve for $\mathbf{J}_s$ directly. Then we can use (4.88) to obtain an approximate value for $\mathbf{M}_s$. One can also easily derive the MFIE equivalence of the above equations.

We shall investigate how good these local approximations are by studying reflection from a flat surface and reflection from a cylindrical surface.

4.7.3   Reflection by a Flat Lossy Ground Plane

Let us consider first a general reflection of a plane wave by a penetrable half-space (Figure 4.9). This problem has closed form solution, and the electric field in region 1 can be written as

$$\mathbf{E} = \hat{x}E_1 \left(e^{-ik_1 z} + R_{12}e^{ik_1 z} \right) \tag{4.93}$$

$$\mathbf{H} = -\hat{y}\frac{E_1}{\eta_1} \left(e^{-ik_1 z} - R_{12}e^{ik_1 z} \right) \tag{4.94}$$

Figure 4.9: Reflection by a conductive half space can be used to derive an impedance boundary condition.

The above is valid even for a general half space. In general, the tangential $\mathbf{E}$ field is not zero at $z = 0$. Also, the extinction theorem for $z < 0$ can be applied exactly for this problem even if the half space is not conductive.

To apply the extinction theorem, we need $\mathbf{M}_s \neq 0$ and $\mathbf{J}_s \neq 0$ in the following equation

$$\mathbf{E}^{inc} + \mathcal{L}_{1E}(\mathbf{J}_s) + \mathcal{K}_{1E}(\mathbf{M}_s) = 0, \qquad z < 0 \tag{4.95}$$

We ignore the $\overline{\mathbf{I}}_t \cdot$ operation because for this 1D problem, the field is naturally tangential. In the above, the equivalence surface current $\mathbf{J}_s$ equals the discontinuity of the magnetic field. Hence, from (4.94)

$$\mathbf{J}_s = \hat{x}\frac{E_1}{\eta_1}(1 - R_{12}) \tag{4.96}$$

Furthermore, $\mathbf{M}_s$ equals the discontinuity in the total tangential $\mathbf{E}$ field, which in this case, from (4.93), is

$$\mathbf{M}_s = -\hat{y}E_1(1 + R_{12}) \tag{4.97}$$

In (4.95), $\mathcal{L}_{1E}(\mathbf{J}_s)$ is the $\mathbf{E}$ field produced by an infinite electric current sheet (4.96) extended over the xy plane, and for $z < 0$, the electric field it produces is

$$\mathcal{L}_{1E}(\mathbf{J}_s) = \frac{1}{2}\hat{x}E_1(1 - R_{12})e^{-ik_1 z} \tag{4.98}$$

Remember that in invoking the extinction theorem, we remove medium 2, and replace it with medium 1; hence, we use the wavenumber k_1, in (4.98) for $z < 0$. In (4.95), $\mathcal{K}_{1E}(\mathbf{M}_s)$ is the electric field produced by an infinite magnetic current sheet (4.97), and for $z < 0$

$$\mathcal{K}_{1E}(\mathbf{M}_s) = \frac{1}{2}\hat{x}E_1(1 + R_{12})e^{-ik_1 z} \tag{4.99}$$

The incident field is

$$\mathbf{E}^{inc} = \hat{x}E_1 e^{-ik_1 z} \tag{4.100}$$

Using (4.98), (4.99), and (4.100) in (4.95), we see the extinction theorem is exactly satisfied for $z < 0$ for this general penetrable half space problem.

Next, we want to see if the equivalent currents $\mathbf{M}_s$ and $\mathbf{J}_s$ are related in a simple fashion when the half-space is metallic or when its conductivity is high. To this end, we make a high conductivity approximation to R_{12} where

$$R_{12} = \frac{\mu_2 k_1 - \mu_1 k_2}{\mu_2 k_1 + \mu_1 k_2} = -\frac{1 - \frac{\mu_2}{\mu_1}\frac{k_1}{k_2}}{1 + \frac{\mu_2}{\mu_1}\frac{k_1}{k_2}} = -\frac{1 - \eta_2 \eta_1^{-1}}{1 + \eta_2 \eta_1^{-1}} \simeq -1 + 2\eta_2 \eta_1^{-1} \tag{4.101}$$

where $\eta_i = \sqrt{\frac{\mu_i}{\epsilon_i}}$. If region 2 is highly conductive, $\eta_2 \simeq \sqrt{\frac{\omega \mu_2}{i \sigma_2}} \to 0, \sigma_2 \to \infty$. Using this in (4.96) and (4.97), we have

$$\mathbf{J}_s \cong \hat{x}2\frac{E_1}{\eta_1}, \qquad \mathbf{M}_s = -\hat{y}2E_1\eta_2\eta_1^{-1} \tag{4.102}$$

It is clear that $\hat{n} \times \mathbf{M}_s = Z_s \mathbf{J}_s$ where $Z_s = \eta_2 = \sqrt{\frac{\omega \mu_2}{i \sigma_2}}$. Hence the surface impedance approximation, the first local approximation indicated by (4.88) is at least good for this problem with the appropriate value of Z_s.

Furthermore, since

$$\mathcal{K}_{1E}(\mathbf{M}_s) = \pm\frac{1}{2}\hat{n} \times \mathbf{M}_s + P.V.\mathcal{K}_{1E}(\mathbf{M}_s) \tag{4.103}$$

We can ignore the $P.V.\mathcal{K}$ term: for a flat surface, it is exactly zero, and for a quasi-planar surface, it should be small. Hence, the second local approximation indicated by (4.90) is also reasonable. Consequently, (4.95) becomes

$$\mathbf{E}^{inc} + \mathcal{L}_{1E}(\mathbf{J}_s) - \frac{1}{2}\hat{n} \times \mathbf{M}_s = 0, \qquad z < 0 \tag{4.104}$$

If we let $\hat{n} \times \mathbf{M}_s = \eta_2 \mathbf{J}_s$, then (4.104) can be readily solved for $\mathbf{J}_s$ as this is a simple 1D problem. We can also ignore the $\mathbf{M}_s$ term in (4.104) and solve for $\mathbf{J}_s$ and show that it is correct to leading order in η_2 when we assume that η_2 is a small parameter.

4.7.4   Reflection by a Lossy Cylinder

Many lossy structures are wire-like or cylinder-like. Hence, it is useful to see how the IBC is affected by the presence of a curvature as in an infinitely long cylinder.

For simplicity of analysis, we consider a single cylindrical harmonic wave with $\partial/\partial z = 0$ incident on the conducting cylinder. The general penetrable circular cylinder scattering can be solved exactly (Figure 4.10). We have

$$\mathbf{E}_1 = \hat{z}E_1\left[J_n(k_1\rho) + R_{12}H_n^{(1)}(k_1\rho)\right]e^{in\phi} \tag{4.105}$$

$$\mathbf{E}_2 = \hat{z}E_1 T_{12} J_n(k_2\rho)e^{in\phi} \tag{4.106}$$

$$H_{1\phi} = \frac{i}{\eta_1}E_1\left[J_n'(k_1\rho) + R_{12}H_n^{(1)'}(k_1\rho)\right]e^{in\phi} \tag{4.107}$$

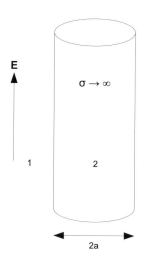

Figure 4.10: Reflection by a lossy cylindrical surface can be used to derive an IBC for a circular wire.

$$H_{2\phi} = \frac{i}{\eta_2} E_1 T_{12} J_n'(k_2\rho) e^{in\phi} \tag{4.108}$$

For $n \neq 0$, the magnetic field may have a nonzero H_ρ component, but only the H_ϕ component is needed for the ensuing discussions.

To show the extinction theorem in action for $\rho < a$, we need the surface equivalence electric and magnetic currents. The surface equivalence electric current $\mathbf{J}_s$ is

$$\mathbf{J}_s = \hat{z} \frac{iE_1}{\eta_1} \left[J_n'(k_1a) + R_{12} H_n^{(1)'}(k_1a) \right] e^{in\phi} = \hat{z} J_0 e^{in\phi} \tag{4.109}$$

The surface equivalence magnetic current $\mathbf{M}_s$ is

$$\mathbf{M}_s = \hat{\phi} E_1 \left[J_n(k_1a) + R_{12} H_n^{(1)}(k_1a) \right] e^{in\phi} = \hat{\phi} M_0 e^{in\phi} \tag{4.110}$$

To invoke the extinction theorem, we remove medium 2, and investigate the fields produced by a current sheet $\mathbf{J}_s$. If we have $\mathbf{J}_s = \hat{z} J_0 e^{in\phi}$, it can be shown that such an electric current sheet produces an electric field which is continuous across the current sheet, and a magnetic field which is discontinuous across the same current sheet. The electric field is given by

$$\mathcal{L}_{1E}(\mathbf{J}_s) = \mathbf{E}_J = -\hat{z} \frac{\eta_1 J_0 \pi k_1 a}{2} e^{in\phi} \begin{cases} J_n(k_1a) H_n^{(1)}(k_1\rho), & \rho > a \\ H_n^{(1)}(k_1a) J_n(k_1\rho), & \rho < a \end{cases} \tag{4.111}$$

The corresponding $H_{J\phi}$ is given by

$$H_{J\phi} = -\frac{iJ_0\pi k_1 a}{2} e^{in\phi} \begin{cases} J_n(k_1 a) H_n^{(1)'}(k_1\rho), & \rho > a \\ H^{(1)}(k_1 a) J_n'(k_1\rho), & \rho < a \end{cases} \tag{4.112}$$

On the other hand, the magnetic current sheet $\mathbf{M}_s = \hat{\phi} M_0 e^{in\phi}$ will give rise to a discontinuous electric field and a continuous magnetic field. Namely, we have

$$\mathcal{K}_{1E}(\mathbf{M}_s) = \mathbf{E}_M = -\hat{z}\frac{M_0\pi k_1 a}{2i} e^{in\phi} \begin{cases} J_n'(k_1 a) H_n^{(1)}(k_1\rho), & \rho > a \\ H_n^{(1)'}(k_1 a) J_n(k_1\rho), & \rho < a \end{cases} \tag{4.113}$$

$$H_{M\phi} = -\frac{M_0\pi k_1 a}{2\eta} e^{in\phi} \begin{cases} J_n'(k_1 a) H_n^{(1)'}(k_1\rho), & \rho > a \\ H_n^{(1)'}(k_1 a) J_n'(k_1\rho), & \rho < a \end{cases} \tag{4.114}$$

The incident field in this problem is given by the first part of (4.105), viz.,

$$\mathbf{E}^{inc} = \hat{z} E_1 J_n(k_1\rho) e^{in\phi} \tag{4.115}$$

To show the validity of the extinction theorem, we need to take terms from (4.111) and (4.113) for $\rho < a$ together with (4.115), and show that

$$E_1 - \frac{\eta_1 J_0 \pi k_1 a}{2} H_n^{(1)}(k_1 a) - \frac{M_0\pi k_1 a}{2i} H_n^{(1)'}(k_1 a) = 0 \quad , \tag{4.116}$$

with J_0 and M_0 given by (4.109) and (4.110). To evaluate them, the reflection coefficient R_{12} is needed, and it can be derived to be [16]

$$R_{12} = -\frac{\eta_1 J_n'(k_2 a) J_n(k_1 a) - \eta_2 J_n(k_2 a) J_n'(k_1 a)}{\eta_1 J_n'(k_2 a) H_n^{(1)}(k_1 a) - \eta_2 J_n(k_2 a) H_n^{(1)'}(k_1 a)} \tag{4.117}$$

As a result, J_0 is given by

$$\begin{aligned} J_0 &= \frac{iE_1}{\eta_1} \left[J_n'(k_1 a) + R_{12} H_n^{(1)'}(k_1 a) \right] \\ &= \frac{iE_1 J_n'(k_2 a) \frac{-2i}{\pi k_1 a}}{\eta_1 J_n'(k_2 a) H_n^{(1)}(k_1 a) - \eta_2 J_n(k_2 a) H_n^{(1)'}(k_1 a)} \end{aligned} \tag{4.118}$$

where the Wronskian of Bessel functions has been used, that is,

$$H_n^{(1)}(x) J_n'(x) - J_n(x) H_n^{(1)'}(x) = -\frac{2i}{\pi x}.$$

Similarly, M_0 is

$$\begin{aligned} M_0 &= E_1 \left[J_n(k_1 a) + R_{12} H_n^{(1)}(k_1 a) \right] \\ &= \frac{E_1 \eta_2 J_n(k_2 a) \frac{-2i}{\pi k_1 a}}{\eta_1 J_n'(k_2 a) H_n^{(1)}(k_1 a) - \eta_2 J_n(k_2 a) H_n^{(1)'}(k_1 a)} \end{aligned} \tag{4.119}$$

Consequently, we can show by substituting (4.118) and (4.119) that (4.116) is true for a general penetrable circular cylinder.

From the above, by comparing (4.118) with (4.119), we deduce that

$$M_0 = -i\eta_2 \frac{J_n(k_2a)}{J_n'(k_2a)} J_0 \qquad (4.120)$$

The above is equivalent to

$$\hat{n} \times \mathbf{M}_s = -i\eta_2 \frac{J_n(k_2a)}{J_n'(k_2a)} \mathbf{J}_s \qquad (4.121)$$

The above is dependent on the the harmonic order n, and hence it is not convenient for use. It should be applied to the dominant harmonic around a cylinder. It can be shown that when the radius of the cylinder is small compared to wavelength, the dominant harmonic is the $n = 0$ harmonic.

In the limit when k_2 corresponds to a highly conductive medium, using the large argument approximations to the Bessel functions in (4.120) or (4.121), the above becomes $M_0 = \eta_2 J_0$ or $\hat{n} \times \mathbf{M}_s = \eta_2 \mathbf{J}_s$. This result is valid for arbitrary n that is finite. Hence, the equivalence of (4.82) for a cylinder can be solved by using the above IBCs.

Next we would like to know if the approximation

$$\mathcal{K}_{1E}(\mathbf{M}_s) \simeq -\frac{1}{2}\hat{n} \times \mathbf{M}_s, \quad \mathbf{r} \in S^-$$

is still good for a cylinder. From (4.113), by setting $\rho = a^-$, it can be seen that when $k_1a \to \infty$ with k_1 having a small amount of loss, and the large argument approximations are used for the Hankel and Bessel functions, or the radius of curvature of the cylinder is large compared to wavelength, this approximation is a good one. This confirms our assertion that this approximation is good for quasi-planar surfaces. The small amount of loss is needed so that the Bessel wave, which is a standing wave, can be approximated by an incoming wave around the cylinder.

Finally, we want to study how the $\mathcal{K}_{1E}(\mathbf{M}_s)$ term can be approximated around a circular wire. Because the monopole or $n = 0$ component is dominant when $k_1a \to 0$, we may approximate (4.113) as

$$\mathcal{K}_{1E}(\mathbf{M}_s) \approx \hat{n} \times \mathbf{M}_s C_0 \qquad (4.122)$$

where

$$C_0 = \frac{\pi k_1 a}{2i} H_0^{(1)'}(k_1a) J_0(k_1a) \qquad (4.123)$$

When $k_1a \to 0$, we can further have the approximation that

$$C_0 \sim 1, \qquad k_1a \to 0 \qquad (4.124)$$

Therefore, we see that the local approximation given by (4.122) for a thin wire is in fact quite different compared to the local approximation of a quasi-planar surface given by (4.90).

To test the idea espoused here, we calculate the input admittance of a copper loop excited by a delta gap source. The copper loop is 1984 μm by 122 μm, and it has a cross section of 4 μm by 4 μm. For copper, the conductivity is chosen as 5.8e7 s/m. The skin depth decreased

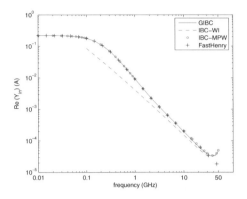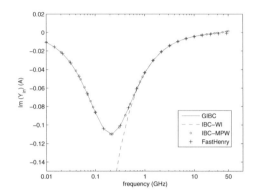

Figure 4.11: Real and imaginary parts of Y_{in} for the copper rectangular loop. In the figure, GIBC is compared with three other results. IBC-WI refers to IBC based on wave impedance of the conductive medium. IBC-MPW refers to IBC based on the modified plane wave model introduced in [52]. FastHenry is the code from J. White's group at MIT [53].

from 20.9 μm at 0.01 GHz to 0.3 μm at 50 GHz. Via the GIBC, the input admittance could be compared to the one from FastHenry [53], IBC-WI based on wave impedance and IBC-MPW based on modified plane wave model [52]. Since empirically FastHenry works for electrical size less than 0.1 wavelength, it will provide accurate results up to about 10 GHz. From the numerical results in Figure 4.11, the GIBC shows accurate solution for conductor analysis over a broadband of frequencies. More of this work including numerical examples can be found in [54].

4.8 Conclusions

In this chapter, we have discussed the formulations of integral equations for penetrable objects. They can be solved with SIEs. We have shown by Gedanken experiments how some of the formulations avoid the internal resonance problem commonly encountered in SIEs. We have also discussed the use of VIEs to solve the inhomogeneous medium penetrable object case. Then we presented various formulations for TDS. Finally, we have discussed the use of impedance boundary condition (IBC) and generalized impedance boundary condition (GIBC) to approximately model lossy penetrable objects.

Bibliography

[1] R. C. Hansen, ed., *Geometric Theory of Diffraction,* New York: IEEE Press, 1981.

[2] P. C. Waterman, "Scattering by dielectric obstacles," *Alta Freq.,* vol. 38, pp. 348–352, 1969.

[3] A. C. Ludwig, "The generalized multipole method," *Comput. Phys. Commun.,* vol. 68, pp. 306–314, 1991.

[4] M. A. Morgan, ed., *Finite Element and Finite Difference Methods in Electromagnetic Scattering,* Amsterdam: Elsevier, 1990.

[5] A. Taflove and K. R. Umashankar, "Review of FDTD numerical modeling of electromagnetic wave scattering and radar cross-section," *Proc. IEEE,* vol. 77, pp. 682–699, 1989.

[6] P. P. Silvester and R. L. Ferrari, *Finite Elements for Electrical Engineers,* Cambridge, MA: Cambridge University Press, 1986.

[7] J. M. Jin, *The Finite Element Method in Electromagnetics,* New York: Wiley, 1993.

[8] J. L. Volakis, A. Chatterjee and L. C. Kempel, *Finite Element Method for Electromagnetics: Antennas, Microwave Circuits, and Scattering Applications,* New York: IEEE Press and Oxford University Press, 1998.

[9] R. F. Harrington, *Field Computation by Moment Methods,* Malabar, FL: Krieger, 1968.

[10] J. Richmond, "Scattering by a dielectric cylinder of arbitrary cross section shape," *IEEE Trans. Antennas Propag.,* vol. 13, no. 3, pp. 334-341, May 1965.

[11] J. Richmond, "TE-wave scattering by a dielectric cylinder of arbitrary crosssection shape," *IEEE Trans. Antennas Propag.,* vol. 14, no. 4, pp. 460-464, Jul. 1966.

[12] A. J. Poggio and E. K. Miller, "Integral equation solutions of three-dimensional scattering problems," in *Computer Techniques for Electromagnetics,* R. Mittra, Ed., 2nd ed., Hemisphere, New York, 1987, pp. 159–264.

[13] E. K. Miller, L. Medgyesi-Mitschang, and E. H. Newman, Ed.,*Computational Electromagnetics: Frequency-Domain Method of Moments,* New York: IEEE Press, 1992.

[14] L. N. Medgyesi-Mitschang, J. M. Putnam, and M. B. Gedera, "Generalized method of moments for three-dimensional penetrable scatterers," *J. Opt. Soc. Am. A*, vol. 11, pp. 1383–1398, Apr. 1994.

[15] R. F. Harrington, *Time-Harmonic Electromagnetic Fields*, New York: McGraw-Hill, 1961.

[16] W. C. Chew, *Waves and Fields in Inhomogeneous Media,* New York: Van Nostrand Reinhold, 1990. Reprinted by Piscataway, NJ: IEEE Press, 1995.

[17] W. C. Chew, J.-M. Jin, E. Michielssen, and J. M. Song, *Fast and Efficient Algorithms in Computational Electromagnetics*, Boston: Artech House, 2001.

[18] J. R. Mautz and R. F. Harrington, "*H*-field, *E*-field, and combined-field solutions for conducting body of revolution," *AEÜ*, vol. 32, pp. 157–164, Apr. 1978.

[19] R. F. Harrington, "Boundary integral formulations for homogenous material bodies," *J. Electromagn. Waves and Appl.*, vol. 3, no. 1, pp. 1-15, 1989.

[20] C. Müller, *Foundation of the Mathematical Theory of Electromagnetic Waves*, Berlin: Springer-Verlag, 1969.

[21] P. Werner "On the exrerior boundary value problem of the perfect reflection for the stationary electromagnetic wave fields." *J. Math. Anal. Appl.*, vol. 7, pp. 348–396, 1963.

[22] H. A. Schenck, "Improved integral formulation for acoustic radiation problems," *J. Acoust. Soc. Am.*, vol. 44, pp. 41–58, 1968.

[23] K. M. Mitzner, "Numerical solution of the exterior scattering problem at the eigenfrequencies of the interior problem." *URSI Meeting Digest*, pp. 75, Boston, 1968.

[24] A. J. Burton and G. F. Miller, "The application of integral equation methods to the numerical solution of some exterior boundary value problems." *Proc. R. Soc., Lond.* A323, pp. 201–210, 1971.

[25] J. C. Bolomey and W. Tabbara, "Numerical aspects on coupling between complementary boundary value problems." *IEEE Trans. Antennas Propag.*, vol. 21, pp. 356–63, 1973.

[26] R. Mittra and C. A. Klein, "Stability and convergence of moment method solutions," In *Numerical and Asymptotic Techniques in Electromagnetics*, Ed. R. Mittra. New York: Springer-Verlag, 1975.

[27] N. Morita, "Resonant solutions involved in the integral equation approach to scattering from conducting and dielectric cylinders." *IEEE Trans. Antennas Propag.*27, pp. 869–871, 1979.

[28] A. D. Yaghjian, "Augmented electric- and magnetic-field integral equations," *Radio Sci.*, vol. 16, pp. 987–1001, 1981.

[29] A. F. Peterson, "The interior resonance problem associated with surface integral equations of electromagnetics: numerical consequences and a survey of remedies," *Electromagn.*, vol. 10, pp. 293–312, 1990.

[30] W. C. Chew and J. M. Song, "Gedanken experiments to understand the internal resonance problems of electromagnetic scattering," *Electromagn.*, vol. 27, no. 8, pp. 457–472, Nov.-Dec. 2007.

[31] J. M. Song, W. W. Shu, and W. C. Chew, "Numerical resonances in method of moments," *IEEE APS Int. Symp. Dig.*, Honolulu, Hawaii, pp. 4861-4864, Jun. 2007.

[32] D. R. Wilton, "Review of current status and trends in the use of integral equations in computational electromagnetics," *Electromagn.*, vol. 12, pp. 287-341, 1992.

[33] J. R. Mautz and R. F. Harrington, "A combined-source solution for radiation and scattering from a perfectly conducting body," *IEEE Trans. Antennas Propag.*, AP-27, no. 4, pp. 445–454, Jul. 1979.

[34] J. C. Monzon and N. J. Damaskos, "A scheme for eliminating internal resonances: the parasitic body technique," *IEEE Trans. Antennas Propag.*, vol. 42, no. 8, pp. 1089–1091, 1994.

[35] K. F. Warnick and W. C. Chew, "Error analysis of surface integral equation methods," in *Fast and Efficient Algorithms in Computational Electromagnetics,* eds. W. Chew, J. Jin, E. Michielssen, J. Song, Boston: Artech House, 2001.

[36] H. Gan and W. C. Chew, "A discrete BCG-FFT algorithm for solving 3D inhomogeneous scatterer problems," *J. Electromagn. Waves and Appl.*, vol. 9, no. 10, pp. 1339-1357, 1995.

[37] J. C. Nedelec, "Mixed finite elements in $\mathbb{R}^3$," *Numer. Math.*, vol. 35, pp. 315–341, 1980.

[38] A. Bossavit: "On Finite Elements for the Electricity Equation," in *The Mathematics of Finite Elements* (J.R. Whiteman, ed.), London: Academic Press, pp. 85–92, 1982.

[39] W. C. Chew and L. E. Sun, "A novel formulation of the volume integral equation for electromagnetic scattering," *Tenty-Third Ann. Rev. Prog. Appl. Comput. Electromag.*, Verona, Italy, Mar. 19–23, 2007.

[40] R. D. Graglia, D. R. Wilton, and A. F. Peterson, "Higher order interpolatory vector bases for computational electromagnetics," *IEEE Trans. Antennas Propag.*, vol. 45, no. 3, pp. 329–342, Mar. 1997.

[41] R. F. Harrington and J. R. Mautz, "An impedance sheet approximation for thin dielectric shells," *IEEE Trans. Antennas Propag.*, vol. AP-23, pp. 531-534, Jul. 1975.

[42] E. H. Newman and M. R. Schrote, "An open surface integral formulation for electromagnetic scattering by material plates," *IEEE Trans. Antennas Propag.*, vol. AP-32, pp. 672-678, Jul. 1984.

[43] E. Bleszynski, M. Bleszynski, and T. Jaroszewicz, "Surface-integral equations for electromagnetic scattering from impenetrable and penetrable sheets," *IEEE Antennas Propag. Mag.*, vol. 35, pp. 14-25, Dec. 1993.

[44] I. T. Chiang and W. C. Chew, "Thin dielectric sheet simulation by surface integral equation using modified RWG and pulse bases," *IEEE Trans. Antennas Propag.*, vol. 54, no. 7, pp. 1927–1934, Jul. 2006.

[45] I. T. Chiang and W. C. Chew, "A Coupled PEC-TDS Surface Integral Equation Approach for Electromagnetic Scattering and Radiation from Composite Metallic and Thin Dielectric Objects," *IEEE Trans. Antennas Propag.*, vol. 54, no. 11, pp. 3511–3516, Nov. 2006.

[46] S. W. Lee and W. Gee, "How good is the impedance boundary condition?" *IEEE Trans. Antennas Propag.*, vol. AP-35, pp. 1313–1315, Nov. 1987.

[47] T. B. A. Senior and J. L. Volakis, "Derivation and application of a class of generalized impedance boundary conditions," *IEEE Trans. Antennas Propag.*, vol. AP-37, pp. 1566–1572, Dec. 1989.

[48] D. J. Hoppe and Y. Rahmat-Samii, "Scattering by superquadric dielectric coated cylinders using high order impedance boundary conditions," *IEEE Trans. Antennas Propag.*, vol. 40, pp. 1513–1523, Dec. 1992.

[49] O. Marceaux and B. Stupfel, "High-order impedance boundary condition for multilayer coated 3-D objects," *IEEE Trans. Antennas Propag.*, vol. 48, pp. 429–436, Mar. 2000.

[50] B. Stupfel and Y. Pion, "Impedance boundary conditions for finite planar and curved frequency selective surfaces," *IEEE Trans. Antennas Propag.*, vol. 53, pp. 1415–1425 Apr. 2005.

[51] J. A. Stratton, *Electromagnetic Theory*, New York: McGraw-Hill, 1941.

[52] B.-T. Lee, "Efficient Series Impedance Extraction Using Effective Internal Impedance," *Ph.D. Thesis*, University of Texas at Austin, Aug. 1996.

[53] M. Kamon, M. J. Tsuk, and J. White, "FastHenry: A multipole-accelerated 3-D inductance extraction program," *IEEE Trans. Microwave Theory Tech.*, vol. 42, no. 9, pp. 1750-58, Sept. 1994.

[54] Z. G. Qian, W. C. Chew, and R. Suaya, "Generalized impedance boundary condition for conductor modeling in surface integral equation," *IEEE Trans. Microwave Theory Tech.*, vol. 55, no. 11, pp. 2354–2364, Nov. 2007.

CHAPTER 5

Low-Frequency Problems in Integral Equations

5.1 Introduction

The development of a number of technologies has called for the solutions of complex structures when the structures are small compared to wavelength, and yet the complexity of these structures cannot be ignored. These are the area of high-speed circuit in computer chip and package designs, the area of small antennas where the antennas can be a fraction of a wavelength, but its complex design is important, as well as the area of nanotechnology and metamaterial. These complex subwavelength structures are richly endowed with physics, especially when they are coupled with nonlinear devices such as transistors and diodes. Their microscopic designs manifest themselves macroscopically in a vastly different manner. An example of such is in radio frequency identification (RFID) tag [1].

One can relate the richness of the physics that is buried in microcircuits as due to the singularity of the dyadic Green's function ([2, Chap. 7]). Such a singularity is not found for instance in the elastodynamic dyadic Green's function [3], marking a contrast between micro devices in the mechanical world and those in the electrical world.

It is important to solve the electromagnetic equations in multiscales, and yet preserve the microscale physics, as it can be manifested on a macroscopic scale. This is unlike some area where the microscale physics can be homogenized for their connection to the macroscale physics. In this chapter, we discuss the solutions of integral equations when the wavelength is much larger than the dimensions of the structures.

5.2 Low-Frequency Breakdown of Electric Field Integral Equation

The electric field integral equation (EFIE) suffers from a low-frequency breakdown problem [4–11]. This problem, however, is not specific to integral equations. It is also germane to differential equation solutions of electromagnetics. In other words, a numerical electromagnetic solver, when written for mid frequency, may not work well for low frequency.

We can foresee this problem by looking at the electromagnetic equations when $\omega \to 0$ [11]. Right at zero frequency, they become

$$\nabla \times \mathbf{E} = 0, \tag{5.1}$$

$$\nabla \times \mathbf{H} = \mathbf{J}, \tag{5.2}$$

$$\nabla \cdot \epsilon \mathbf{E} = \rho = \lim_{\omega \to 0} \frac{\nabla \cdot \mathbf{J}}{i\omega}, \tag{5.3}$$

$$\nabla \cdot \mu \mathbf{H} = 0. \tag{5.4}$$

Notice that from (5.2), the current $\mathbf{J}$ that produces the magnetic field must be divergence free. However, the current $\mathbf{J}$ that produces the charge ρ in (5.3) cannot be divergence free, but in order to keep the right-hand side of (5.3) bounded, this $\mathbf{J} \sim O(\omega)$, when $\omega \to 0$. Hence we see the total $\mathbf{J}$ naturally decomposes into a solenoidal (divergence free) part and a irrotational (curl-free) part. Namely

$$\mathbf{J} = \mathbf{J}_{sol} + \mathbf{J}_{irr} \tag{5.5}$$

This is also known as the Helmholtz decomposition [12]. Moreover, the curl-free part, $\mathbf{J}_{irr} \sim O(\omega)$ when $\omega \to 0$ as alluded to above, and hence,

$$|\mathbf{J}_{irr}| \ll |\mathbf{J}_{sol}|, \qquad \omega \to 0 \tag{5.6}$$

If the total $\mathbf{J}$ is the unknown current to be solved for, the above poses a severe numerical problem.[1] One may think that perhaps $\mathbf{J}_{irr}$ is not important when $\omega \to 0$, since it is so much smaller than $\mathbf{J}_{sol}$. However, this is not the case. The electromagnetic equations separate into the magnetoquasistatic equations and the electroquasistatic equations. In circuit theory, the magnetoquasistatic equations are important for describing the world of inductors, whereas the electroquasistatic equations are important for the world of capacitors. Both are equally important in capturing circuit physics. Hence, the accuracy of one cannot be compromised over the other. The solenoidal current $\mathbf{J}_{sol}$ represents eddy currents that produce primarily the magnetic field at low frequency, whereas the irrotational current $\mathbf{J}_{irr}$ represents charge currents that produce primarily the electric field at low frequency.

This problem can also be seen when we look at the EFIE operator (see the previous chapter),

$$\mathcal{L}_E \mathbf{J}_s = i\omega\mu \int_S dS' g(\mathbf{r}, \mathbf{r}')\mathbf{J}_s(\mathbf{r}') + \frac{1}{i\omega\epsilon}\nabla \int_S dS' g(\mathbf{r}, \mathbf{r}')\nabla' \cdot \mathbf{J}_s(\mathbf{r}') \tag{5.7}$$

where $\mathbf{J}_s$ is the surface current on the surface S. The first term is due to the vector potential or it corresponds to the electric field generated by a time-varying magnetic field. We shall call it the induction term. The second term is due to the scalar potential, or it corresponds to the electric field produced by the charge in the system. We shall call it the charge term.

---

[1]The disparity between the magnitudes of $\mathbf{J}_{irr}$ and $\mathbf{J}_{sol}$ remains true if the above argument is repeated for a conductive medium.

When $\omega \to 0$, the second term dominates over the first term, and we have

$$\mathcal{L}_E \mathbf{J}_s \simeq \frac{1}{i\omega\epsilon} \nabla \int_S dS' g(\mathbf{r}, \mathbf{r}') \nabla' \cdot \mathbf{J}_s(\mathbf{r}') \tag{5.8}$$

Since there exists $\mathbf{J}_s(\mathbf{r}')$ such that $\nabla' \cdot \mathbf{J}_s(\mathbf{r}') = 0$, that is, when $\mathbf{J}_s = \mathbf{J}_{sol}$, the above integral operator has a null space. The eigenvalues of the operator $\mathcal{L}_E$ in (5.7) flip-flop between very large values for $\mathbf{J}_s = \mathbf{J}_{irr}$ when the charge term dominates, and very small eigenvalues for $\mathbf{J}_s = \mathbf{J}_{sol}$ when the induction term dominates. Hence, the matrix representation of the above operator becomes extremely ill-conditioned when $\omega \to 0$. This is the reason for the low-frequency breakdown of the EFIE operator.

Another point of view of the low-frequency breakdown is to look at the differential equation. One can convert the electromagnetic equations to a vector wave equation for a lossless inhomogeneous medium as

$$\nabla \times \mu^{-1} \nabla \times \mathbf{E} - \omega^2 \epsilon \mathbf{E} = i\omega \mathbf{J} \tag{5.9}$$

When $\omega \to 0$, the above becomes

$$\nabla \times \mu^{-1} \nabla \times \mathbf{E} \simeq i\omega \mathbf{J} \tag{5.10}$$

Eq. (5.10) has no unique solution because $\nabla \times \mathbf{A} = \mathbf{b}$ does not guarantee a unique $\mathbf{A}$ because $\mathbf{A}' = \mathbf{A} + \nabla\phi$ will equally satisfy $\nabla \times \mathbf{A}' = \mathbf{b}$. In other words, the curl operator has a null space. Hence, the matrix representation of the differential operator in (5.10) yields a badly conditioned system. The breakdown of the electromagnetic equations when $\omega \to 0$ is also because that $\omega = 0$ is mathematically a singular perturbation point for solutions to the electromagnetic equations [13, p. 159].

The fact that $\omega = 0$ is a singular perturbation point is both a curse as well as a blessing. A curse because it makes the development of robust numerical solvers in this regime (which is referred to as the twilight zone [14, 15]) difficult. It is a blessing because it is this difficulty that enriches the world of circuit physics. In other words, it is not simple to transition from the world of wave physics to the world of circuit physics. If one transitions smoothly from the world of wave physics to the world of the circuit physics, however, the richness in the world of circuit physics will be revealed. This is almost like the Alice in Wonderland story!

5.3 Remedy—Loop-Tree Decomposition and Frequency Normalization

The remedy to this low-frequency breakdown is to solve the problem in such a way so that the world of inductors is separated from the world of the capacitors. That is, we have to forcefully decompose the current into two parts by performing a Helmholtz decomposition.

In the preceding discussion, it is seen that the charge term is responsible for producing instability of the EFIE operator depending on if it is acting on a divergence-free current or nondivergence-free current. To stabilize it, it is necessary to decompose the current into a divergence-free part and the nondivergence-free part [4, 5]. This is not a full Helmholtz decomposition, but a quasi-Helmholtz decomposition current. The Rao-Wilton-Glisson (RWG) basis[2] is amenable to the aforementioned quasi-Helmholtz decomposition, but not a full

---

[2]We use the word basis here to mean a set of functions that spans a space.

Helmholtz decomposition.[3]

Instead of using RWG functions as a basis to span the approximate linear vector space, we decompose the RWG basis into the loop basis whose members have zero divergence, and the tree basis (or star basis) whose members have nonzero divergence. This is known as the loop-tree or loop-star decomposition, which is a quasi-Helmholtz decomposition because the tree or the star expansion functions are not curl-free. We shall call the space spanned by RWG basis the RWG space, the subspace spanned by the tree (or star) basis the tree (or star) space, and that spanned by loop basis the loop space.[4]

Consequently, we expand [11]

$$\mathbf{J}_s(\mathbf{r}') = \sum_{n=1}^{N_L} I_{Ln}\mathbf{J}_{Ln}(\mathbf{r}') + \sum_{n=1}^{N_C} I_{Cn}\mathbf{J}_{Cn}(\mathbf{r}') \tag{5.11}$$

where $\mathbf{J}_{Ln}(\mathbf{r}')$ is a loop expansion function such that $\nabla \cdot \mathbf{J}_{Ln}(\mathbf{r}') = 0$, and $\mathbf{J}_{Cn}(\mathbf{r}')$ is a tree expansion (or a star expansion) function such that $\nabla \cdot \mathbf{J}_{Cn}(\mathbf{r}') \neq 0$ and it is used to model the charge in the system. We write the above compactly as

$$\mathbf{J}_s(\mathbf{r}') = \overline{\mathbf{J}}_L^t(\mathbf{r}') \cdot \mathbf{I}_L + \overline{\mathbf{J}}_C^t(\mathbf{r}') \cdot \mathbf{I}_C \tag{5.12}$$

where

$$\left[\overline{\mathbf{J}}_L(\mathbf{r}')\right]_n = \mathbf{J}_{Ln}(\mathbf{r}'), \qquad [\mathbf{I}_L]_n = I_{Ln} \tag{5.13a}$$

$$\left[\overline{\mathbf{J}}_C(\mathbf{r}')\right]_n = \mathbf{J}_{Cn}(\mathbf{r}'), \qquad [\mathbf{I}_C]_n = I_{Cn} \tag{5.13b}$$

When the above is substituted into the EFIE,

$$\mathcal{L}_E \mathbf{J}_s = -\mathbf{E}^{inc} \tag{5.14}$$

and testing the above with the same set of functions as in Galerkin's method, we have a matrix equation

$$\begin{bmatrix} \overline{\mathbf{Z}}_{LL} & \overline{\mathbf{Z}}_{LC} \\ \overline{\mathbf{Z}}_{CL} & \overline{\mathbf{Z}}_{CC} \end{bmatrix} \cdot \begin{bmatrix} \mathbf{I}_L \\ \mathbf{I}_C \end{bmatrix} = \begin{bmatrix} \mathbf{V}_L \\ \mathbf{V}_C \end{bmatrix} \tag{5.15}$$

where

$$\mathbf{V}_L = -\langle \overline{\mathbf{J}}_L(\mathbf{r}), \mathbf{E}^{inc}(\mathbf{r}) \rangle \tag{5.16a}$$

$$\mathbf{V}_C = -\langle \overline{\mathbf{J}}_C(\mathbf{r}), \mathbf{E}^{inc}(\mathbf{r}) \rangle \tag{5.16b}$$

$$\overline{\mathbf{Z}}_{LL} = i\omega\mu \langle \overline{\mathbf{J}}_L(\mathbf{r}), g(\mathbf{r}, \mathbf{r}'), \overline{\mathbf{J}}_L^t(\mathbf{r}') \rangle \tag{5.16c}$$

$$\overline{\mathbf{Z}}_{LC} = i\omega\mu \langle \overline{\mathbf{J}}_L(\mathbf{r}), g(\mathbf{r}, \mathbf{r}'), \overline{\mathbf{J}}_C^t(\mathbf{r}') \rangle \tag{5.16d}$$

---

[3]It is interesting to note that the stabilizing of the differential equation (5.9) is done via a quasi-Helmholtz decomposition where the field is decomposed into a curl-free and the noncurl-free part [4].

[4]This is also known as the tree-cotree decomposition in some work [16].

$$\overline{\mathbf{Z}}_{CL} = i\omega\mu\langle \overline{\mathbf{J}}_C(\mathbf{r}), g(\mathbf{r}, \mathbf{r}'), \overline{\mathbf{J}}_L^t(\mathbf{r}')\rangle = \overline{\mathbf{Z}}_{LC}^t \tag{5.16e}$$

$$\overline{\mathbf{Z}}_{CC} = i\omega\mu\langle \overline{\mathbf{J}}_C(\mathbf{r}), g(\mathbf{r}, \mathbf{r}'), \overline{\mathbf{J}}_C^t(\mathbf{r}')\rangle$$
$$- \tfrac{i}{\omega\epsilon}\langle \nabla \cdot \mathbf{J}_C(\mathbf{r}), g(\mathbf{r}, \mathbf{r}'), \nabla' \cdot \mathbf{J}_C(\mathbf{r}')\rangle \tag{5.16f}$$

In this case, the integral is over the supports of the expansion and testing functions.

When $\omega \to 0$, the frequency-scaling behavior of the matrix system is as follows:

$$\begin{bmatrix} \overline{\mathbf{Z}}_{LL}(O(\omega)) & \overline{\mathbf{Z}}_{LC}(O(\omega)) \\ \overline{\mathbf{Z}}_{CL}(O(\omega)) & \overline{\mathbf{Z}}_{CC}(O(\omega^{-1})) \end{bmatrix} \cdot \begin{bmatrix} \mathbf{I}_L \\ \mathbf{I}_C \end{bmatrix} = \begin{bmatrix} \mathbf{V}_L \\ \mathbf{V}_C \end{bmatrix} \tag{5.17}$$

where $O(\omega^\alpha)$ in the parentheses represents the scaling properties of the matrix elements with respect to frequency ω when $\omega \to 0$. Hence, even after the loop-tree decomposition, the matrix system is still ill conditioned because of the imbalance of the matrix elements; some blocks have small values, whereas the lower right block is inordinately large when $\omega \to 0$. One still needs to frequency-normalize the matrix system in order to obtain a stable system. However, one should note that this determinate frequency scaling is because of the quasi-Helmholtz decomposition, which isolates the frequency-scaling behavior into different blocks. This makes the block matrices amenable to frequency normalization.

To this end, we multiply the first row of the equation by ω^{-1}, and rearrange the matrix system such that

$$\begin{bmatrix} \omega^{-1}\overline{\mathbf{Z}}_{LL}(O(1)) & \overline{\mathbf{Z}}_{LC}(O(\omega)) \\ \overline{\mathbf{Z}}_{CL}(O(\omega)) & \omega\overline{\mathbf{Z}}_{CC}(O(1)) \end{bmatrix} \begin{bmatrix} \mathbf{I}_L \\ \omega^{-1}\mathbf{I}_C \end{bmatrix} = \begin{bmatrix} \omega^{-1}\mathbf{V}_L \\ \mathbf{V}_C \end{bmatrix} \tag{5.18}$$

As can be seen, now the diagonal blocks are of $O(1)$ and the off diagonal blocks are of $O(\omega)$ when $\omega \to 0$. Now the system is diagonally dominant and hence is better conditioned.

Frequency normalization can be regarded as a kind of post- and prediagonal preconditioning such that

$$\begin{bmatrix} \omega^{-1} & 0 \\ 0 & 1 \end{bmatrix} \cdot \overline{\mathbf{Z}} \cdot \begin{bmatrix} 1 & 0 \\ 0 & \omega \end{bmatrix} \cdot \begin{bmatrix} 1 & 0 \\ 0 & \omega^{-1} \end{bmatrix} \cdot \mathbf{I} = \begin{bmatrix} \omega^{-1} & 0 \\ 0 & 1 \end{bmatrix} \cdot \mathbf{V} \tag{5.19}$$

or

$$\overline{\mathbf{P}}_{PR} \cdot \overline{\mathbf{Z}} \cdot \overline{\mathbf{P}}_{PO} \cdot \overline{\mathbf{P}}_{PO}^{-1} \cdot \mathbf{I} = \overline{\mathbf{P}}_{PR} \cdot \mathbf{V} \tag{5.20}$$

where $\overline{\mathbf{P}}_{PR}$ and $\overline{\mathbf{P}}_{PO}$, the pre- and the postpreconditioning matrices, can be identified by comparing (5.19) with (5.20).

The above frequency normalization boosts the importance of the vector potential term, or the $\overline{\mathbf{Z}}_{LL}$ matrix elements, and suppresses the divergence of the $\overline{\mathbf{Z}}_{CC}$ matrix elements. It also boosts the importance of the $\mathbf{I}_C$ and $\mathbf{V}_L$. These two terms tend to be vanishingly small from our discussion in Eq. (5.6) when $\omega \to 0$.

In the original EFIE, as $\omega \to 0$, it becomes an electroquasistatic system where the electric field dominates over the magnetic field. Frequency normalization makes the magnetic field equally important compared to the electric field, and hence, the magnetoquasistatic field is solved to the same accuracy as the electroquasistatic field. Consequently, the physics of inductors is captured with the same accuracy as the physics of capacitors.

To elucidate the physics of the magnetoquasistatic part in the above testing procedure, since $\nabla \cdot \mathbf{J}_{Li} = 0$ for the loop expansion function, we can express $\mathbf{J}_{Li}(\mathbf{r}) = \nabla \times \hat{n}\psi_i(\mathbf{r})$. Then an element of the $\overline{\mathbf{Z}}_{LL}$ matrix in (5.18) becomes

$$\left[\overline{\mathbf{Z}}_{LL}\right]_{ij} = i\omega\mu\langle\nabla \times \hat{n}\psi_i(r), g(\mathbf{r}, \mathbf{r}'), \mathbf{J}_{Lj}(\mathbf{r}')\rangle$$
$$= i\omega\mu\langle\hat{n}\psi_i(\mathbf{r}), \nabla \times \mathbf{I}g(\mathbf{r}, \mathbf{r}'), \mathbf{J}_{Lj}(\mathbf{r}')\rangle$$
$$= i\omega\mu\langle\hat{n}\psi_i(\mathbf{r}), \mathbf{H}_j(\mathbf{r})\rangle \tag{5.21}$$

where integration is over the $\mathbf{r}$ variable, and $\mathbf{H}_j(\mathbf{r})$ is the magnetic field produced by $\mathbf{J}_{Lj}(\mathbf{r})$. Hence, testing by the loop function is equivalent to testing the normal component of the magnetic field. In solving a magnetoquasistatic problem, it is sufficient to impose the boundary condition that the normal component of the magnetic field be zero on a metallic surface. Consequently, the upper left block reduces to a magnetoquasistatic problem for low frequency.

Similarly, from (5.16f), it is seen that the lower right block reduces to an electroquasistatic problem when $\omega \rightarrow 0$. Moreover, the electrostatic problem is decoupled from the magnetostatic problem at $\omega = 0$. This point is underscored in the Eqs. (5.1), (5.2), (5.3), and (5.4) in the beginning of this chapter.

It is to be noted that right when $\omega = 0$, the problem decomposes into two uncoupled electrostatic and magnetostatic problems. These problems are important for describing the worlds of capacitors and inductors respectively, and hence are important for circuit physics. When the frequency increases, these two worlds start to couple to each other, and it often is important to include the physics of these two worlds concurrently. We shall call this the regime of full-field physics, since both the $\mathbf{E}$ and $\mathbf{H}$ fields are involved, but the regime of full-wave physics has not fully set in yet. It is in fact found that the loop-tree decomposition is deleterious in modeling full-wave physics [17]. When the regime of full-wave physics sets in, the RWG functions are better suited to model this regime.

5.3.1   Loop-Tree Decomposition

The loop-tree decomposition of a subspace spanned by RWG expansion functions or rooftop expansion functions can be found quite easily [18–21]. We will discuss the case whose surface is triangulated and that the approximate subspace for the current is spanned by the RWG basis.

There are two cases to consider, the open surface and the closed surface. In the open surface case, we first introduce cuts on the triangulated surface (see Figure 5.1). The cuts should be such that the triangulated surface remains simply connected. Namely, if one thinks of the cuts as being introduced on a sheet of paper, the paper should not be broken into two or more pieces. Also, the number of cuts should be taken such that no loops can be formed from the remaining RWG expansion functions on the surface. An RWG function is then removed from the edges that have been cut. Then associated with each cut, one introduces a loop expansion function. Hence, for every RWG function removed, a loop function is introduced. For a simple open surface, the number of cuts is equal to the number of interior nodes of the surface. Hence, the number of loops is equal to the number of interior nodes N_{Nint}.

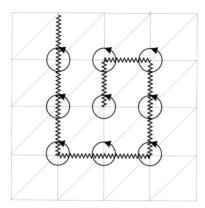

Figure 5.1: Loop-tree decomposition for an open surface. The serrated cuts represent RWG expansion functions removed from the edges, and the loops represent loop expansion function around a node.

For a simple closed surface, every node is an interior node. However, there are only $N_{node} - 1$ loops where N_{node} is number of nodes on the triangulated geometry (see Figure 5.2). If one adds a loop to every node on a closed surface, then the loop added to the last (final) node is dependent on the previous loops added. One can see this for a tetrahedron, which is the simplest triangulated closed surface (see Figure 5.2). A loop is dependent on other loops if it can be formed by a linear superposition of these other loops.

In addition to simple loops described above, there are global loops. Global loops are distinguished by their irreducibility to a loop around a single node. For instance, if we have a ribbon, there will be a loop around a ribbon, which cannot be reduced to a loop around a single node. A ribbon is topologically the same as an open surface with a hole in it (see Figure 5.3). For a surface with two holes, the number of global loops is two. The number of global loops is proportional to the number of holes in an open surface.

A closed surface with a hole is equivalent to a doughnut (see Figure 5.4). For a doughnut-shape object, the number of global loops is two, as can be seen from the figure. If we add one more hole to the doughnut, the number of global loops increases to four, and so on. When more than one global loop exist, they may be entangled and special care has to be taken to account for them [22].

5.3.2   The Electrostatic Problem

Even though the lower right block reduces to the electrostatic problem when $\omega \to 0$, the matrix is unlike the capacitance matrix from solving an electrostatic problem. When iterative solvers are applied to the above equation, convergence is still slow. To gain further insight, we will review the electrostatic problem here.

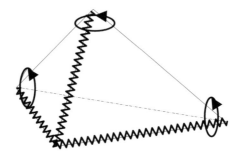

Figure 5.2: Loop-tree decomposition for a closed surface. A tetrahedron represents the simplest triangulated closed surface. The number of loops is one less than the number of nodes.

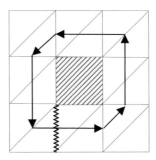

Figure 5.3: When an open surface has a hole, a global loop is formed.

An electrostatic boundary value problem can be cast as [23]

$$\nabla^2 \phi(\mathbf{r}) = s(\mathbf{r}) \tag{5.22}$$
$$\phi(\mathbf{r}) = 0, \quad \mathbf{r} \in S$$

If $s(\mathbf{r})$ is not charge neutral, we may assume that there are charges at infinity that neutralize the net charge of $s(\mathbf{r})$. The integral equation for the above can be written as

$$\phi(\mathbf{r}) = \phi_0(\mathbf{r}) + \int_S d\mathbf{r}' g_0(\mathbf{r}, \mathbf{r}') \sigma(\mathbf{r}') = 0, \qquad \mathbf{r} \in S \tag{5.23}$$

where $\sigma(\mathbf{r}')$ is the surface charge on S, $g_0(\mathbf{r}, \mathbf{r}') = \frac{1}{4\pi|\mathbf{r}-\mathbf{r}'|}$ is the static Green's function, and $\phi_0(\mathbf{r})$ is the potential generated by the source $s(\mathbf{r})$. The above can be converted into a matrix equation by letting

$$\sigma(\mathbf{r}') = \sum_{n=1}^{N} Q_n \sigma_n(\mathbf{r}') \tag{5.24}$$

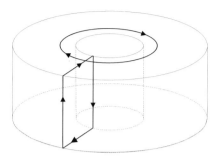

Figure 5.4: When a closed surface has a hole in it forming a doughnut, two global loops are formed.

where $\sigma_n(\mathbf{r}')$ is an expansion function for expanding the charge and N is the number of expansion functions used. Using (5.24) in (5.23), and testing the result by $\sigma_m(\mathbf{r}), m = 1, \ldots, N$, and integrating over S, and finally, imposing that the total $\phi = 0$ on S gives

$$-\langle \sigma_m, \phi_0 \rangle = \sum_{n=1}^{N} Q_n \langle \sigma_m, g_0, \sigma_n \rangle \tag{5.25}$$

The above is a matrix equation

$$\overline{\mathbf{A}} \cdot \mathbf{Q} = \mathbf{V} \tag{5.26}$$

Since in a capacitive system, $\mathbf{Q} = \overline{\mathbf{C}} \cdot \mathbf{V}$, the matrix $\overline{\mathbf{A}} = \overline{\mathbf{C}}^{-1}$ is the inverse of the capacitance matrix, and is given by

$$\left[\mathbf{A} \right]_{mn} = \langle \sigma_m, g_0, \sigma_n \rangle \tag{5.27}$$

$$[\mathbf{Q}]_n = Q_n, \quad [\mathbf{V}]_m = -\langle \sigma_m, \phi_0 \rangle \tag{5.28}$$

The above matrix equation is easily proven to be Hermitian positive definite, with good convergence property when solved with iterative solvers.

When $\sigma_m(\mathbf{r})$ is chosen to be a pulse expansion function, we expect Eq. (5.26) to be related to the frequency normalized $\overline{\mathbf{Z}}_{CC}$. However, we find that

$$\lim_{\omega \to 0} \left[\omega \overline{\mathbf{Z}}_{CC} \right]_{mn} = -\frac{i}{\epsilon} \langle \nabla \cdot \mathbf{J}_m(\mathbf{r}), g_0(\mathbf{r}, \mathbf{r}'), \nabla \cdot \mathbf{J}_n(\mathbf{r}') \rangle \tag{5.29}$$

When $\mathbf{J}_m(\mathbf{r})$ is chosen to be an RWG basis, $\nabla \cdot \mathbf{J}_m(\mathbf{r})$ does not become a pulse function, but a pulse pair of opposite signs (see Figure 5.5). We shall call this the pulse-pair function. Hence, the matrix in (5.29) is the matrix representation of the integral operator in the subspace spanned by pulse-pair functions. Moreover, each of those pulse-pair functions is charge neutral, namely, the total integral sum of the charge is zero.

Expanding $\sigma(\mathbf{r})$ in terms of pulse expansion functions, we have

$$\sigma(\mathbf{r}) = \sum_{n=1}^{N_p} Q_n P_n(\mathbf{r}) \tag{5.30}$$

Figure 5.5: A pulse-pair function resulting from taking the divergence of the RWG function.

where N_p is the number of pulse functions used. We can connect this with the charge that is expanded in terms of the pulse-pair functions, namely,

$$\sigma(\mathbf{r}) = \sum_{n=1}^{N_C} I_n \nabla \cdot \mathbf{\Lambda}_n(\mathbf{r}) \tag{5.31}$$

where $\mathbf{\Lambda}_n(\mathbf{r})$ represents an RWG function, I_n is the unknown coefficient, and N_C is the number of the subset of the RWG functions that span the tree space, the subspace of non-divergence-free functions complement to the loop space. In (5.31), $\int_S \sigma(\mathbf{r})dS = 0$, or it is charge neutral because $\int_S \nabla \cdot \Lambda_n(\mathbf{r})dS = 0$. Hence, if we impose charge neutrality on (5.30), we can make the connection between Q_n and I_n, because $\nabla \cdot \mathbf{\Lambda}_n(\mathbf{r})$ is the superposition of two pulse functions. This connection is necessary to improve on the convergence property of the frequency normalized equation (5.18).

5.3.3 Basis Rearrangement

The above method so far gives rise to a matrix system with no null space when $\omega \to 0$, but the matrix system is still not well-conditioned enough for iterative solvers. The basis rearrangement method [11] has to be invoked to make the solution of the matrix system more amenable to iterative solvers. The basis rearrangement relates the electroquasistatic part of (5.18) to (5.26), the matrix equation of an electrostatic problem.

If we assume that $P_n(\mathbf{r})$ is normalized such that $\int_S dS P_n(\mathbf{r}) = 1$, then the charge neutrality condition for (5.30) is

$$\sum_{n=1}^{N_p} Q_n = 0 \tag{5.32}$$

or

$$Q_{N_p} = - \sum_{n=1}^{N_p-1} Q_n \tag{5.33}$$

Hence, using (5.33), (5.30) can be written as

$$\sigma(\mathbf{r}) = \sum_{n=1}^{N_p-1} Q_n \left[P_n(\mathbf{r}) - P_{N_F}(\mathbf{r}) \right] \tag{5.34}$$

where now $\sigma(\mathbf{r})$ is charge neutral, that is, $\int_S \sigma(\mathbf{r})dS = 0$. Charge neutrality is true, because the pulse functions are the consequence of splitting an RWG function into two, but each RWG function is charge neutral, namely, $\int_S \nabla \cdot \Lambda_n(\mathbf{r})dS = 0$. We can express the above as

$$\sigma(\mathbf{r}) = \mathbf{N}^t(\mathbf{r}) \cdot \mathbf{Q} \tag{5.35}$$

where $\mathbf{N}^t(\mathbf{r})$ and $\mathbf{Q}$ are vectors of length $N_p - 1$.

$$\left[\mathbf{N}^t(\mathbf{r})\right]_n = P_n(\mathbf{r}) - P_{N_F}(\mathbf{r}), \quad n = 1, \ldots, N_p - 1 \tag{5.36}$$

$$[\mathbf{Q}(\mathbf{r})]_n = Q_n, \quad n = 1, \ldots, N_p - 1 \tag{5.37}$$

From the continuity equation, we note that

$$\nabla \cdot \mathbf{J}_s = i\omega\sigma(\mathbf{r}) \tag{5.38}$$

We can apply this to the surface current and surface charge density on the surface of a perfect electric conduction (PEC).

At low frequency, the solenoidal part of $\mathbf{J}_s$ does not play a role in generating charges, and hence, only the tree current plays a role in (5.38). The tree or charge current is related to the N_C tree RWG function as

$$\mathbf{J}_C = \overline{\mathbf{J}}_C^t(\mathbf{r}) \cdot \mathbf{I}_C \tag{5.39}$$

where $\overline{\mathbf{J}}_C^t(\mathbf{r})$ is a $3 \times N_C$ matrix and $\mathbf{I}_C$ is an $N_C \times 1$ vector. Using (5.39) in (5.38), we have

$$\nabla \cdot \overline{\mathbf{J}}_C^t(\mathbf{r}) \cdot \mathbf{I}_C = i\omega\mathbf{N}^t(\mathbf{r}) \cdot \mathbf{Q} \tag{5.40}$$

where $\nabla \cdot \overline{\mathbf{J}}_C^t(\mathbf{r})$ is a $1 \times N_C$ vector. Taking the inner product of (5.40) with $\mathbf{P}(\mathbf{r})$, where $[\mathbf{P}(\mathbf{r})]_n = P_n(\mathbf{r}), n = 1, \ldots, N_p - 1$, we have

$$\langle \mathbf{P}(\mathbf{r}), \nabla \cdot \overline{\mathbf{J}}_C^t(\mathbf{r}) \rangle \cdot \mathbf{I}_C = i\omega\langle \mathbf{P}(\mathbf{r}), \mathbf{N}^t(\mathbf{r}) \rangle \cdot \mathbf{Q} \tag{5.41}$$

In this manner, $\langle \mathbf{P}(\mathbf{r}), \mathbf{N}^t(\mathbf{r}) \rangle$ is the identity matrix because $P_n(\mathbf{r})$ is normalized. Consequently, (5.41) becomes

$$\overline{\mathbf{K}} \cdot \mathbf{I}_C = i\omega\mathbf{Q} \tag{5.42}$$

where $\overline{\mathbf{K}}$ is a square matrix given by

$$\overline{\mathbf{K}} = \langle \mathbf{P}(\mathbf{r}), \nabla \cdot \overline{\mathbf{J}}_C^t(\mathbf{r}) \rangle \tag{5.43}$$

and more explicitly,

$$\left[\overline{\mathbf{K}}\right]_{mn} = \langle P_m(\mathbf{r}), \nabla \cdot \Lambda_n(\mathbf{r}) \rangle \tag{5.44}$$

$\overline{\mathbf{K}}$ is an $(N_C - 1) \times (N_C - 1)$ sparse matrix dense to $O(N_C)$ and can be inverted in $O(N_C)$ operations. Eq. (5.42) implies that $\mathbf{I}_C \sim O(\omega)$, when $\omega \to 0$, as is required in our earlier discussion.

Consequently, the original method of moments (MOM) matrix (5.18) can be written in a form more suggestive of circuit theory, namely,

$$\begin{bmatrix} \overline{\mathbf{L}}_{LL} & i\omega\overline{\mathbf{L}}_{LC} \\ i\omega\overline{\mathbf{L}}_{CL} & \overline{\mathbf{C}}_{CC}^{-1} \end{bmatrix} \begin{bmatrix} \mathbf{I}_L \\ \mathbf{Q} \end{bmatrix} = \begin{bmatrix} \frac{1}{i\omega}\mathbf{V}_L \\ \left(\overline{\mathbf{K}}^t\right)^{-1} \cdot \mathbf{V}_C \end{bmatrix} \tag{5.45}$$

where

$$\overline{\mathbf{L}}_{LL} = \frac{1}{i\omega}\overline{\mathbf{Z}}_{LL}, \tag{5.46}$$

$$\overline{\mathbf{L}}_{LC} = \frac{1}{i\omega}\overline{\mathbf{Z}}_{LC} \cdot \overline{\mathbf{K}}^{-1}, \tag{5.47}$$

$$\overline{\mathbf{L}}_{CL} = \frac{1}{i\omega}(\overline{\mathbf{K}}^t)^{-1} \cdot \overline{\mathbf{Z}}_{CL} = \overline{\mathbf{L}}_{LC}^t, \tag{5.48}$$

$$\overline{\mathbf{C}}_{CC}^{-1} = i\omega \left(\overline{\mathbf{K}}^t\right)^{-1} \cdot \overline{\mathbf{Z}}_{CC} \cdot \overline{\mathbf{K}}^{-1} \tag{5.49}$$

and $\overline{\mathbf{L}}_{LL}$ is the self-inductance matrix represented in the loop basis, $\overline{\mathbf{L}}_{LC}$ has a unit of inductance, and $i\omega\overline{\mathbf{L}}_{LC}$ can be regarded as the mutual impedance matrix between the inductance and the capacitance, and $\overline{\mathbf{C}}_{CC}$ is the self-capacitance matrix represented in the tree or charge basis.

Because inductance is proportional to μ and capacitance is proportional to ϵ, and in MKS units, μ and ϵ are vastly different in numerical values, the corresponding diagonal blocks in (5.45) are vastly different. A remedy is to express (5.45) in dimensionless form, or multiply the top row in (5.45) by μ^{-1}, and the bottom row by ϵ to further normalize the system.

For acceleration purposes, the final equation may be written in terms of a sequence of matrix-vector products. For instance, we can rewrite (5.45) as

$$\begin{bmatrix} \overline{\mathbf{I}} & 0 \\ 0 & \left(\overline{\mathbf{K}}^t\right)^{-1} \end{bmatrix} \cdot \overline{\mathbf{F}}^t \cdot \overline{\mathbf{Z}}_{\mathrm{RWG}} \cdot \overline{\mathbf{F}} \cdot \begin{bmatrix} \overline{\mathbf{I}} & 0 \\ 0 & \overline{\mathbf{K}}^{-1} \end{bmatrix} \cdot \begin{bmatrix} \mathbf{I}_L \\ \mathbf{Q} \end{bmatrix} = \begin{bmatrix} \frac{1}{i\omega}\mathbf{V}_L \\ \left(\overline{\mathbf{K}}^t\right)^{-1} \cdot \mathbf{V}_C \end{bmatrix} \tag{5.50}$$

When an iterative solver is used to solve the matrix equation, the matrix-vector product can be decomposed into a sequence of matrix-vector products as above. A fast matrix-vector product can be written for $\overline{\mathbf{Z}}_{\mathrm{RWG}} \cdot \mathbf{a}$, and be used to accelerate the total matrix-vector product.

In the above, $\overline{\mathbf{Z}}_{\mathrm{RWG}}$ is a conventional impedance matrix constructed from using RWG functions as a basis, but it can represent a matrix system obtained by using other basis set. $\overline{\mathbf{F}}$ is a transformation matrix that changes from the conventional basis to the loop-tree basis. It is a sparse matrix with $O(N)$ elements and hence, its associated matrix-vector product can be performed in $O(N)$ operations. The frequency and matrix normalization can be incorporated in $\overline{\mathbf{F}}$. The construction of the $\overline{\mathbf{F}}$ matrix requires the knowledge of the loop-tree decomposition of a structure. This can be formed by loop-tree search algorithms [18–21]. The matrix-vector product associated with $\overline{\mathbf{Z}}_{\mathrm{RWG}} \cdot \mathbf{a}$ can be accelerated by a low-frequency fast multipole algorithm (LF-FMA) [24] requiring $O(N)$ operations (for simplicity, we use the acronym LF-FMA to replace LF-MLFMA). Hence, if $\overline{\mathbf{K}}^{-1} \cdot \mathbf{Q}$ can be effected in $O(N)$ operations, the whole matrix vector product can be effected in $O(N)$ operations.

5.3.4   Computation of $\overline{\mathbf{K}}^{-1} \cdot \mathbf{Q}$ in $O(N)$ Operations

As discussed above, it is imperative to effect $\overline{\mathbf{K}}^{-1} \cdot \mathbf{Q}$ in $O(N)$ operations. Otherwise it will be the bottleneck in the entire matrix-vector product. Applying $\overline{\mathbf{K}}^{-1} \cdot \mathbf{Q}$ in $O(N)$ operations is equivalent to solving

$$\overline{\mathbf{K}} \cdot \mathbf{I} = i\omega\mathbf{Q} \tag{5.51}$$

in $O(N)$ operations. We regroup $\mathbf{I}_C$, the nondivergence-free current into a part that involves junctions, $\mathbf{I}_{C1}$, and the rest that does not involve junctions, $\mathbf{I}_{C2}$. Hence, (5.51) can be written as

$$i\omega \begin{bmatrix} \mathbf{Q}_1 \\ \mathbf{Q}_2 \end{bmatrix} = \overline{\mathbf{K}} \cdot \begin{bmatrix} \mathbf{I}_{C1} \\ \mathbf{I}_{C2} \end{bmatrix} = \begin{bmatrix} \overline{\mathbf{A}}_1 & \overline{\mathbf{B}}_1 \\ \overline{\mathbf{A}}_2 & \overline{\mathbf{B}}_2 \end{bmatrix} \cdot \begin{bmatrix} \mathbf{I}_{C1} \\ \mathbf{I}_{C2} \end{bmatrix} \tag{5.52}$$

Then by the matrix partitioning method,

$$\overline{\mathbf{K}}^{-1} = \begin{bmatrix} \overline{\mathbf{A}}_1^{-1} + \overline{\mathbf{A}}_1^{-1} \cdot \overline{\mathbf{B}}_1 \cdot \overline{\mathbf{D}}_2^{-1} \cdot \overline{\mathbf{A}}_2 \cdot \overline{\mathbf{A}}_1^{-1}, & -\overline{\mathbf{A}}_1^{-1} \cdot \overline{\mathbf{B}}_1 \cdot \overline{\mathbf{D}}_1^{-1} \\ -\overline{\mathbf{D}}_2^{-1} \cdot \overline{\mathbf{A}}_2 \cdot \overline{\mathbf{A}}_1^{-1}, & \overline{\mathbf{D}}_2^{-1} \end{bmatrix} \tag{5.53}$$

In the above, the dimension of $\overline{\mathbf{A}}_1$ is much smaller than that of $\overline{\mathbf{B}}_2$, and $\overline{\mathbf{D}}_2$ is a small matrix given by

$$\overline{\mathbf{D}}_2 = \overline{\mathbf{B}}_2 - \overline{\mathbf{A}}_2^{-1} \cdot \overline{\mathbf{A}}_1^{-1} \cdot \overline{\mathbf{B}}_2 \tag{5.54}$$

Hence, the main cost of $\overline{\mathbf{K}}^{-1} \cdot \mathbf{Q}$ is the action of $\overline{\mathbf{A}}_1^{-1}$ on a vector $\mathbf{y}$ on finding

$$\mathbf{x} = \overline{\mathbf{A}}_1^{-1} \cdot \mathbf{y} \tag{5.55}$$

which is similar to solving

$$\mathbf{y} = \overline{\mathbf{A}}_1 \cdot \mathbf{x} \tag{5.56}$$

Similarly, when we need to calculate the action of $\left(\overline{\mathbf{K}}^t\right)^{-1}$ on a vector, we need to calculate the action of

$$\left(\overline{\mathbf{A}}_1^t\right)^{-1} \cdot \mathbf{x} = \mathbf{y} \tag{5.57}$$

5.3.5   Motivation for Inverting $\overline{\mathbf{K}}$ Matrix in $O(N)$ Operations

The connection matrix $\overline{\mathbf{K}}$ that connects charge due to current basis such as RWG to the pulse basis for a triangle is a sparse matrix system. Many sparse matrices, such as those in finite element method (FEM), cannot be inverted in $O(N)$ operation. The cost of inverting sparse matrices arising from FEM is $O(NB_w^2)$ where B_w is the bandwidth of the matrix. Surprisingly, the $\overline{\mathbf{K}}$ matrix here can be inverted in $O(N)$ operations.

We can motivate such an inversion because for one dimensional structure, it can be done in $O(N)$ operations. For the $\overline{\mathbf{K}}$ matrix involving a tree basis, it can be thought of as a string of one dimensional problems hooked up like branches of a tree.

A Simple 1D Problem

Before discussing the general problem, it is pedagogical to study the one dimensional (1D) problem, as it can be more lucidly described. In a simple 1D problem regarding a strip, the strip current can be expanded in a simple rooftop function

$$\mathbf{J}_n(x) = \hat{x} J_n(x) = \hat{x} \begin{cases} \left(1 - \frac{x - x_n}{l_{n+1}}\right), & x_n < x < x_{n+1}, \\ \left(1 - \frac{x_n - x}{l_n}\right), & x_{n-1} < x < x_n, \\ 0, & \text{otherwise.} \end{cases} \tag{5.58}$$

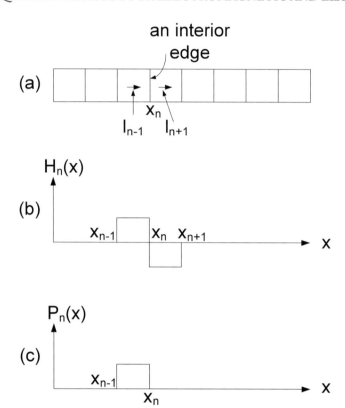

Figure 5.6: Illustration of the 1D tree problem.

The divergence of this current expansion function yields a charge that is proportional to the simple Haar function [25] shown in Figure 5.6. The surface current can be expanded in terms of the simple rooftop function, viz.,

$$\mathbf{J}_s(x) = \sum_{n=1}^{N} I_n \mathbf{J}_n(x) \tag{5.59}$$

Taking the divergence, we have the surface charge

$$i\omega\sigma(x) = \sum_{n=1}^{N} I_n H_n(x) \tag{5.60}$$

where the Haar function is defined as

$$H_n(x) = \begin{cases} -l_{n+1}^{-1}, & x_n < x < x_{n+1}, \\ l_n^{-1}, & x_{n-1} < x < x_n, \\ 0, & \text{otherwise.} \end{cases} \tag{5.61}$$

But we can also expand the surface charge in terms of the pulse basis

$$\sigma(x) = \sum_{n=1}^{n+1} Q_n P_n(x), \tag{5.62}$$

where

$$P_n(x) = \begin{cases} l_{n-1}^{-1/2}, & x_{n-1} < x < x_n, \\ 0, & \text{otherwise.} \end{cases} \tag{5.63}$$

There is one more pulse (patch) function than rooftop functions as can be seen from the figure. However, we can assign one of the patches to be a charge-neutralizing patch so that the number of unknowns in (5.60) and (5.62) is the same. Charge neutrality is necessary because the charge in Eq. (5.60) is naturally neutral, and also the Haar function in (5.61) has zero net area. For simplicity, we choose the last patch in the figure to be the neutralizing function.

Multiplying (5.62) by $i\omega$, equating the right hand sides of (5.62) and (5.60), and testing with a pulse function $P_m(x), m = 1, \ldots, N$, we have

$$i\omega Q_m = \sum_{n=1}^{N} \langle P_m(x), H_n(x) \rangle I_n$$

$$= \sum_{n=1}^{N} K_{mn} I_n, \quad m = 1, \ldots, N \tag{5.64}$$

or

$$i\omega \overline{Q} = \overline{K} \cdot I \tag{5.65}$$

It is quite clear that when the pulse function is used to test the left-most patch of the structure, only one unknown from the right hand side of (5.64) is involved. In other words, the first row of $\overline{K}$ has only one element, and hence, one can solve for I_1 in terms of Q_1. The subsequent row involves two elements, but one of the two unknowns is already solved for from the previous row. Therefore, one can recursively solve for all the I_n's in terms of the Q_n's, $n = 1, \ldots, N$. Here, Q_{N+1} is picked for charge neutrality.

A General Tree Problem

For a general surface whose spanning tree (which is the set of RWG basis complementary to the loop basis) has been found, the tree graph will be shown in Figure 5.7. Each node in the tree represents the center of a triangular patch, which supports half of the spanning tree RWG function, or the quad patch which supports half of a general rooftop. Each segment between two nodes can be associated with the current that flows between two patches. If we go through the same process of constructing the connection matrix $\overline{K}$ as in the 1D problem, then the patch unknown that is associated with a tip of the dendritic branch is only related to one current unknown. We can recursively solve for the unknowns starting from the branch tips until a junction node is reached. For RWG basis, a junction node cannot have more than three segments connected to it, even though some junction nodes may be associated with

Figure 5.7: Schematic of a general tree. Each node represents the center of an RWG expansion function, and the line indicates the possibility of current flowing from one center to another center.

more than three current unknowns, for example, when a quad patch is used. Eventually, for the RWG basis case, after the unknowns on two of the open branches are solved for, the third current unknown can be solved for. Finally all current unknowns, I_n's, can be solved for in terms of Q_n's, in $O(N)$ operation, because the CPU cost of recursion is proportional to the number of nodes or N.

Figure 5.8 shows the loop-tree decomposition of a conical antenna [26]. The conical antenna consists of two circular, annular surfaces. It is driven from the inner edges which are connected by a circular cylinder. The outer edges are connected by four wires which can be shorted or inductively loaded.

The current on the triangulated conical antenna is expanded in terms of RWG functions. The center of each triangle is assigned a node, and the centers are connected to form loops (see Figure 5.8(b)). Hence, an RWG function, which is associated with an inner edge on a triangulated surface, lives between two nodes. The RWG basis is then decomposed into the loop basis and the tree basis. In the case of Figure 5.8(b), when a group of RWG functions form closed loops, they form a loop function that belongs to the loop basis. In the case of Figure 5.8(c), the RWG functions do not form closed loops. Each RWG function living on a segment of the tree belongs to the tree basis.

5.3.6   Reason for Ill-Convergence Without Basis Rearrangement

Without basis rearrangement, the charge part of the integral operator is acting on an unknown current through the divergence operator, viz.,

$$\mathcal{L}_C \mathbf{J}_s = \frac{\nabla}{i\omega\epsilon} \int_S d\mathbf{r}' g(\mathbf{r}, \mathbf{r}') \nabla' \cdot \mathbf{J}_s(\mathbf{r}') \tag{5.66}$$

If we restrict $\mathbf{J}_s(\mathbf{r}')$ to nondivergence-free basis, such as the tree basis, it turns out that there are currents in the tree basis where $\nabla' \cdot \mathbf{J}_s(\mathbf{r}') \simeq 0$ implying the existence of small eigenvalues even when the above is applied to the tree space [27].

A way to avoid the divergence operator is via basis rearrangement so that the fundamental unknown is the charge, which is related to the current via $\nabla \cdot \mathbf{J}_s(\mathbf{r}) = i\omega\sigma(\mathbf{r})$, and hence Eq.

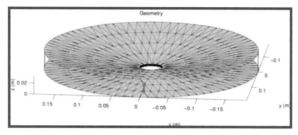

(a) Triangulated conical antenna.

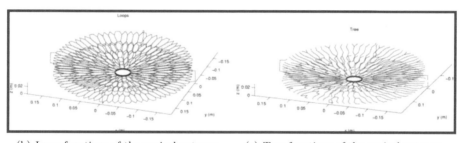

(b) Loop functions of the conical antenna.　　(c) Tree functions of the conical antenna.

Figure 5.8: The decomposition of (a) a conical antenna into (b) a set of loop functions plus (c) a set of tree functions.

(5.66) becomes

$$\mathcal{L}_C \mathbf{J}_s = \frac{\nabla}{\epsilon} \int d\mathbf{r}' g(\mathbf{r}, \mathbf{r}') \sigma(\mathbf{r}') = \hat{\mathcal{L}}_C \sigma \qquad (5.67)$$

In the above, $\overline{\mathbf{K}}$ is the matrix representation of the divergence operator in the tree space [see (5.42)], while $\overline{\mathbf{K}}^{-1}$ is a preconditioner that removes the effect of the divergence operator.

5.4 Testing of the Incident Field with the Loop Function

When the frequency is very low, we need to test the right-hand side of the EFIE, which consists of the incident fields, with the loop functions. This is computed as

$$V_n^L = -\langle \mathbf{J}_{Ln}(\mathbf{r}), \mathbf{E}^{inc}(\mathbf{r}) \rangle \qquad (5.68)$$

When the frequency is low, the above integral in the inner product can yield numerical inaccuracies [28]. To remedy this problem, we use the same trick as in Eq. (5.21), and express $\mathbf{J}_{Ln}(\mathbf{r}) = \nabla \times \hat{n} \psi_n(\mathbf{r})$. By using integration by parts, we convert (5.68) to

$$V_n^L = i\omega\mu\langle \psi_n(\mathbf{r}), \hat{n} \cdot \mathbf{H}^{inc}(\mathbf{r}) \rangle \qquad (5.69)$$

so that V_n^L drops linearly with ω explicitly when $\omega \to 0$, whereas this explicit linear dependence on ω is not in (5.68).

The use of low-frequency MOM (LF-MOM) with loop-tree decomposition to solve small problems is demonstrated in the modeling of a Hertzian dipole in Figure 5.9(a) [11]. The input impedance is calculated with LF-MOM where the equation is solved using conjugate gradient (CG) or low-frequency multilevel fast multipole algorithm (LF-MLFMA). The capacitive model where the input impedance is modeled solely by the capacitance between the two spheres is also shown.

It is seen that at low frequencies, the dipole is a capacitive structure, in agreement with the capacitive model. At higher frequencies, the capacitive model starts to depart from the full-field model using LF-MOM where the magnetic field is solved concurrently with the electric field. The reason is that the input feed wires of the Hertzian dipole start to become inductive. At even higher frequencies as shown in Figure 5.9(b), where a nonloop-tree model is used, the structure starts to resonate like an LC tank circuit.

Figure 5.10 shows the use of this technique to calculate the input impedance of the conical antenna shown in Figure 5.8 [26]. With this technique, the input impedance can be calculated to very low frequencies without low-frequency breakdown, but without this technique as in the finite element radiation method (FERM) code developed at Lincoln Lab [29], the calculation at low frequency is impossible.

5.5 The Multi-Dielectric-Region Problem

The problem involving a single dielectric region can be solved in terms of the PMCHWT (Poggio-Miller-Chang-Harrington-Wu-Tsai) formulation (see previous chapter). Even the

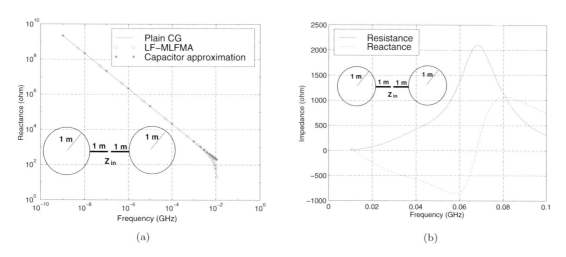

(a)   (b)

Figure 5.9: Input impedance of a Hertzian dipole from (a) very low frequency to (b) microwave frequency.

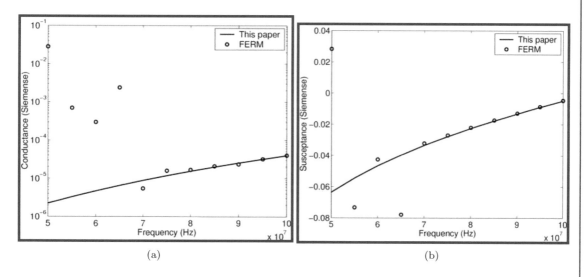

(a)   (b)

Figure 5.10: Input impedance of a conical antenna at very low frequencies indicating (a) the conductance, and (b) the susceptance. Data are taken from [26].

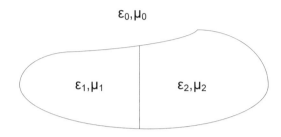

Figure 5.11: The multiple dielectric region problem.

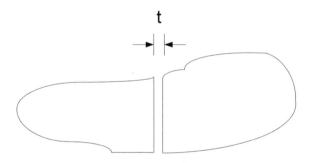

Figure 5.12: The multidielectric region problem with contacting regions can be solved as the limiting case of the noncontact region problem.

multidielectric-region problem can be solved in terms of PMCHWT formulation, or the composite region problem as shown in Figure 5.11 [30]. The PMCHWT formulation involves the use of the EFIE operator which becomes unstable or ill-conditioned when $\omega \to 0$. The MFIE operator or the $\mathcal{K}$ operator does not exhibit low-frequency breakdown as in the EFIE operator, but because of the disparate amplitude of the eddy current with respect to the charge current, numerical inaccuracy still persists [31]. Hence, it is still necessary to use a loop-tree decomposition method to stabilize the EFIE operator [28, 32–34]. However, performing a loop-tree decomposition for a composite surface shown in Figure 5.11 is a difficult task. Instead, we break the composite object into two nontouching objects with closed surfaces as shown in Figure 5.12. The loop-tree decomposition can then be preformed for the individual closed surfaces, and the corresponding integral operators can be stabilized as such. As a result, the solution for Figure 5.11 can be obtained by letting $t \to 0$. We call this the contact-region modeling method [33]. The disadvantage of such a method is that the number of unknowns needed to solve the problem is increased, but the loop-tree decomposition on the surface currents can be performed.

5.5.1 Numerical Error with Basis Rearrangement

A simple way to rearrange the basis so that the charges from the RWG currents are connected to the pulse basis yields numerical instability. This is because of the numerical cancelation that happens. So when a large problem is solved, the error becomes unacceptable, and iterative solution method does not converge [28].

One can appreciate this problem as follows. Originally, when we take the divergence of the RWG basis to obtain the charge, a pair of pulse bases of opposite signs are formed. We shall call this the short patch pair (SPP) function. This is the expansion function for the charge before basis rearrangement.

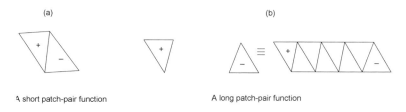

Figure 5.13: A short patch-pair function and a long patch-pair function.

After basis rearrangement, the fundamental expansion function is a patch pair which is shown in Figure 5.13(b). A patch pair can be formed by linearly superposing a number of SPP functions, where there are charge cancelations between the patch pair. Before basis rearrangement, these SPPs are producing electric field of opposite signs, where supposedly after basis rearrangement, they are to cancel exactly, but in a numerical setting, they do not cancel exactly, giving rise to numerical errors. Hence, when the total field is tested at a field point where exact cancelation should have occurred, numerical error persists. This is especially true when large problems are solved where the separation between the patch pair can be very large, and many SPPs are involved before they cancel to form the long patch pair (LPP).

As a remedy to this problem, we first form a matrix representation of the Green's operator using a single-patch expansion function, that is to find

$$A_{nm} = \langle \sigma_n(\mathbf{r}), g(\mathbf{r}, \mathbf{r}'), \sigma_m(\mathbf{r}') \rangle \qquad (5.70)$$

directly where $\sigma_n(\mathbf{r})$ is a single-patch expansion function. Then a linear transformation converts the single-patch basis to the patch-pair basis, because a patch pair function is just a linear sum of two single-patch expansion function. Since the transformation is the change of basis, a corresponding similarity transform can be performed on the matrix in Eq. (5.70) to obtain the matrix representation based on patch-pair basis. The transformation is not full rank because for each object where charge neutrality is maintained, the rank is reduced by one. For N_b independent noncontacting objects, the rank is reduced by N_b.

In this alternative implementation, the LF-FMA [24] and other low-frequency algorithms [35–38] can still be used to accelerate the matrix-vector product associated with A_{mn}. The

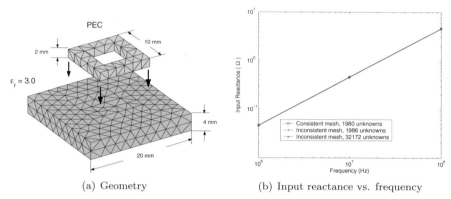

(a) Geometry   (b) Input reactance vs. frequency

Figure 5.14: The input reactance of an inductance structure using contact region modeling.

transformation matrix involved with the similarity transform is sparse with $O(N)$ elements. Hence, a matrix-vector product can be performed in $O(N)$ operations.

Figures 5.14(a) and 5.14(b) show the use of contact-region modeling to model an inductive ring on top of a dielectric substrate [34]. It is shown that the meshes on the two contacting surfaces need not be the same, and yet the result can be reasonable. The inductor-like PEC-dielectric composite structure as shown in Figure 5.14(a), where the top part is a square PEC ring located on the dielectric substrate. For convenience of display, the top part is shifted upwards in Figure 5.14(a). A delta-gap source is placed peripherally in the middle of one arm of the PEC part. The input reactance is shown in Figure 5.14(b), which demonstrates the inductor-like character of the structure. Again the input resistance is negligible compared to the reactance. Good agreement among different meshes is observed.

Figures 5.15(a), 5.15(b), 5.15(c) show the use of LF-MOM for the solution of some large structures using LF-MLFMA acceleration with the stabilization of the patch pair described above [28]. It is only with stabilization that such a large problem can be solved. This example involves two layers of wires and a ground plane. The width of each wire is 5 μm and the thickness is 1 μm. The distance between two adjacent parallel wires is 4 μm. The ground plane is $210 \times 210 \ \mu m^2$. The first layer of wires is 4 μm above the ground plane and the second layer is 8 μm above. Two ports are defined as in the figure. The conductivity values of the wires and the ground plane is 3.7×10^7 S/m. The entire structure is assumed to be in a silicon dioxide background with $\epsilon_r = 3.8$.

5.6   Multiscale Problems in Electromagnetics

The solution to the multiscale problems in electromagnetics is important, especially so in the area of computer design. Inside a computer chassis, electromagnetic problems of different length-scales occur. Moreover, unlike some problems, homogenization method cannot be used to simplify the problems of small length-scale so that they can be interfaced with problems of larger length-scales.

Recently, the mixed-form fast multipole algorithm (MF-FMA) [17] and broadband fast algorithm [39, 40] have been developed so that integral equation solvers can be accelerated

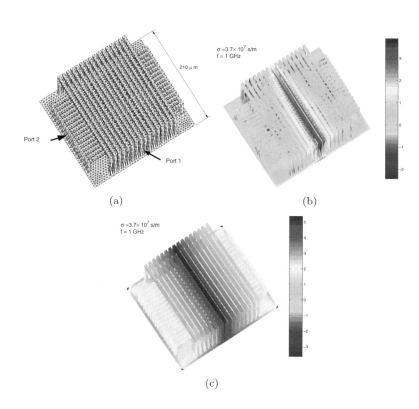

Figure 5.15: (a) Two layers of wires with a ground plane. (b) Charge magnitude on the three-layer structure. (c) Current magnitude on the three-layer structure.

all the way from very low frequency for objects with very tiny length-scale compared to wavelength to mid length-scale where the objects are of the order of wavelength or larger. These methods will be suitable for accelerating integral equation solvers for problems pertinent to multiscale problems.

However, because of the disparate nature of the physics of small length-scale and physics of mid-length-scale, it is necessary to decouple these two physical regimes when solutions are sought. The equivalence principle algorithm (EPA) [41,42] will be a good method to decouple the solutions of these two regimes. In this manner, solvers can be used separately to solve for solutions in the regime of small length-scales, and the regime of mid length-scales. This is still an active area of research.

5.7 Conclusions

This chapter introduces the low-frequency problems and solutions in solving integral equations. It suggests using the loop-tree decomposition to overcome the low-frequency breakdown problem. The basis rearrangement method is introduced to arrive at a better conditioned matrix system, and the physical reason is also given. The low-frequency technique is very important in solving problems associated with computer technology where wave physics meets circuit physics. It is also important in modeling nanostructures where the wavelength is much longer than the size of the structures. However, most structures, as those in computer technology (see end of Chapter 1), are multiscale. The multiscale problems presage new solution techniques. We are currently investigating a stable low-frequency solution without the need for loop-tree decomposition [43].

Bibliography

[1] J. Blau, "Supermarket's futuristic outlet," *IEEE Spectrum*, vol. 41, no. 4, pp. 21–22, 25, Apr. 2004.

[2] W. C. Chew *Waves and Fields in Inhomogeneous Media*, New York: Van Nostrand Reinhold, 1990. Reprinted by Piscataway, NJ: IEEE Press, 1995.

[3] M. S. Tong and W. C. Chew, "Unified boundary integral equation for the scattering of elastic wave and acoustic wave: solution by the method of moments," *Waves Random Complex Media*, vol. 18, pp. 303–324, May 2008.

[4] D. R. Wilton and A. W. Glisson, "On improving the electric field integral equation at low frequencies," *URSI Radio Sci. Meet. Dig.*, pp. 24, Los Angeles, CA, Jun. 1981.

[5] J. R. Mautz and R. F. Harrington, "An E-field solution for a conducting surface small or comparable to the wavelength," *IEEE Trans. Antennas Propag.*, vol. 32, pp. 330–339, Apr. 1984.

[6] E. Arvas, R. F. Harrington, and J. R. Mautz, "Radiation and scattering from electrically small conducting bodies of arbitrary shape," *IEEE Trans. Antennas Propag.*, vol. 34, pp. 66–77, Jan. 1986.

[7] W. Wu, A. W. Glisson, and D. Kajfez, "A comparison of two low-frequency formulations for the electric field integral equation," *Tenth Ann. Rev. Prog. Appl. Comput. Electromag.*, vol. 2, pp. 484–491, Monterey, California, Mar. 1994.

[8] M. Burton and S. Kashyap, "A study of a recent, moment-method algorithm that is accurate to very low frequencies," *Appl. Comput. Electromagn. Soc. J.*, vol. 10, no. 3, pp. 58–68, Nov. 1995.

[9] F. Yuan, "Analysis of power/ground noise and decoupling capacitors in printed circuit board systems," *IEEE Electromagn. Compat. Symp. Proc.*, pp. 425–430, Austin, TX, Aug. 1997.

[10] G. Vecchi, "Loop-star decomposition of basis functions in the discretization of EFIE," *IEEE Trans. Antennas Propag.*, vol. 47, pp. 339–346, Feb. 1999.

[11] J. S. Zhao and W. C. Chew, "Integral equation solution of maxwell's equations from zero frequency to microwave frequencies," *IEEE Trans. Antennas Propag., James R. Wait Memorial Special Issue*, vol. 48. no. 10, pp. 1635–1645, Oct. 2000.

[12] R. E. Collins, *Foundations for Microwave Engineering*, New York: McGraw-Hill, 1966.

[13] D. Colton and R. Kress, *Integral Equation Methods in Scattering Theory*, Melbourne, FL: Krieger Publ., 1983.

[14] W. C. Chew, B. Hu, Y. C. Pan and L. J. Jiang, "Fast algorithm for layered medium," *C. R. Phys.*, vol. 6, pp. 604–617, 2005.

[15] W. C. Chew, L. J. Jiang, Y. H. Chu, G. L. Wang, I. T. Chiang, Y. C. Pan, and J. S. Zhao, "Toward a more robust and accurate cem fast integral equation solver for IC applications," *IEEE Trans. Adv. Packag.*, vol. 28, no. 3, pp. 449–464, Aug. 2005.

[16] R. Albanese and G. Rubinacci, "Integral formulation for 3D eddy-current computation using edge elements," *IEE Proc.*, vol. 135, pt. A, no. 7, pp. 457–462, Sept. 1988.

[17] L. J. Jiang and W. C. Chew, "A mixed-form fast multipole algorithm," *IEEE Trans. Antennas Propag.*, vol. AP-53, no. 12, pp. 4145–4156, Dec. 2005.

[18] L. Bai, "An efficient algorithm for finding the minimal loop basis of a graph and its applications in computational electromagnetics," *M.S. Thesis*, Dept. ECE, U. Illinois, Urbana-Champaign, Sept. 2000.

[19] V. I. Okhmatovski, "An efficient algorithm for generation of loop-tree basis in 2.5D interconnect models," *Elect. Perf. Electron. Packag.*, Austin, TX, 2005.

[20] S. Chakraborty, D. Gope, G. Ouyang, and V. Jandhyala, "A three-stage preconditioner for geometries with multiple holes and handles in integral equation-based electromagnetic simulation of integrated packages," *Elect. Perf. Electron. Packag.*, Austin, TX, 2005.

[21] H. Y. Chao and W. C. Chew, "A linear-time algorithm for extracting tree and loop bases in computational electromagnetics," *IEEE Antennas Propag. Soc. Int. Symp. Proc.*, Albuquerque, NM, pp. 4115–4118, 2006.

[22] Y. Liu, Y. Chu and W. C. Chew, "LFFMA and induction well-logging modeling," *IEEE Antennas Propag. Soc. Int. Symp. Proc.*, Washington, DC, vol. 3A, pp. 224–227, Jul. 2005.

[23] J. D. Jackson, *Classical Electrodynamics*, 3rd ed., New York: Wiley, 1999.

[24] J. S. Zhao and W. C. Chew, "Three dimensional multilevel fast multipole algorithm from static to electrodynamic," *Microwave Opt. Tech. Lett.*, vol. 26, no. 1, pp. 43–48, Jul. 2000.

[25] I. Daubechies, "Orthonormal bases of compactly supported wavelets," *Commun. Pure Appl. Math.*, vol XLI, pp. 909–996, 1988.

[26] J. S. Zhao, W. C. Chew, and P. E. Mayes, "Accurate analysis of electrically small conical antennas by using the low-frequency method," *Antennas Appl. Symp.*, Monticello, IL, pp. 135–151, Sept. 19–21, 2001.

[27] K. F. Warnick and W. C. Chew, "Error analysis of surface integral equation methods," in *Fast and Efficient Algorithms in Computational Electromagnetics,* eds. W. Chew, J. Jin, E. Michielssen, J. Song, Boston: Artech House, 2001.

[28] Y. H. Chu and W. C. Chew, "Large-scale computation for electrically small structures using surface integral equation method," *Microwave Opt. Tech. Lett.*, vol. 47, no. 6, pp. 525–530, Dec. 2005.

[29] D. A. Shnidman and S. Lee, "The finite element radiation model (FERM) program," *Third Ann. Rev. Prog. Appl. Comput. Electromag.*, Monterey, CA, Mar. 24–26, 1987.

[30] L. N. Medgyesi-Mitschang, J. M. Putnam, and M. B. Gedera, "Generalized method of moments for three-dimensional penetrable scatterers," *J. Opt. Soc. Am. A*, vol. 11, pp. 1383–1398, Apr. 1994.

[31] Y. Zhang, T. J. Cui, W. C. Chew and J. S. Zhao, "Magnetic field integral equation at very low frequencies," *IEEE Trans. Antennas Propag.*, vol. 51, no. 8, pp. 1864–1871, Aug. 2003.

[32] S. Y. Chen, W. C. Chew, J. M. Song, and J. S. Zhao, "Analysis of low frequency scattering from penetrable scatterers," *IEEE Geosci. Remote Sens.*, vol. 39, no. 4, pp. 726–735, Apr. 2001.

[33] Y. Chu, W. C. Chew, S. Chen, and J. Zhao, "A surface integral equation method for low-frequency scattering from a composite object," *IEEE Trans. Antennas Propag.*, vol. 51, no. 10, pp. 2837-2844, Oct. 2003.

[34] Y. Chu and W. C. Chew, "A multi-level fast multipole algorithm for electrically small composite structures," *Microwave Opt. Tech. Lett.*, vol. 43, no. 3, pp. 202–207, Nov. 2004.

[35] L. Greengard, J. F. Huang, V. Rokhlin, and S. Wandzura, "Accelerating fast multipole methods for the Helmholtz equation at low frequencies," *IEEE Comput. Sci. Eng.*, vol. 5, no. 3, pp. 32–38, 1998.

[36] L. Xuan and R. J. Adams, "Computing the translation operator by sampling the Green function at low frequencies," *URSI Radio Sci. Meet. Dig.*, Columbus, OH, pp. 683, Jun. 2003.

[37] L. J. Jiang, and W. C. Chew, "Low frequency inhomogeneous plane wave algorithm–LF-FIPWA," *Microwave Opt. Tech. Lett.*, vol. 40, no. 2, pp. 117–122, Jan. 2004.

[38] E. Darve and P. Havè, "Efficient fast multipole method for low-frequency scattering," *J. Comput. Phys.*, vol. 197, no. 1 pp. 341–363, 2004.

[39] H. Wallén and J. Sarvas, "Translation procedures for broadband MLFMA," *Prog. Electromagn. Res.*, PIER 55, pp. 47-78, 2005.

[40] H. Cheng, W. Y. Crutchfield, Z. Gimbutasa, L. F. Greengard, J. F. Ethridgea, J. Huang, V. Rokhlin, N. Yarvin, and J. Zhao, "A wideband fast multipole method for the Helmholtz equation in three dimensions," *J. Comput. Phys.*, vol. 216, no. 1, pp. 300–325, 2006.

[41] M. K. Li and W. C. Chew, "A domain decomposition scheme based on equivalence theorem," *Microwave Opt. Tech. Lett.*, vol. 48, no. 9, pp. 1853–1857, Sept. 2006.

[42] M. K. Li and W. C. Chew, "Wave-Field Interaction With Complex Structures Using Equivalence Principle Algorithm," *IEEE Trans. Antennas Propag.*, vol. 55, no. 1, pp. 130–138, Jan. 2007.

[43] Z. G. Qian and W. C. Chew, "An augmented EFIE for high speed interconnect analysis," *Microwave Opt. Tech. Lett.*, scheduled for October 2008.

CHAPTER 6

Dyadic Green's Function for Layered Media and Integral Equations

6.1 Introduction

In computational electromagnetics (CEM), modeling with layered medium is quite common. This problem can be formulated in terms of integral equations using the layered medium Green's function. As such, the dyadic Green's function needs to be derived. When the layered medium Green's function is used, unknowns are only associated with the embedded scatterers or objects, but not with the interfaces of the layers. Hence, it greatly reduces the number of unknowns required to model the problem accurately. The integral equation thus formulated is then converted into a matrix equation by the method of moments (MOM) [1] described in the previous chapters.

In this chapter, we introduce an elegant way of deriving the matrix representation of the dyadic Green's function for layered media (DGLM) [2]. The derivation is based on the pilot vector potential approach where two vector potentials in terms of TE and TM fields are derived. The derivation is succinct and the resulting Sommerfeld integrals only involve zeroth order Bessel function and two scalar potentials.

The problem of a dipole over a half space was first solved by Sommerfeld [3] in 1909 using Hertz vector potential. Kong in 1972 suggested using the E_z and H_z formulation to solve such problems, extending them to layered media [4]. The layered medium dyadic Green's function can only be written in terms of Fourier integrals which are expressed in terms of Sommerfeld integrals [5, 6]. The evaluation of these integrals is time consuming, especially if they are associated with high order coordinate-space singularities. Because these integrals are spectral integrals, the coordinate-space singularities translate into higher spectral components in the Fourier space, causing their slow convergence.

The filling of these matrix elements involves evaluation of Sommerfeld integrals for singular functions in the coordinate space. These singular functions are due to the spatial derivatives of the Green's function. To evaluate these matrix elements, it is important that the matrix representations be manipulated into a form whose coordinate-space singularities are as weak as possible. We shall call such matrix elements the matrix-friendly forms. By this, we mean that their associated Sommerfeld integrals are more rapidly convergent.

6.2 Dyadic Green's Function for Layered Media

The DGLM has been derived in various forms [5–10]. Michalski and Zheng [7] used the mixed potential formulation to derive a dyadic Green's function. The Michalski-Zheng formulation is vastly popular and has been adopted by many workers[1] [11–18]. We will start with a dyadic Green's function that is succinctly derived in [6], and then manipulate it into a form that is matrix-friendly for evaluating elements of its matrix representation.

To derive the dyadic Green's function for a layered medium, first we derive that for a homogeneous medium. The dyadic Green's function for a homogeneous medium (or free space) exists in many different forms. However, we will use the form that is expressed in terms of a linear superposition of plane waves. This is with the hindsight that the interaction of plane waves with layered medium can be easily solved for.

Using the z-directed pilot vector potential approach [19], the free space dyadic Green's function can be found using the vector wave function expansion approach. These are the $\mathbf{M}$, $\mathbf{N}$ and $\mathbf{L}$ functions where $\mathbf{M}$, $\mathbf{N}$ are divergence free but not curl free (solenoidal) whereas $\mathbf{L}$ is curl free (irrotational) but not divergence free. In Cartesian coordinates, when $\hat{z}$ is chosen as the pilot vector, they are defined in terms of plane waves as

$$\mathbf{M}(\mathbf{k}, \mathbf{r}) = \nabla \times \hat{z} e^{i\mathbf{k} \cdot \mathbf{r}} = i\mathbf{k} \times \hat{z} e^{i\mathbf{k} \cdot \mathbf{r}} \tag{6.1}$$

$$\mathbf{N}(\mathbf{k}, \mathbf{r}) = \frac{1}{k} \nabla \times \mathbf{M}(\mathbf{k}, \mathbf{r}) = -\frac{1}{k} \mathbf{k} \times \mathbf{k} \times \hat{z} e^{i\mathbf{k} \cdot \mathbf{r}} \tag{6.2}$$

$$\mathbf{L}(\mathbf{k}, \mathbf{r}) = \nabla e^{i\mathbf{k} \cdot \mathbf{r}} = i\mathbf{k} e^{i\mathbf{k} \cdot \mathbf{r}} \tag{6.3}$$

Here, $\mathbf{M}$ and $\mathbf{N}$ are transverse waves because they are orthogonal to the direction of the plane wave propagation or the $\mathbf{k}$ vector, whereas $\mathbf{L}$ is a longitudinal wave because it points in the direction of the plane wave propagation or the $\mathbf{k}$ vector.

The dyadic Green's function, when expressed in terms of the vector wave functions, is

$$\overline{\mathbf{G}}(\mathbf{r}, \mathbf{r}') = \frac{1}{(2\pi)^3} \iiint\limits_{-\infty}^{\infty} d\mathbf{k} \left[\frac{\mathbf{M}(\mathbf{k}, \mathbf{r}) \, \mathbf{M}(-\mathbf{k}, \mathbf{r}')}{(k^2 - k_m^2) k_s^2} \right.$$
$$\left. + \frac{\mathbf{N}(\mathbf{k}, \mathbf{r}) \, \mathbf{N}(-\mathbf{k}, \mathbf{r}')}{(k^2 - k_m^2) k_s^2} - \frac{\mathbf{L}(\mathbf{k}, \mathbf{r}) \, \mathbf{L}(-\mathbf{k}, \mathbf{r}')}{k_m^2 k^2} \right]. \tag{6.4}$$

where $k_s^2 = k_x^2 + k_y^2$, $k^2 = k_x^2 + k_y^2 + k_z^2$, $k_m = \omega \sqrt{\mu_m \epsilon_m}$ is the wavenumber of the homogeneous medium, and $d\mathbf{k} = dk_x dk_y dk_z$ is a three-dimensional Fourier integral. Hence, k is not a constant at this point. A continuum of Fourier wave number is needed because the above corresponds to a Fourier expansion in an infinitely large box in the x, y, and z directions.

---

[1]This list is by no means complete.

The above choice of pilot vector $\hat{z}$ makes the vector $\mathbf{M}$ function transverse to $\hat{z}$. If the above represents the electric field dyadic Green's function, $\mathbf{M}$ corresponds to a TE to z wave, whereas $\mathbf{N}$ resembles the electric field of a TM to z wave. The $\mathbf{L}$ function is needed for completeness. This follows from Helmholtz theorem that a vector field can be decomposed into the sum of a solenoidal (divergence-free) field and an irrotational (curl-free) field.

A contour integration technique can be applied to the above to evaluate the dk_z integration in closed form. Consequently, the dyadic Green's function can be decomposed into TE_z and TM_z components. Hence, the dyadic Green's function for a homogeneous medium can be derived in terms of vector wave functions as [6, p. 411]

$$\overline{\mathbf{G}}(\mathbf{r}, \mathbf{r}') = \frac{i}{8\pi^2} \iint_{-\infty}^{\infty} \frac{d\mathbf{k}_s}{k_{mz} k_s^2} \left[\mathbf{M}(\mathbf{k}_s, \mathbf{r}) \mathbf{M}(-\mathbf{k}_s, \mathbf{r}') \right.$$
$$\left. + \mathbf{N}(\mathbf{k}_s, \mathbf{r}) \mathbf{N}(-\mathbf{k}_s, \mathbf{r}') \right] - \frac{\hat{z}\hat{z}}{k_m^2} \delta(\mathbf{r} - \mathbf{r}'), \tag{6.5}$$

where

$$\mathbf{M}(\mathbf{k}_s, \mathbf{r}) \mathbf{M}(-\mathbf{k}_s, \mathbf{r}') = (\mathbf{k}_s \times \hat{z})(\mathbf{k}_s \times \hat{z}) e^{i\mathbf{k}_s \cdot (\mathbf{r}_s - \mathbf{r}'_s)} e^{ik_{mz}|z - z'|} \tag{6.6}$$

$$\mathbf{N}(\mathbf{k}_s, \mathbf{r}) \mathbf{N}(-\mathbf{k}_s, \mathbf{r}') = \frac{1}{k_m^2} (\mathbf{k}_{m\pm} \times \mathbf{k}_s \times \hat{z})(\mathbf{k}_{m\mp} \times \mathbf{k}_s \times \hat{z}) \cdot e^{i\mathbf{k}_s \cdot (\mathbf{r}_s - \mathbf{r}'_s)} e^{ik_{mz}|z - z'|} \tag{6.7}$$

$\mathbf{k}_{m\pm} = \mathbf{k}_s \pm \hat{z} k_{mz}$, $\mathbf{k}_s = \hat{x} k_x + \hat{y} k_y$, and the upper sign is chosen when $z > z'$ and the lower sign is chosen when $z < z'$. Here, $k_{mz} = \sqrt{k_m^2 - k_s^2}$ and $k_s^2 = \mathbf{k}_s \cdot \mathbf{k}_s$.

In (6.4), the dyadic Green's function represents the electric field produced by a point current source at $\mathbf{r}'$. Hence, this field must be divergence-free when $\mathbf{r} \neq \mathbf{r}'$. The $\mathbf{L}$ function is generally nondivergence free, and seemingly, it cannot be nonzero outside the source region. But because the $\nabla \cdot \mathbf{L} = -k^2 e^{i\mathbf{k} \cdot \mathbf{r}}$ which is divergence free at $k = 0$, the $\mathbf{L}$ function can be nonzero outside the source point for $k = 0$. In fact, $k = 0$ with $\mathbf{k} \neq 0$ is possible, where the three components of the $\mathbf{k}$ vector are related by $k_z = \pm\sqrt{k_x^2 + k_y^2}$, so that $\mathbf{L}$ defined in (6.3) can be nonzero. Consequently, looking at (6.4), for the $\mathbf{L}$ term, a pole exists at $k = 0$ giving rise to a pole contribution for the $\mathbf{L}$ term when the contour integration technique is used to evaluate the dk_z integration. This field is nonzero outside the source point, but it is Laplacian (static) in nature, and hence, is unphysical for a time-harmonic field. Fortunately, because $\mathbf{N} \sim \frac{1}{k}$, $k \to 0$, a pole contribution from the $\mathbf{N}$ term produces a static field that exactly cancels this unphysical field. The last term in (6.5) is the remnant of this cancelation.

The above integrals are ill-convergent when the source point and the field point are coplanar, viz., $z = z'$. This is because this plane contains the source point singularity whose Fourier transform is ill-convergent (see [13, 14, 18] for details). However, numerical evaluation of them can be avoided because they have closed form in coordinate space for the homogeneous medium case, that is,

$$\overline{\mathbf{G}}(\mathbf{r}, \mathbf{r}') = \left(\overline{\mathbf{I}} + \frac{1}{k_m^2} \nabla\nabla \right) \frac{e^{ik_m|\mathbf{r} - \mathbf{r}'|}}{4\pi|\mathbf{r} - \mathbf{r}'|} \tag{6.8}$$

In the above, the $\mathbf{M}$ function corresponds to TE_z waves whereas the $\mathbf{N}$ function corresponds to TM_z waves. For a layered medium, we can replace the above by [6]

$$\mathbf{M}(\mathbf{k}_s, \mathbf{r})\mathbf{M}(-\mathbf{k}_s, \mathbf{r}') = (\nabla \times \hat{z})(\nabla' \times \hat{z})e^{i\mathbf{k}_s \cdot (\mathbf{r}_s - \mathbf{r}'_s)}F^{TE}(k_s, z, z'), \tag{6.9}$$

$$\mathbf{N}(\mathbf{k}_s, \mathbf{r})\mathbf{N}(-\mathbf{k}_s, \mathbf{r}') = -\left(\frac{\nabla \times \nabla \times \hat{z}}{i\omega\epsilon_n}\right)\left(\frac{\nabla' \times \nabla' \times \hat{z}}{i\omega\mu_m}\right) \cdot e^{i\mathbf{k}_s \cdot (\mathbf{r}_s - \mathbf{r}'_s)}F^{TM}(k_s, z, z'), \tag{6.10}$$

where we have assumed that the source point is in region m whereas the field point is in region n. The function $F^\alpha(k_s, z, z')$ above, where α is either TE or TM, satisfies the following ordinary differential equation [6, 20]

$$\left[\frac{d}{dz}\frac{1}{p(z)}\frac{d}{dz} + \frac{1}{p(z)}k_z^2(z)\right]F^\alpha(k_s, z, z') = \frac{2i}{p(z)}k_z(z)\delta(z, z') \tag{6.11}$$

where $p = \mu$ for TE waves, and $p = \epsilon$ for TM waves. Moreover, $k_z = \sqrt{k^2(z) - k_s^2}$. When $k(z)$ is piecewise constant denoted by k_m for region m, F^α can be found in closed form using a recursive procedure [6, p. 76]. Furthermore, F^α describes the propagation of B_z or D_z fields for TE and TM fields respectively in an inhomogeneous layer. For a homogeneous medium, $F^\alpha(k_s, z, z') = \exp(ik_{mz}|z - z'|)$.

After using Eqs. (6.9) and (6.10) in Eq. (6.5), and exchanging the order of derivatives and integrals, we can rewrite the dyadic Green's function for layered media as

$$\overline{\mathbf{G}}_e(\mathbf{r}, \mathbf{r}') = (\nabla \times \hat{z})(\nabla' \times \hat{z})g^{TE}(\mathbf{r}, \mathbf{r}')$$
$$+ \frac{1}{k_{nm}^2}(\nabla \times \nabla \times \hat{z})(\nabla' \times \nabla' \times \hat{z})g^{TM}(\mathbf{r}, \mathbf{r}'), \qquad |z - z'| > 0, \tag{6.12}$$

where $k_{nm}^2 = \omega^2\epsilon_n\mu_m$. The subscript e is used to denote that this is an electric dyadic Green's function that generates electric field when acting on an electric current source. The above representation is strictly correct only when $|z - z'| > 0$, because when the derivative operators act on $\exp(ik_{mz}|z - z'|)$, a singularity is produced at $z = z'$ which is not the same as that in Eq. (6.5). Here, $g^{TE}(\mathbf{r}, \mathbf{r}')$ and $g^{TM}(\mathbf{r}, \mathbf{r}')$ consist of the primary (direct) field terms and the secondary (reflected) field terms. In a layer that does not contain the source point, the field is entirely secondary. The primary field term contains the source singularity, and hence, its representation in the Fourier space is inefficient. Because it is the same as the homogeneous medium case for which closed form exists, it is more expedient to express it in closed form in coordinate space (see Eq. (6.8)).

Another point to be noted is that in (6.12), the derivative operators are taken out of the spectral integrals. Hence, $g^{TE}(\mathbf{r}, \mathbf{r}')$ and $g^{TM}(\mathbf{r}, \mathbf{r}')$ in their coordinate-space representations, are less singular than the dyadic Green's function itself. As such, their spectral integrals are more rapidly convergent compared to the spectral integrals in (6.5).

The explicit expressions for $g^{TE}(\mathbf{r}, \mathbf{r}')$ and $g^{TM}(\mathbf{r}, \mathbf{r}')$ are

$$g^{TE}(\mathbf{r}, \mathbf{r}') = \frac{i}{8\pi^2}\iint_{-\infty}^{\infty}\frac{d\mathbf{k}_s}{k_{mz}k_s^2}e^{i\mathbf{k}_s \cdot (\mathbf{r}_s - \mathbf{r}'_s)}F^{TE}(k_s, z, z') \tag{6.13}$$

$$g^{TM}(\mathbf{r}, \mathbf{r}') = \frac{i}{8\pi^2}\iint_{-\infty}^{\infty}\frac{d\mathbf{k}_s}{k_{mz}k_s^2}e^{i\mathbf{k}_s \cdot (\mathbf{r}_s - \mathbf{r}'_s)}F^{TM}(k_s, z, z') \tag{6.14}$$

The above Fourier integrals can be expressed in terms of Sommerfeld integrals involving a single integral and Bessel function (where $k_\rho = k_s$) using the techniques described in [6], namely

$$g^{\text{TE}}(\mathbf{r}, \mathbf{r}') = \frac{i}{4\pi} \int_0^\infty \frac{dk_\rho}{k_{mz}k_\rho} J_0(k_\rho |\mathbf{r}_s - \mathbf{r}'_s|) F^{\text{TE}}(k_\rho, z, z'), \tag{6.15}$$

$$g^{\text{TM}}(\mathbf{r}, \mathbf{r}') = \frac{i}{4\pi} \int_0^\infty \frac{dk_\rho}{k_{mz}k_\rho} J_0(k_\rho |\mathbf{r}_s - \mathbf{r}'_s|) F^{\text{TM}}(k_\rho, z, z'). \tag{6.16}$$

To arrive at the above, we have made use of the identity that

$$\iint_{-\infty}^\infty d\mathbf{k}_s = \int_0^{2\pi} d\alpha \int_0^\infty k_\rho dk_\rho \tag{6.17}$$

$$J_0(k_\rho |\mathbf{r}_s - \mathbf{r}'_s|) = \frac{1}{2\pi} \int_0^{2\pi} e^{i\mathbf{k}_\rho \cdot (\mathbf{r}_s - \mathbf{r}'_s)} d\alpha \tag{6.18}$$

where $\mathbf{k}_\rho = \hat{x} k_\rho \cos\alpha + \hat{y} k_\rho \sin\alpha$. Because of the $1/k_s^2$ or the $1/k_\rho$ term in the integrands, these spectral integrals are more rapidly convergent compared to those in (6.5). When F^{TM} and F^{TE} are defined in (6.15)–(6.16), the generalized reflection coefficients $\tilde{R}^{\text{TE}}$ and $\tilde{R}^{\text{TM}}$ [6, Chap. 2] embody polarization effects of the layered medium.

Notice that in (6.15) and (6.16), poles exist at $k_\rho = 0$, but these poles are unphysical causing the integrals to diverge. This dilemma is nonexistent because g^{TE} and g^{TM} are not the final physical quantities, but $\overline{\mathbf{G}}_e$ in (6.12). In fact, the poles disappear because they cancel each other when (6.15) and (6.16) are substituted in (6.12).

6.3   Matrix Representation

The primary field term is best expressed in coordinate space in closed form as in Eq. (6.8). Hence, its matrix representation is straightforward and can be done following conventional methods. The secondary field terms do not contain singularities, and are regular, except when both the field point and source point are at the interface—a near singularity exists when the source point and field points are close to the interface.

In the following, the manipulation to obtain the matrix representation of the Green's function operator is intended for the secondary fields where the image source point and the field point do not coincide. Focussing on the TM wave term for the dyadic Green's function, we have

$$\overline{\mathbf{G}}_e^{\text{TM},S}(\mathbf{r}, \mathbf{r}') = \frac{1}{k_{nm}^2} (\nabla \times \nabla \times \hat{z})(\nabla' \times \nabla' \times \hat{z}) g^{\text{TM},S}(\mathbf{r}, \mathbf{r}')$$

$$= \frac{1}{k_{nm}^2} (\nabla\nabla \cdot \hat{z} + k_n^2 \hat{z})(\nabla'\nabla' \cdot \hat{z} + k_m^2 \hat{z}) g^{\text{TM},S}(\mathbf{r}, \mathbf{r}')$$

$$= \frac{1}{k_{nm}^2} (\nabla\nabla' \partial_z \partial_{z'} + k_n^2 \nabla'_z \nabla' + k_m^2 \nabla\nabla_z + k_n^2 k_m^2 \hat{z}\hat{z}) g^{\text{TM},S}(\mathbf{r}, \mathbf{r}'), \tag{6.19}$$

where the superscript S indicates that these are secondary field terms, and $\nabla_z = \hat{z} \partial_z$. In the above, we have assumed that $(\nabla^2 + k_n^2) g^{\text{TM},S}(\mathbf{r}, \mathbf{r}') = 0$ and $(\nabla'^2 + k_m^2) g^{\text{TM},S}(\mathbf{r}, \mathbf{r}') = 0$, because we assume that the field point and the image source point do not coincide. Likewise, the TE wave term for the DGLM is

$$\overline{\mathbf{G}}_e^{\mathrm{TE},S}(\mathbf{r},\mathbf{r}') = (\nabla \times \hat{z})(\nabla' \times \hat{z})g^{\mathrm{TE},S}(\mathbf{r},\mathbf{r}'). \qquad (6.20)$$

The electric field due to an electric current can be expressed with the dyadic Green's function for an inhomogeneous medium as [6, p. 411]

$$\mathbf{E}(\mathbf{r}) = i\omega \int_V d\mathbf{r}' \overline{\mathbf{G}}_e(\mathbf{r},\mathbf{r}')\mu(\mathbf{r}') \cdot \mathbf{J}(\mathbf{r}'). \qquad (6.21)$$

Next, we evaluate the matrix element for the impedance matrix as

$$\begin{aligned} Z_{ij} &= i\omega\mu_m \langle \mathbf{J}_{Ti}(\mathbf{r}), \overline{\mathbf{G}}_e(\mathbf{r},\mathbf{r}'), \mathbf{J}_j(\mathbf{r}') \rangle \\ &= i\omega\mu_m \int d\mathbf{r} \mathbf{J}_{Ti}(\mathbf{r}) \cdot \int d\mathbf{r}' \overline{\mathbf{G}}_e(\mathbf{r},\mathbf{r}') \cdot \mathbf{J}_j(\mathbf{r}'), \end{aligned} \qquad (6.22)$$

where we have assumed that the source expansion function $\mathbf{J}_j(\mathbf{r}')$ is entirely in the m-th region. When a source expansion function straddles two or more regions, the above expression has to be modified accordingly.

In the above, the matrix representation of the primary field term is obtained in the conventional method. For the secondary field, the term associated with TM waves is

$$\begin{aligned} \langle \mathbf{J}_{Ti}(\mathbf{r}), \overline{\mathbf{G}}_e^{\mathrm{TM},S}(\mathbf{r},\mathbf{r}'), \mathbf{J}_j(\mathbf{r}') \rangle = \frac{1}{k_{nm}^2} &\left[\langle \nabla \cdot \mathbf{J}_{Ti}(\mathbf{r}), \ \partial_z \partial_{z'} g^{\mathrm{TM},S}(\mathbf{r},\mathbf{r}'), \ \nabla' \cdot \mathbf{J}_j(\mathbf{r}') \rangle \right. \\ &- k_n^2 \langle \mathbf{J}_{Ti}(\mathbf{r}) \cdot \hat{z}, \ \partial_{z'} g^{\mathrm{TM},S}(\mathbf{r},\mathbf{r}'), \ \nabla' \cdot \mathbf{J}_j(\mathbf{r}') \rangle \\ &- k_m^2 \langle \nabla \cdot \mathbf{J}_{Ti}(\mathbf{r}), \ \partial_z g^{\mathrm{TM},S}(\mathbf{r},\mathbf{r}'), \ \hat{z} \cdot \mathbf{J}_j(\mathbf{r}') \rangle \\ &\left. + k_n^2 k_m^2 \langle \hat{z} \cdot \mathbf{J}_{Ti}(\mathbf{r}), \ g^{\mathrm{TM},S}(\mathbf{r},\mathbf{r}'), \ \hat{z} \cdot \mathbf{J}_j(\mathbf{r}') \rangle \right]. \end{aligned} \qquad (6.23)$$

In deriving the above, we have made use of integration by parts as much as possible to move the ∇ operators away from $g^{\mathrm{TM},S}$. Notice that $\partial_z \partial_{z'} g^{\mathrm{TM},S}$ is the most singular function since derivatives enhance the order of the singularity. If the z derivative operators are brought into the spectral integral of $g^{\mathrm{TM},S}$, the resultant integral will be slowly convergent. The other integrals are more benign and are rapidly convergent when numerically integrated.

For the term associated with TE waves, using (6.20), we can express it as

$$\langle \mathbf{J}_{Ti}(\mathbf{r}), \overline{\mathbf{G}}_e^{\mathrm{TE},S}(\mathbf{r},\mathbf{r}'), \mathbf{J}_j(\mathbf{r}') \rangle = \langle \mathbf{J}_{Ti}(\mathbf{r}), \left[(\nabla_s \times \hat{z})(\nabla'_s \times \hat{z})g^{\mathrm{TE},S}(\mathbf{r},\mathbf{r}') \right], \mathbf{J}_j(\mathbf{r}') \rangle, \qquad (6.24)$$

where $\nabla_s = \hat{x}\partial_x + \hat{y}\partial_y$. If $\mathbf{J}_{Ti}$ and $\mathbf{J}_j$ are curl-conforming volume expansion function, integration by parts can be used to move the curl operators to the expansion and testing functions. We manipulate the expression by first showing that

$$\begin{aligned} (\hat{z} \times \nabla_s)(\hat{z} \times \nabla'_s) &= k_s^2 \overline{\mathbf{I}}_s - \nabla_s \nabla'_s \\ &= \overline{\mathbf{I}}_s k_s^2 - (\nabla \nabla' - \nabla_z \nabla' - \nabla \nabla'_z + \nabla_z \nabla'_z). \end{aligned} \qquad (6.25)$$

(The first equality above can be easily proved in the spectral domain.) Then we can rewrite (6.24) as

$$
\begin{aligned}
\langle \mathbf{J}_{Ti}(\mathbf{r}), \overline{\mathbf{G}}_e^{\text{TE},S}(\mathbf{r},\mathbf{r}'), \mathbf{J}_j(\mathbf{r}')\rangle =&\langle \mathbf{J}_{TSi}(\mathbf{r}), g_s^{\text{TE},S}(\mathbf{r},\mathbf{r}'), \mathbf{J}_{Sj}(\mathbf{r}')\rangle \\
& - \langle \nabla \cdot \mathbf{J}_{Ti}(\mathbf{r}),\ g^{\text{TE},S}(\mathbf{r},\mathbf{r}'), \nabla' \cdot \mathbf{J}_j(\mathbf{r}')\rangle \\
& - \langle \mathbf{J}_{Ti}(\mathbf{r}) \cdot \hat{z},\ \partial_z g^{\text{TE},S}(\mathbf{r},\mathbf{r}'), \nabla' \cdot \mathbf{J}_j(\mathbf{r}')\rangle \\
& - \langle \nabla \cdot \mathbf{J}_{Ti}(\mathbf{r}),\ \partial_{z'} g^{\text{TE},S}(\mathbf{r},\mathbf{r}'), \hat{z} \cdot \mathbf{J}_j(\mathbf{r}')\rangle \\
& - \langle \mathbf{J}_{Ti}(\mathbf{r}) \cdot \hat{z},\ \partial_z \partial_{z'} g^{\text{TE},S}(\mathbf{r},\mathbf{r}'), \hat{z} \cdot \mathbf{J}_j(\mathbf{r}')\rangle.
\end{aligned}
\tag{6.26}
$$

In the above,

$$
g_s^{\text{TE},S}(\mathbf{r},\mathbf{r}') = \frac{i}{4\pi} \int_0^\infty \frac{dk_\rho k_\rho}{k_{mz}} J_0(k_\rho |\mathbf{r}_s - \mathbf{r}'_s|) F^{\text{TE},S}(k_\rho, z, z'),
\tag{6.27}
$$

where the primary field term has been removed in $F^{\text{TE},S}$, and $\mathbf{J}_{TSi}$ is the part of $\mathbf{J}_{Ti}$ that is transverse to $\hat{z}$, and similarly for $\mathbf{J}_{Sj}$.

We can combine the integrals in (6.26) with those in (6.23) to arrive at five basic terms. By combining the TM and the TE Green's functions together, and collecting like terms, for the secondary field, we get

$$
\begin{aligned}
&\langle \mathbf{J}_{Ti}(\mathbf{r}), \overline{\mathbf{G}}_e^{S}(\mathbf{r},\mathbf{r}'), \mathbf{J}_j(\mathbf{r}')\rangle \\
&= \left\langle \nabla \cdot \mathbf{J}_{Ti}(\mathbf{r}),\ \frac{\partial_z \partial_{z'}}{k_{nm}^2} g^{\text{TM},S}(\mathbf{r},\mathbf{r}') - g^{\text{TE},S}(\mathbf{r},\mathbf{r}'),\ \nabla' \cdot \mathbf{J}_j(\mathbf{r}') \right\rangle \\
&\quad - \left\langle \mathbf{J}_{Ti}(\mathbf{r}) \cdot \hat{z},\ \frac{\mu_n}{\mu_m}\partial_{z'} g^{\text{TM},S}(\mathbf{r},\mathbf{r}') + \partial_z g^{\text{TE},S}(\mathbf{r},\mathbf{r}'),\ \nabla' \cdot \mathbf{J}_j(\mathbf{r}') \right\rangle \\
&\quad - \left\langle \nabla \cdot \mathbf{J}_{Ti}(\mathbf{r}),\ \frac{\epsilon_m}{\epsilon_n}\partial_z g^{\text{TM},S}(\mathbf{r},\mathbf{r}') + \partial_{z'} g^{\text{TE},S}(\mathbf{r},\mathbf{r}'),\ \hat{z} \cdot \mathbf{J}_j(\mathbf{r}') \right\rangle \\
&\quad + \left\langle \hat{z} \cdot \mathbf{J}_{Ti}(\mathbf{r}),\ k_{mn}^2 g^{\text{TM},S}(\mathbf{r},\mathbf{r}') - \partial_z \partial_{z'} g^{\text{TE},S}(\mathbf{r},\mathbf{r}'),\ \hat{z} \cdot \mathbf{J}_j(\mathbf{r}') \right\rangle \\
&\quad + \left\langle \mathbf{J}_{TSi}(\mathbf{r}),\ g_s^{\text{TE},S}(\mathbf{r},\mathbf{r}'),\ \mathbf{J}_{Sj}(\mathbf{r}') \right\rangle.
\end{aligned}
\tag{6.28}
$$

Evaluating (6.28) involves the computation of two basic Green's functions $g^{\text{TM},S}$, $g^{\text{TE},S}$ and their z derivatives. The $g_s^{\text{TE},S}$ Green's function can be obtained from $g^{\text{TE},S}$ by taking R_s derivatives, where $R_s = |\mathbf{r}_s - \mathbf{r}'_s|$. It can be shown by using the property of the Bessel equation that

$$
\frac{1}{R_s} \frac{\partial}{\partial R_s} R_s \frac{\partial}{\partial R_s} g^{\text{TE},S} + g_s^{\text{TE},S} = 0
\tag{6.29}
$$

these Green's functions, $g^{\text{TM},S}$, $g^{\text{TE},S}$, can be computed and tabulated via Sommerfeld integrals, and their derivatives can be approximated by finite difference on tabulated values to evaluate the above five terms.

Alternatively, for higher accuracy, one may want to evaluate the z derivatives exactly by bringing them inside the spectral integrals to avoid the need for finite difference. To this end, the Green's functions sandwiched between the source expansion function and the test function can be combined into one integral, and tabulated for efficiency. Also, in the special case when the source point and the field point are in the uppermost layer or the lowermost layer, the second and the third terms in (6.28) can be further combined, leaving only four integrals.

We consider the matrix representation when Rao-Wilton-Glisson (RWG) functions are used for expansion and testing. Using the vector notation shown in Figure 6.1, the matrix elements for the impedance matrix can be explicitly expressed as

$$Z_{jk} = i\omega\mu_m \sum_{p,q=1}^{2} \frac{l_j l_k}{4 S_j^p S_k^q} \int_{S_j^p} d\mathbf{r} \int_{S_k^q} d\mathbf{r}' \left\{ Z_{ss,jk}^{pq} + Z_{zz,jk}^{pq} + Z_{z_1,jk}^{pq} + Z_{z_2,jk}^{pq} \right.$$
$$\left. + Z_{\Phi,jk}^{pq} \right\} \tag{6.30}$$

where

$$Z_{ss,jk}^{pq} = \left(\hat{z} \times \hat{z} \times \boldsymbol{\xi}_j^p \right) \cdot \left(\hat{z} \times \hat{z} \times \boldsymbol{\xi}_k^q \right) g_s^{\mathrm{TE}} \left(\mathbf{r}, \mathbf{r}' \right) \tag{6.31}$$

$$Z_{zz,jk}^{pq} = \left(\hat{z} \cdot \boldsymbol{\xi}_j^p \right) \left(\hat{z} \cdot \boldsymbol{\xi}_k^q \right) \left[k_{mn}^2 g^{\mathrm{TM}} - \partial_z \partial_{z'} g^{\mathrm{TE}} \right] \tag{6.32}$$

$$Z_{z_1,jk}^{pq} = - 2\epsilon_q \left(\boldsymbol{\xi}_j^p \cdot \hat{z} \right) \left[\frac{\mu_n}{\mu_m} \partial_{z'} g^{\mathrm{TM}} + \partial_z g^{\mathrm{TE}} \right] \tag{6.33}$$

$$Z_{z_2,jk}^{pq} = - 2\epsilon_p \left(\boldsymbol{\xi}_k^q \cdot \hat{z} \right) \left[\frac{\epsilon_m}{\epsilon_n} \partial_z g^{\mathrm{TM}} + \partial_{z'} g^{\mathrm{TE}} \right] \tag{6.34}$$

$$Z_{\Phi,jk}^{pq} = 4\epsilon_{pq} \left[\frac{\partial_z \partial_{z'}}{k_{nm}^2} g^{\mathrm{TM}} - g^{\mathrm{TE}} \right] \tag{6.35}$$

where $\epsilon_{p,q} = \begin{cases} 1, & p = q \\ -1, & p \neq q \end{cases}$, $\epsilon_s = \begin{cases} 1, & s = 1 \\ -1, & s = 2 \end{cases}$, $s = p, q$. The subscripts ss, z_1, z_2, zz, Φ of $Z_{\cdot,jk}^{pq}$ follow [7] for ease of comparison.

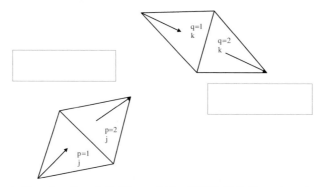

Figure 6.1: The notation for the definition of the RWG [21] functions in this Chapter. The function is also used as a testing function.

6.4 The $\nabla \times \overline{\mathbf{G}}_e$ Operator

In addition to calculating the electric field due to an electric current source, we need to find the magnetic field due to an electric current source. Then the other operators can be found by invoking duality.

By taking the curl of the $\overline{\mathbf{G}}_e$ operator, the primary field term from (6.4), is given by

$$\nabla \times \overline{\mathbf{G}}_e^P(\mathbf{r}, \mathbf{r}') = \nabla \times \overline{\mathbf{I}} \frac{e^{ik_m|\mathbf{r}-\mathbf{r}'|}}{4\pi|\mathbf{r}-\mathbf{r}'|} \tag{6.36}$$

By taking the curl of the secondary field term, from (6.19), we have

$$\nabla \times \overline{\mathbf{G}}_e^{\mathrm{TM},S}(\mathbf{r}, \mathbf{r}') = \frac{\mu_n}{\mu_m}(\nabla \times \hat{z})(\nabla' \times \nabla' \times \hat{z})g^{\mathrm{TM},S}(\mathbf{r}, \mathbf{r}') \tag{6.37}$$

From (6.20), we have

$$\nabla \times \overline{\mathbf{G}}_e^{\mathrm{TE},S}(\mathbf{r}, \mathbf{r}') = (\nabla \times \nabla \times \hat{z})(\nabla' \times \hat{z})g^{\mathrm{TE},S}(\mathbf{r}, \mathbf{r}') \tag{6.38}$$

As

$$\mathbf{H}(\mathbf{r}) = \int_V \nabla \times \overline{\mathbf{G}}_e(\mathbf{r}, \mathbf{r}') \cdot \mathbf{J}(\mathbf{r}')d\mathbf{r}' \tag{6.39}$$

Equations (6.37) and (6.38) will generate the magnetic field when operating on the electric current. In fact (6.37) and (6.38) are related by duality.

When we need to find the matrix representation of the above operators, we rewrite

$$\nabla \times \overline{\mathbf{G}}_e^{\mathrm{TM},S}(\mathbf{r}, \mathbf{r}') = \frac{\mu_n}{\mu_m}(\nabla \times \hat{z})(\nabla'\nabla' \cdot \hat{z} + k_m^2\hat{z})g^{\mathrm{TM},S}(\mathbf{r}, \mathbf{r}') \tag{6.40}$$

Consequently, we have

$$\langle \mathbf{J}_{Ti}(\mathbf{r}), \nabla \times \overline{\mathbf{G}}_e^{\mathrm{TM},S}(\mathbf{r}, \mathbf{r}'), \mathbf{J}_j(\mathbf{r}')\rangle$$
$$= \frac{\mu_n}{\mu_m}\left[-\langle \mathbf{J}_{Ti}(\mathbf{r}), \nabla \times \hat{z}\partial_{z'}g^{\mathrm{TM},S}(\mathbf{r}, \mathbf{r}'), \nabla' \cdot \mathbf{J}_j(\mathbf{r}')\rangle \right.$$
$$\left. +k_m^2\langle \mathbf{J}_{Ti}(\mathbf{r}), \nabla \times \hat{z}g^{\mathrm{TM},S}(\mathbf{r}, \mathbf{r}'), \hat{z} \cdot \mathbf{J}_j(\mathbf{r}')\rangle\right] \tag{6.41}$$

Similarly, for $\nabla \times \overline{\mathbf{G}}_e^{\mathrm{TE},S}(\mathbf{r}, \mathbf{r}')$, we have

$$\nabla \times \overline{\mathbf{G}}_e^{\mathrm{TE},S}(\mathbf{r}, \mathbf{r}') = (\nabla\nabla \cdot \hat{z} + k_m^2\hat{z})(\nabla' \times \hat{z})g^{\mathrm{TE},S}(\mathbf{r}, \mathbf{r}') \tag{6.42}$$

$$\langle \mathbf{J}_{Ti}(\mathbf{r}), \nabla \times \overline{\mathbf{G}}_e^{\mathrm{TE},S}(\mathbf{r}, \mathbf{r}'), \mathbf{J}_j(\mathbf{r}')\rangle = -\langle \nabla \cdot \mathbf{J}_{Ti}(\mathbf{r}), \nabla' \times \hat{z}\partial_z g^{\mathrm{TE},S}(\mathbf{r}, \mathbf{r}'), \mathbf{J}_j(\mathbf{r}')\rangle$$
$$+k_m^2\langle \mathbf{J}_{Ti}(\mathbf{r}) \cdot \hat{z}, \nabla' \times \hat{z}g^{\mathrm{TE},S}(\mathbf{r}, \mathbf{r}'), \mathbf{J}_j(\mathbf{r}')\rangle \tag{6.43}$$

When $\mathbf{J}_{Ti}$ and $\mathbf{J}_j$ are curl-conforming volume expansion functions, an additional integration by parts can be used to move the $\nabla\times$ operators onto them. However, for surface currents, which are distributed over a thin sheet in space, their curls are singular, and hence, it is prudent to leave the curl operators the way they are in the above.

6.5   The $\mathcal{L}$ and $\mathcal{K}$ Operators

The $\overline{\mathbf{G}}_e$ and $\nabla \times \overline{\mathbf{G}}_e$ operators above allow us to formulate the EFIE and MFIE for impenetrable objects embedded in layered media. To formulate integral equations for penetrable objects in a layered medium, it is necessary to derive the full suite of $\mathcal{L}$ and $\mathcal{K}$ operators as have been done for the homogeneous medium case, which are

$$
\begin{aligned}
\mathbf{E}(\mathbf{r}) &= i\omega \int \overline{\mathbf{G}}_e(\mathbf{r}, \mathbf{r}')\mu(\mathbf{r}') \cdot \mathbf{J}(\mathbf{r}') \\
&= \mathcal{L}_e(\mathbf{r}, \mathbf{r}') \cdot \mathbf{J}(\mathbf{r}')
\end{aligned}
\tag{6.44}
$$

where

$$
\mathcal{L}_e(\mathbf{r}, \mathbf{r}') = i\omega \overline{\mathbf{G}}_e(\mathbf{r}, \mathbf{r}')\mu(\mathbf{r}'),
\tag{6.45}
$$

and

$$
\begin{aligned}
\mathbf{H}(\mathbf{r}) &= \mu^{-1}(\mathbf{r}) \int \nabla \times \overline{\mathbf{G}}_e(\mathbf{r}, \mathbf{r}')\mu(\mathbf{r}') \cdot \mathbf{J}(\mathbf{r}') \\
&= \mathcal{K}_m(\mathbf{r}, \mathbf{r}') \cdot \mathbf{J}(\mathbf{r}')
\end{aligned}
\tag{6.46}
$$

where

$$
\mathcal{K}_m(\mathbf{r}, \mathbf{r}') = \mu^{-1}(\mathbf{r})\nabla \times \overline{\mathbf{G}}_e(\mathbf{r}, \mathbf{r}')\mu(\mathbf{r}')
\tag{6.47}
$$

and in the above, integration is implied over repeated variables.

By invoking duality, it follows that

$$
\begin{aligned}
\mathbf{H}(\mathbf{r}) &= i\omega \int \overline{\mathbf{G}}_m(\mathbf{r}, \mathbf{r}')\epsilon(\mathbf{r}') \cdot \mathbf{M}(\mathbf{r}') \\
&= \mathcal{L}_m(\mathbf{r}, \mathbf{r}') \cdot \mathbf{M}(\mathbf{r}')
\end{aligned}
\tag{6.48}
$$

$$
\begin{aligned}
\mathbf{E}(\mathbf{r}) &= -\epsilon^{-1}(\mathbf{r}) \int \nabla \times \overline{\mathbf{G}}_m(\mathbf{r}, \mathbf{r}')\epsilon(\mathbf{r}') \cdot \mathbf{M}(\mathbf{r}') \\
&= \mathcal{K}_e(\mathbf{r}, \mathbf{r}') \cdot \mathbf{J}(\mathbf{r}')
\end{aligned}
\tag{6.49}
$$

where

$$
\mathcal{L}_m(\mathbf{r}, \mathbf{r}') = i\omega \overline{\mathbf{G}}_m(\mathbf{r}, \mathbf{r}')\epsilon(\mathbf{r}')
\tag{6.50}
$$

$$
\mathcal{K}_e(\mathbf{r}, \mathbf{r}') = -\epsilon^{-1}(\mathbf{r})\nabla \times \overline{\mathbf{G}}_m(\mathbf{r}, \mathbf{r}')\epsilon(\mathbf{r}')
\tag{6.51}
$$

In the above,

$$
\begin{aligned}
\overline{\mathbf{G}}_m(\mathbf{r}, \mathbf{r}') = {} & (\nabla \times \hat{z})(\nabla' \times \hat{z})g^{\mathrm{TM}}(\mathbf{r}, \mathbf{r}') \\
& + \frac{1}{k_{mn}^2}(\nabla \times \nabla \times \hat{z})(\nabla' \times \nabla' \times \hat{z})g^{\mathrm{TE}}(\mathbf{r}, \mathbf{r}'), \qquad |z - z'| > 0,
\end{aligned}
\tag{6.52}
$$

6.6   The E_z-H_z Formulation

The E_z-H_z formulation was used in some of our earlier works [8–10, 22–25]. For completeness, we will describe it here. In this formulation, the E_z and H_z components of the electromagnetic fields are first derived in the layered medium due to a point source, and other components can then be derived from them [4]. But in this section, we will derive the fields so that there will be less derivative operators operating on these spectral integrals. This will enable the easier evaluations of the spectral integrals.

Consider an arbitrary Hertzian dipole point source $\mathbf{J}(\mathbf{r}) = \hat{\alpha}' I_0 \ell \delta(\mathbf{r} - \mathbf{r}')$, carrying a current I_0 with effective length ℓ, and located in region m in a layered medium. For simplicity, we can assume that $I_0 \ell = 1$. The electric field it produces can be derived, and from these expressions, we can glean off the dyadic Green's function. By breaking an arbitrary point source into vertical and horizontal components, the solution due to this point source can be written in the spectral domain as

$$\tilde{E}_{mz} = \hat{z} \cdot \hat{\alpha}' \left(1 + \frac{1}{k_m^2} \frac{\partial^2}{\partial z^2} \right) \tilde{g}_m^{TM} - \frac{1}{k_m^2} \frac{\partial}{\partial z'} \nabla_s \cdot \hat{\alpha}' \tilde{g}_m^{TM}, \tag{6.53}$$

where the first term is due to the vertical component of the source whereas the second term is due to its horizontal component. In general,

$$\tilde{g}_m^{TM} = \tilde{g}_m^{TM,P} + \tilde{g}_m^{TM,R} \tag{6.54}$$

where

$$\tilde{g}_m^{TM,P} = \frac{-\omega \mu_m}{8\pi^2} \frac{e^{i\mathbf{k}_s \cdot (\mathbf{r}_s - \mathbf{r}_s') + ik_{mz}|z-z'|}}{k_{mz}}, \tag{6.55}$$

$$\tilde{g}_m^{TM,R} = \frac{-\omega \mu_m}{8\pi^2} \frac{e^{i\mathbf{k}_s \cdot (\mathbf{r}_s - \mathbf{r}_s')}}{k_{mz}} \left[B_m^{TM} e^{-ik_{mz}z} + D_m^{TM} e^{ik_{mz}z} \right]. \tag{6.56}$$

B_m^{TM} and D_m^{TM} can be found as in [6]. Here, $\tilde{g}_m^{TM,P}$ is the primary field contribution whereas $\tilde{g}_m^{TM,R}$ is due to reflections at boundaries at $z = -d_{m-1}$ and $z = -d_m$. Because $\tilde{g}_m^{TM,P}$ is the same as the homogeneous medium case, we can focus on $\tilde{g}_m^{TM,R}$. Moreover, $\tilde{E}_{mz}$ due to $\tilde{g}_m^{TM,R}$ can be further simplified as using $\frac{\partial^2}{\partial z^2} \to -k_{mz}^2$ to obtain

$$\tilde{E}_{mz}^R = \left(\hat{z} \cdot \hat{\alpha}' \frac{k_s^2}{k_m^2} - \frac{1}{k_m^2} \frac{\partial}{\partial z'} \nabla_s \cdot \hat{\alpha}' \right) \tilde{g}_m^{TM,R}. \tag{6.57}$$

Using [6, Eq. (2.3.17)] and from the above, we can derive the other components of the TM electric field to obtain

$$\tilde{\mathbf{E}}_{ms}^{TM,R} = \frac{1}{k_s^2} \nabla_s \frac{\partial}{\partial z} \tilde{E}_{mz}^R = \frac{1}{k_s^2} \nabla_s \frac{\partial}{\partial z} \left(\hat{z} \cdot \hat{\alpha}' \frac{k_s^2}{k_m^2} - \frac{1}{k_m^2} \frac{\partial}{\partial z'} \nabla_s \cdot \hat{\alpha}' \right) \tilde{g}_m^{TM,R}. \tag{6.58}$$

Adding the z component to the above, we have

$$\tilde{\mathbf{E}}_m^{TM,R} = \left(\frac{1}{k_m^2} \nabla_s \nabla_z \cdot \hat{\alpha}' - \frac{1}{k_m^2 k_s^2} \frac{\partial}{\partial z} \frac{\partial}{\partial z'} \nabla_s \nabla_s \cdot \hat{\alpha}' \right.$$
$$\left. + \hat{z}\hat{z} \cdot \hat{\alpha}' \frac{k_s^2}{k_m^2} - \frac{1}{k_m^2} \nabla_z' \nabla_s \cdot \hat{\alpha}' \right) \tilde{g}_m^{TM,R}. \tag{6.59}$$

We can write

$$\frac{1}{k_m^2}\nabla_s\nabla_z = \frac{1}{k_m^2}\left(\nabla\nabla_z - \nabla_z\nabla_z\right) = \frac{1}{k_m^2}\nabla\nabla_z + \hat{z}\hat{z}\frac{k_z^2}{k_m^2} = \frac{1}{k_m^2}\nabla\nabla_z + \hat{z}\hat{z} - \hat{z}\hat{z}\frac{k_s^2}{k_m^2}. \qquad (6.60)$$

Using (6.60) in (6.59), we have

$$\tilde{\mathbf{E}}_m^{TM,R} = \left(\frac{1}{k_m^2}\nabla\nabla_z - \frac{1}{k_m^2 k_s^2}\frac{\partial}{\partial z}\frac{\partial}{\partial z'}\nabla_s\nabla_s + \hat{z}\hat{z} - \frac{1}{k_m^2}\nabla_z'\nabla_s\right)\cdot\hat{\alpha}'\tilde{g}_m^{TM,R}. \qquad (6.61)$$

We can always define a new function $g_{mh}^{TM,R}$ such that

$$\frac{\partial}{\partial z'}\tilde{g}_m^{TM,R} = -\frac{\partial}{\partial z}\tilde{g}_{mh}^{TM,R}. \qquad (6.62)$$

It can be shown that such a $g_{mh}^{TM,R}$ can always be found. In this case, (6.61) can be written as

$$\tilde{\mathbf{E}}_m^{TM,R} = \left(\frac{1}{k_m^2}\nabla\nabla_z + \hat{z}\hat{z}\right)\cdot\hat{\alpha}'\tilde{g}_m^{TM,R} + \frac{1}{k_m^2}\left(\frac{1}{k_s^2}\frac{\partial^2}{\partial z^2}\nabla_s\nabla_s + \nabla_z\nabla_s\right)\cdot\hat{\alpha}'\tilde{g}_{mh}^{TM,R}. \qquad (6.63)$$

Using $\frac{\partial^2}{\partial z^2} = -k_{mz}^2 = -k_m^2 + k_s^2$, the above can be rewritten as

$$\tilde{\mathbf{E}}_m^{TM,R} = \left(\frac{1}{k_m^2}\nabla\nabla_z + \hat{z}\hat{z}\right)\cdot\hat{\alpha}'\tilde{g}_m^{TM,R} + \left(\frac{1}{k_m^2}\nabla\nabla_s - \frac{1}{k_s^2}\nabla_s\nabla_s\right)\cdot\hat{\alpha}'\tilde{g}_{mh}^{TM,R}. \qquad (6.64)$$

For the TE wave, we can focus again on the reflected wave to obtain the spectral domain solution as

$$\tilde{H}_{mz}^R = \hat{z}\cdot\nabla_s\times\hat{\alpha}'\frac{1}{i\omega\mu_m}\tilde{g}_m^{TE,R} = \hat{\alpha}'\cdot\hat{z}\times\nabla_s\frac{1}{i\omega\mu_m}\tilde{g}_m^{TE,R}, \qquad (6.65)$$

where

$$\tilde{g}_m^{TE,R} = -\frac{\omega\mu_m}{8\pi^2}\frac{e^{i\mathbf{k}_s\cdot(\mathbf{r}_s - \mathbf{r}_s')}}{k_{mz}}\left[B_m^{TE}e^{-ik_{mz}z} + D_m^{TE}e^{ik_{mz}z}\right]. \qquad (6.66)$$

Then, the TE electric field in region m is

$$\hat{\alpha}\cdot\tilde{\mathbf{E}}_{ms}^{TE,R} = \hat{\alpha}\cdot\tilde{\mathbf{E}}_m^{TE,R} = -\frac{1}{k_s^2}\left(\hat{\alpha}\cdot\hat{z}\times\nabla_s\right)\left(\hat{\alpha}'\cdot\hat{z}\times\nabla_s\right)\tilde{g}_m^{TE,R}. \qquad (6.67)$$

Using the fact that (an identity easily derived in the spectral domain),

$$\frac{1}{k_s^2}\left(\hat{\alpha}\cdot\hat{z}\times\nabla_s\right)\left(\hat{\alpha}'\cdot\hat{z}\times\nabla_s\right) = -\hat{\alpha}\cdot\hat{z}\hat{z}\cdot\hat{\alpha}' - \frac{1}{k_s^2}\hat{\alpha}\cdot\nabla_s\nabla_s\cdot\hat{\alpha}' - \hat{\alpha}\cdot\hat{\alpha}'$$

$$= -\hat{\alpha}\cdot\bar{\mathbf{I}}_s\cdot\hat{\alpha}' - \frac{1}{k_s^2}\hat{\alpha}\cdot\nabla_s\nabla_s\cdot\hat{\alpha}', \qquad (6.68)$$

where $\bar{\mathbf{I}}_s = \hat{x}\hat{x} + \hat{y}\hat{y}$, we can rewrite (6.67) as

$$\tilde{\mathbf{E}}_m^{TE,R} = \left(\bar{\mathbf{I}}_s + \frac{1}{k_s^2}\nabla_s\nabla_s\right)\cdot\hat{\alpha}'\tilde{g}_m^{TE,R}. \qquad (6.69)$$

Combining (6.64) and (6.69), we have the total reflected electric field as

$$\tilde{\mathbf{E}}_m^R = \left(\frac{1}{k_m^2}\nabla\nabla_z + \hat{z}\hat{z}\right)\cdot\hat{\alpha}'\tilde{g}_m^{TM,R} + \left(\frac{1}{k_m^2}\nabla\nabla_s - \frac{1}{k_s^2}\nabla_s\nabla_s\right)\cdot\hat{\alpha}'\tilde{g}_{mh}^{TM,R}$$
$$+ \left(\bar{\mathbf{I}}_s + \frac{1}{k_s^2}\nabla_s\nabla_s\right)\cdot\hat{\alpha}'\tilde{g}_m^{TE,R}. \tag{6.70}$$

Because the electric field due to a point current source $\mathbf{J}(\mathbf{r}) = \hat{\alpha}\delta(\mathbf{r} - \mathbf{r}')$ is [see (6.21)]

$$\tilde{\mathbf{E}}_m^R(\mathbf{r}) = i\omega\tilde{\overline{\mathbf{G}}}^R(\mathbf{r},\mathbf{r}')\mu(\mathbf{r}')\cdot\hat{\alpha}' \tag{6.71}$$

one can extract an expression for the dyadic Green's function in the spectral domain for the reflected wave term from the above.

The field in region $n < m$ due to a source in region m is of the form

$$\tilde{E}_{nz} = \left(\hat{z}\cdot\hat{\alpha}'\frac{k_s^2}{k_m^2} - \frac{1}{k_m^2}\frac{\partial}{\partial z'}\nabla_s\cdot\hat{\alpha}'\right)\tilde{g}_n^{TM}, \tag{6.72}$$

where

$$\tilde{g}_n^{TM} = -\frac{\omega\mu_m}{8\pi^2}\frac{e^{i\mathbf{k}_s\cdot(\mathbf{r}_s-\mathbf{r}_s')}}{k_{mz}}\frac{\epsilon_m}{\epsilon_n}A_n^{TM}\left[e^{ik_{nz}z} + \tilde{R}_{n,n-1}^{TM}e^{-2ik_{nz}d_{n-1}-ik_{nz}z}\right]. \tag{6.73}$$

A_n^{TE} can be found using the technique described in [6, pp. 53, 78]. The transverse components of the field can be found to be

$$\tilde{\mathbf{E}}_{ns}^{TM} = \frac{1}{k_s^2}\nabla_s\frac{\partial}{\partial z}\tilde{E}_{nz} = \frac{1}{k_m^2}\left(\nabla_s\frac{\partial}{\partial z}\hat{z} - \frac{1}{k_s^2}\nabla_s\frac{\partial^2}{\partial z\partial z'}\nabla_s\right)\cdot\hat{\alpha}'\tilde{g}_n^{TM}. \tag{6.74}$$

Adding the z component to the above, we have

$$\tilde{\mathbf{E}}_n^{TM} = \left(\frac{1}{k_m^2}\nabla_s\nabla_z - \frac{1}{k_m^2 k_s^2}\frac{\partial^2}{\partial z\partial z'}\nabla_s\nabla_s + \hat{z}\hat{z}\frac{k_s^2}{k_m^2} - \frac{1}{k_m^2}\nabla_z'\nabla_s\right)\cdot\hat{\alpha}'\tilde{g}_n^{TM}. \tag{6.75}$$

By using the same trick as before, the first term and third term can be combined, viz,

$$\frac{1}{k_m^2}\nabla_s\nabla_z = \frac{1}{k_m^2}\nabla\nabla_z - \frac{1}{k_m^2}\nabla_z\nabla_z = \frac{1}{k_m^2}\nabla\nabla_z + \frac{k_{nz}^2}{k_m^2}\hat{z}\hat{z} = \frac{1}{k_m^2}\nabla\nabla_z + \frac{k_n^2}{k_m^2}\hat{z}\hat{z} - \frac{k_s^2}{k_m^2}\hat{z}\hat{z}. \tag{6.76}$$

Similar to before, we can define

$$\frac{\partial}{\partial z'}\tilde{g}_n^{TM} = -\frac{\partial}{\partial z}\tilde{g}_{nh}^{TM}, \tag{6.77}$$

to combine the second and the fourth terms, viz,

$$-\frac{1}{k_s^2}\frac{\partial^2}{\partial z\partial z'}\nabla_s\nabla_s - \nabla_z'\nabla_s = \frac{1}{k_s^2}\frac{\partial^2}{\partial z^2}\nabla_s\nabla_s + \nabla_z\nabla_s = -\frac{k_{nz}^2}{k_s^2}\nabla_s\nabla_s + \nabla_z\nabla_s$$
$$= -\frac{k_n^2}{k_s^2}\nabla_s\nabla_s + \nabla\nabla_s. \tag{6.78}$$

Consequently, we have

$$\tilde{\mathbf{E}}_n^{TM} = \frac{1}{k_m^2} \left(\nabla \nabla_z + k_n^2 \hat{z}\hat{z} \right) \cdot \hat{\alpha}' \tilde{g}_n^{TM} - \frac{1}{k_m^2} \left(\frac{k_n^2}{k_s^2} \nabla_s \nabla_s - \nabla \nabla_s \right) \cdot \hat{\alpha}' \tilde{g}_{nh}^{TM}, \tag{6.79}$$

The TE field in region $n < m$ in the spectral domain is

$$\tilde{H}_{nz} = (\hat{\alpha}' \cdot \hat{z} \times \nabla_s) \frac{1}{i\omega\mu_n} \tilde{g}_n^{TE},$$

where

$$\tilde{g}_n^{TE} = -\frac{\omega\mu_m}{8\pi^2} \frac{e^{i\mathbf{k}_s \cdot (\mathbf{r}_s - \mathbf{r}_s')}}{k_{mz}} \frac{\mu_m}{\mu_n} A_n^{TE} \left[e^{ik_{nz}z} + \tilde{R}_{n,n-1}^{TE} e^{-2ik_{nz}d_{n-1} - ik_{nz}z} \right] \tag{6.80}$$

The TE field can be shown to be

$$\tilde{\mathbf{E}}_n^{TE} = \left(\bar{\mathbf{I}}_s + \frac{\nabla_s \nabla_s}{k_s^2} \right) \cdot \hat{\alpha}' \tilde{g}_n^{TE}. \tag{6.81}$$

The total electric field is then

$$\tilde{\mathbf{E}}_n = \frac{1}{k_m^2} \left(\nabla \nabla_z + k_n^2 \hat{z}\hat{z} \right) \cdot \hat{\alpha}' \tilde{g}_n^{TM} - \frac{1}{k_m^2} \left(\frac{k_n^2}{k_s^2} \nabla_s \nabla_s - \nabla \nabla_s \right) \cdot \hat{\alpha}' \tilde{g}_{nh}^{TM}$$
$$+ \left(\bar{\mathbf{I}}_s + \frac{\nabla_s \nabla_s}{k_s^2} \right) \cdot \hat{\alpha}' \tilde{g}_n^{TE}. \tag{6.82}$$

Similar to (6.70), one can extract an expression for the dyadic Green's function in the spectral domain by noticing that

$$\tilde{\mathbf{E}}_n(\mathbf{r}) = i\omega \tilde{\bar{\mathbf{G}}}(\mathbf{r}, \mathbf{r}') \mu(\mathbf{r}') \cdot \hat{\alpha}' \tag{6.83}$$

When the above is used to compute the MOM matrices, they need to be integrated to yield the spatial domain fields. It is to be noted that when the spectral domain integral is performed, the derivative operators are taken out of the spectral integrals. Integration by parts are then used to move the derivative operators to the expansion and testing functions.

6.6.1   Example 1: Half Space

Let us assume that $m = N = 2$, so that it corresponds to two half spaces (see Figure 6.2). In this case, $D_m^{TM} = 0$ in (6.56) and

$$B_m^{TM} = B_2^{TE} = R_{12}^{TM} e^{-ik_{2z}(d_1 + z') - ik_{2z}d_1}. \tag{6.84}$$

$$\tilde{g}_2^{TM,R} = -\frac{\omega\mu_2}{8\pi^2} \frac{e^{i\mathbf{k}_s \cdot (\mathbf{r}_s - \mathbf{r}_s')}}{k_{2z}} R_{12}^{TM} e^{-ik_{2z}(d_1 + z') - ik_{2z}d_1 - ik_{2z}z}. \tag{6.85}$$

Also, $\tilde{g}_{2h}^{TM,R}$ defined in (6.62) is

$$\tilde{g}_{2h}^{TM,R} = -\tilde{g}_2^{TM,R}. \tag{6.86}$$

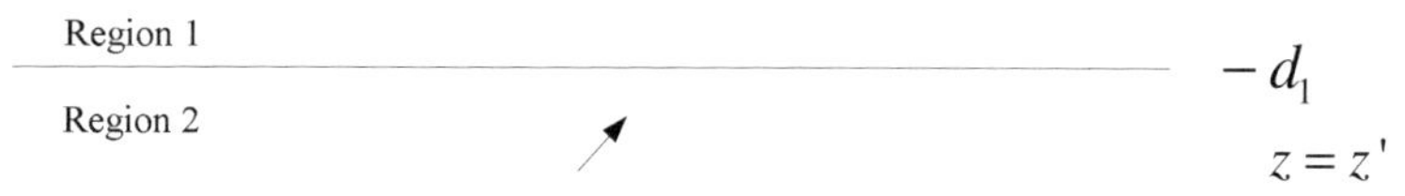

Figure 6.2: A point source in a lower half space.

$\tilde{g}_2^{TE,R}$ is similar to (6.85) with R_{12}^{TE} replacing R_{12}^{TM}. Consequently, (6.70) for this example becomes

$$
\begin{aligned}
\tilde{\mathbf{E}}_2^R &= \left(\frac{1}{k_s^2} \nabla \nabla_z + \hat{z}\hat{z} \right) \cdot \hat{\alpha} \tilde{g}_2^{TM,R} - \frac{1}{k_2^2} \nabla \nabla_s \cdot \hat{\alpha}' \tilde{g}_2^{TM,R} + \bar{\mathbf{I}}_s \cdot \hat{\alpha}' \tilde{g}_2^{TE,R} \\
&\quad + \frac{1}{k_s^2} \nabla_s \nabla_s \cdot \hat{\alpha}' (\tilde{g}_2^{TE,R} + \tilde{g}_2^{TM,R}) \\
&= -\frac{1}{k_2^2} \nabla \hat{\nabla} \cdot \hat{\alpha}' \tilde{g}_2^{TM,R} + \hat{z}\hat{z} \cdot \hat{\alpha}' \tilde{g}_2^{TM,R} + \bar{\mathbf{I}}_s \cdot \hat{\alpha}' \tilde{g}_2^{TE,R} \\
&\quad + \frac{1}{k_s^2} \nabla_s \nabla_s \cdot \hat{\alpha}' (\tilde{g}_2^{TE,R} + \tilde{g}_2^{TM,R}),
\end{aligned}
\tag{6.87}
$$

where $\hat{\nabla} = \nabla_s - \nabla_z$.

In region 1, $\tilde{g}_1^{TM}$ of (6.73) becomes

$$
\tilde{g}_1^{TM} = -\frac{\omega \mu_2}{8\pi^2} \frac{e^{i\mathbf{k}_s \cdot (\mathbf{r}_s - \mathbf{r}_s')}}{k_{2z}} \cdot \frac{\epsilon_2}{\epsilon_1} A_1^{TM} e^{ik_{1z}z},
\tag{6.88}
$$

where

$$
A_1^{TM} = e^{-ik_{2z}(d_1 + z')} T_{21}^{TM} e^{ik_{1z}d_1}.
\tag{6.89}
$$

In this case $\tilde{g}_{1h}^{TM}$ defined in (6.77) becomes,

$$
\tilde{g}_{1h}^{TM} = \frac{k_{2z}}{k_{1z}} \tilde{g}_1^{TM},
\tag{6.90}
$$

and (6.82) becomes

$$
\tilde{\mathbf{E}}_1 = \frac{1}{k_1^2} \left(\nabla \nabla_z + k_1^2 \hat{z}\hat{z} \right) \cdot \hat{\alpha}' \tilde{g}_1^{TM} - \frac{1}{k_2^2} \left(\frac{k_2^2}{k_s^2} \nabla_s \nabla_s - \nabla \nabla_s \right) \cdot \hat{\alpha}' \frac{k_{2z}}{k_{1z}} \tilde{g}_1^{TM}
$$
$$
+ \left(\bar{\mathbf{I}}_s + \frac{\nabla_s \nabla_s}{k_s^2} \right) \cdot \hat{\alpha}' \tilde{g}_1^{TE},
\tag{6.91}
$$

where $\tilde{g}_1^{TE}$ is similar to $\tilde{g}_1^{TM}$ in (6.88), but with $\left(\frac{\epsilon_2}{\epsilon_1}\right)A_1^{TM}$ replaced by $\left(\frac{\mu_2}{\mu_1}\right)A_1^{TE}$, and A_1^{TE} is similar to A_1^{TM} in (6.89) but with T_{21}^{TM} replaced by T_{21}^{TE}, or

$$\tilde{g}_1^{TE} = -\frac{\omega\mu_2}{8\pi^2}\frac{e^{i\mathbf{k}_s\cdot(\mathbf{r}_s-\mathbf{r}_s')}}{k_{2z}}\frac{\mu_2}{\mu_1}e^{-ik_{2z}(d_1+z')}T_{21}^{TE}e^{ik_{1z}d_1+ik_{1z}z}. \tag{6.92}$$

6.6.2   Example 2: Three-Layer Medium

Figure 6.3: A point source embedded in a three-layer medium.

For field in region 2, B_2^{TM} and D_2^{TM} of (6.56) can be obtained from [6, Eq. (2.4.5)].

$$B_2^{TM} = e^{-ik_{2z}d_1}R_{21}^{TM}\left[e^{-ik_{2z}(d_1+z')}+e^{ik_{2z}(d_2-d_1)}R_{23}^{TM}e^{ik_{2z}(d_2+z')}\right]\tilde{M}_2, \tag{6.93}$$

$$D_2^{TM} = e^{ik_{2z}d_2}R_{23}^{TM}\left[e^{ik_{2z}(d_2+z')}+e^{ik_{2z}(d_2-d_1)}R_{21}^{TM}e^{-ik_{2z}(d_1+z')}\right]\tilde{M}_2, \tag{6.94}$$

where

$$\tilde{M}_2 = \left[1-R_{23}^{TM}R_{21}^{TM}e^{2ik_{2z}(d_2-d_1)}\right]^{-1}, \tag{6.95}$$

and

$$\tilde{g}_2^{TM,R} = -\frac{\omega\mu_2}{8\pi^2}\frac{e^{i\mathbf{k}_s\cdot(\mathbf{r}_s-\mathbf{r}_s')}}{k_{2z}}\left[B_2^{TM}e^{-ik_{2z}z}+D_2^{TM}e^{ik_{2z}z}\right]. \tag{6.96}$$

Moreover, $\tilde{g}_{2h}^{TM,R}$ as defined in (6.62) is

$$\tilde{g}_{2h}^{TM,R} = \frac{\omega\mu_2}{8\pi^2}\frac{e^{i\mathbf{k}_s\cdot(\mathbf{r}_s-\mathbf{r}_s')}}{k_{2z}}\frac{k_{1z}}{k_{2z}}\left[-\hat{B}_2^{TM}e^{-ik_{2z}z}+\hat{D}_2^{TM}e^{ik_{2z}z}\right]. \tag{6.97}$$

where

$$\hat{B}_2^{TM} = e^{-ik_{2z}d_1}R_{21}^{TM}\left[-e^{-ik_{2z}(d_1+z')}+e^{-ik_{2z}(d_2-d_1)}R_{23}^{TM}e^{ik_{2z}(d_2+z')}\right]\tilde{M}_2, \tag{6.98}$$

$$\hat{D}_2^{TM} = e^{ik_{2z}d_2}R_{23}^{TM}\left[e^{ik_{2z}(d_2+z')}-e^{ik_{2z}(d_2-d_1)}R_{21}^{TM}e^{-ik_{2z}(d_1+z')}\right]\tilde{M}_2. \tag{6.99}$$

Consequently, (6.70) becomes

$$\tilde{\mathbf{E}}_2^R = \left(\frac{1}{k_2^2}\nabla\nabla_z + \hat{z}\hat{z}\right)\cdot\hat{\alpha}'\tilde{g}_2^{TM,R} + \left(\frac{1}{k_2^2}\nabla\nabla_s - \frac{1}{k_s^2}\nabla_s\nabla_s\right)\cdot\hat{\alpha}'\tilde{g}_{2h}^{TM,R}$$
$$+ \left(\bar{\mathbf{I}}_s + \frac{\nabla_s\nabla_s}{k_s^2}\right)\cdot\hat{\alpha}'\tilde{g}_2^{TE,R}. \tag{6.100}$$

In region 1, $\tilde{g}_1^{TM}$ of (6.73) is similar to (6.88), but with

$$A_1^{TM} = T_{21}^{TM}e^{-ik_{2z}d_1}\left[e^{-ik_{2z}z'} + e^{ik_{2z}(z'+2d_1)}R_{23}^{TM}\right]\tilde{M}_2 e^{ik_{1z}d_1} \tag{6.101}$$

which can be gathered from (2.4.8) and (2.4.9) of [6]. In this case, $\tilde{g}_{1h}^{TM}$ defined in (6.77) becomes

$$\tilde{g}_{1h}^{TM} = -\frac{\omega\mu_2}{8\pi^2}\frac{e^{i\mathbf{k}_s\cdot(\mathbf{r}_s-\mathbf{r}_s')}}{k_{2z}}\frac{k_{2z}}{k_{1z}}\frac{\epsilon_2}{\epsilon_1}\hat{A}_1^{TM}e^{ik_{1z}z}, \tag{6.102}$$

where

$$\hat{A}_1^{TM} = T_{21}^{TM}e^{-ik_{2z}d_1}\left[-e^{-ik_{2z}z'} + e^{ik_{2z}(z'+2d_1)}R_{23}^{TM}\right]\tilde{M}_2 e^{ik_{1z}d_1}. \tag{6.103}$$

Consequently, (6.82) becomes

$$\tilde{\mathbf{E}}_1 = \frac{1}{k_2^2}\left(\nabla\nabla_z + k_1^2\hat{z}\hat{z}\right)\cdot\hat{\alpha}'\tilde{g}_1^{TM} - \frac{1}{k_2^2}\left(\frac{k_1^2}{k_s^2}\nabla_s\nabla_s - \nabla\nabla_s\right)\cdot\hat{\alpha}'\tilde{g}_{1h}^{TM}$$
$$+ \left(\bar{\mathbf{I}}_s + \frac{\nabla_s\nabla_s}{k_s^2}\right)\cdot\hat{\alpha}'\tilde{g}_1^{TE}, \tag{6.104}$$

where $\tilde{g}_1^{TE}$ is similar to $\tilde{g}_1^{TM}$ defined in (6.88), but with $\frac{\mu_2}{\mu_1}A_1^{TE}$ replacing $\frac{\epsilon_2}{\epsilon_1}A_1^{TM}$, and also A_1^{TE} is similar to (6.101) but with TE replacing TM.

6.7   Validation and Result

From these integrals, this new approach gives rise to matrix elements which agree with the formulation derived in [10] term by term. It also verifies its validity, because the results in [10] have been derived differently using E_z and H_z formulation. We have also checked the expressions in (6.28) against those derived by the E_z-H_z formulation for a general layered medium, and they are proved to be equivalent to each other [26].

Another validation is by comparing the radar cross section (RCS) of a sphere above a three-layered medium when calculated by the approach of Michalski-Zheng formulation [7] and this one. The incident wave is at an elevation angle of $30°$ and an azimuth angle of $0°$, with the operating frequency at 300 MHz. The layer parameters are μ_0, $\epsilon_1 = 1.0$, $\epsilon_2 = 2.56$, $\epsilon_3 = 6.5+0.6i$. The first layer is the upper half-space of infinite size, second one with thickness of 0.3 m, and the third one is the lower half-space of infinite size. The PEC sphere has a radius of 1m and is placed in layer 1, 10 m above the first interface. Figure 6.4 demonstrates the excellent agreement of the two formulations.

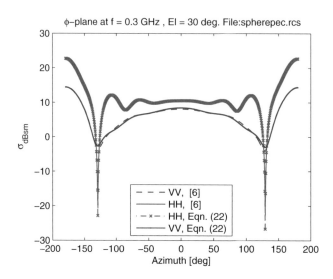

Figure 6.4: The radar cross section of a PEC sphere above a three-layered medium.

6.8 Conclusions

We have arrived at an alternative and elegant way of deriving the matrix representation of the dyadic Green's operator in the electric field integral equation (EFIE). The derivation is based on the pilot vector potential approach where the two vector potentials in terms of TE and TM fields are derived.

Starting with the dyadic Green's function formulas in [6], we manipulate them into a form that is suitable for matrix-element evaluation for CEM, for example, in the MOM. All integrals involve only Bessel function of zeroth order, and they can be reduced to two basic integrals for the Green's function g^{TM} and g^{TE}. When these two Green's functions are tabulated, the other integrals needed can be obtained by finite difference on the tabulated values. This derivation is more succinct than previous derivations.

Appendix: Singularity Subtraction for Scalar Potentials

The dyadic Green's function introduced in Section 6.2 requires the evaluation of two canonical scalar potentials given by (6.15) and (6.16). But the integrands of these integrals have pole singularities at the origin, which disappear (or cancel each other) when they are substituted into the equation for the dyadic Green's function. Hence, these singularities do not contributed to the final value of the dyadic Green's function, and they can be extracted and ignored. This shall be discussed here.

A typical integral for the scalar potential is a divergent integral of the form

$$I = \int_0^\infty dk_\rho \frac{F(k_\rho, z, z', \rho')}{k_\rho} \tag{A-1}$$

for a layered medium Green's function. The above can be rewritten as

$$
\begin{aligned}
I &= \int_0^\infty dk_\rho \frac{1}{k_\rho} \left[F(k_\rho, z, z', \rho') - F(0, z, z', \rho') \frac{a^2}{k_\rho^2 + a^2} \right] \\
&\quad + F(0, z, z', \rho') a^2 \int_0^\infty dk_\rho \frac{1}{k_\rho \left(k_\rho^2 + a^2 \right)} \\
&= I_R + I_S
\end{aligned}
\tag{A-2}
$$

where

$$I_R = \int_0^\infty dk_\rho \frac{1}{k_\rho} \left[F(k_\rho, z, z', \rho') - F(0, z, z', \rho') \frac{a^2}{k_\rho^2 + a^2} \right] \tag{A-3}$$

$$I_S = F(0, z, z', \rho') a^2 \int_0^\infty dk_\rho \frac{1}{k_\rho \left(k_\rho^2 + a^2 \right)} \tag{A-4}$$

I_R is now a convergent integral which may be tabulated. I_S is singular, but can be integrated in closed form. By partial fraction

$$
\begin{aligned}
I_S &= \lim_{\epsilon \to 0} F(0, z, z', \rho') \int_\epsilon^\infty dk_\rho \left[\frac{1}{k_\rho} - \frac{k_\rho}{k_\rho^2 + a^2} \right] \\
&= F(0, z, z', \rho') \lim_{\epsilon \to 0} \left[\ln k_\rho - \frac{1}{2} \ln \left(k_\rho^2 + a^2 \right) \right]_\epsilon^\infty
\end{aligned}
\tag{A-5}
$$

or

$$
\begin{aligned}
I_S &= F(0, z, z', \rho') \lim_{\epsilon \to 0} \left[- \ln \epsilon + \frac{1}{2} \ln \left(\epsilon^2 + a^2 \right) \right] \\
&= -F(0, z, z', \rho') \lim_{\epsilon \to 0} \ln \epsilon + F(0, z, z', \rho') \ln a
\end{aligned}
\tag{A-6}
$$

The singular term in the above will cancel each other in the full dyadic Green's function or else it will contribute to a nonphysical field. Hence, it can be ignored *a priori* and

$$I_S = F(0, z, z', \rho') \ln a \tag{A-7}$$

The above singularity subtraction technique is nonunique, and other ways of subtraction that yield better convergence for the integral are possible.

Bibliography

[1] R. F. Harrington, *Field Computation by Moment Methods,* Melbourne, FL: Krieger Publ., 1982.

[2] W. C. Chew, J. L. Xiong, and M. A. Saville, "A matrix-friendly formulation of layered medium Green's function," *Antennas Wirel. Propag. Lett.*, vol. 5, no. 1, pp. 490–494, Dec. 2006.

[3] A. Sommerfeld, "Uber die Ausbreitung der Wellen in der drahtlosen Telegraphie." *Ann. Phys.*, vol. 28, pp. 665–737, 1909.

[4] J. A. Kong, "Electromagnetic field due to dipole antennas over stratified anisotropic media." *Geophys.*, vol. 38, pp. 985–996, 1972.

[5] L. Tsang, J. A. Kong, and R. T. Shin, *Theory of Microwave Remote Sensing,* Melbourne, FL: Wiley Interscience, 1985.

[6] W. C. Chew, *Waves and Fields in Inhomogeneous Media,* New York: Van Nostrand Reinhold, 1990. Reprinted by Piscataway, NJ: IEEE Press, 1995.

[7] K. A. Michalski and D. Zheng, "Electromagnetic scattering and radiation by surfaces of arbitrary shape in layered media, Part I: Theory," *IEEE Trans. Antennas Propag.*, vol. 38, pp. 335–344, 1990.

[8] J. S. Zhao, W. C. Chew, C. C. Lu, E. Michielssen, and J. M. Song, "Thin-stratified medium fast-multipole algorithm for solving microstrip structures," *IEEE Trans. Microwave Theory Tech.*, vol. 46, no. 4, pp. 395–403, Apr. 1998.

[9] T. J. Cui and W. C. Chew, "Fast evaluation of sommerfield integrals for em scattering and radiation by three-dimensional buried objects," *IEEE Geosci. Remote Sens.*, vol. 37, no. 2, pp. 887–900, Mar. 1999.

[10] W. C. Chew, J. S. Zhao, and T. J. Cui, "The layered medium Green's function—A new look," *Microwave Opt. Tech. Lett.*, vol. 31, no. 4, pp. 252–255, 2001.

[11] G. Dural and M. I. Aksun, "Closed-form Green's functions for general sources and stratified media," *IEEE Trans. Microwave Theory Tech.*, vol. 43, pp. 1545–1552, Jul. 1995.

[12] Y. L. Chow, N. Hojjat, S. Safavi-Naeini, "Spectral Green's functions for multilayer media in a convenient computational form," *IEE Proc.-H.*, vol. 145, pp. 85–91, Feb. 1998.

[13] K. A. Michalski, "Extrapolation methods for Sommerfeld integral tails (Invited Review Paper)," *IEEE Trans. Antennas Propag.*, vol. 46, pp. 1405–1418, Oct. 1998.

[14] W. Cai and T. Yu, "Fast calculation of dyadic Green's functions for electromagnetic scattering in a multi-layered medium," *J. Comput. Phys.*, vol. 165, pp. 1–21, 2000.

[15] F. Ling and J. M. Jin, "Discrete complex image method for Green's Functions of general multilayer media," *IEEE Microwave Guided Wave Lett.*, vol. 10, no. 10, pp. 400–402, Oct. 2000.

[16] J. Y. Chen, A. A. Kishk, and A. W. Glisson, "Application of new MPIE formulation to the analysis of a dielectric resonator embedded in a multilayered medium coupled to a microstrip circuit," *IEEE Trans. Microwave Theory Tech.*, vol. 49, pp. 263–279, Feb. 2001.

[17] T. M. Grzegorczyk and J. R. Mosig, "Full-wave analysis of antennas containing horizontal and vertical metallizations embedded in planar multilayered media," *IEEE Trans. Antennas Propag.*, vol. 51, no. 11, pp. 3047–3054, Nov. 2003.

[18] E. Simsek, Q. H. Liu, and B. Wei, "Singularity subtraction for evaluation of Green's functions for multilayer media," *IEEE Trans. Microwave Theory Tech.*, vol. 54, no. 1, pp. 216–225, Jan. 2006.

[19] W. W. Hansen, "Transformations useful in certain antenna calculations," *J. Appl. Phys.*, vol. 8, pp. 282–286, Apr. 1937.

[20] W. C. Chew and S. Y. Chen, "Response of a point source embedded in a layered medium," *Antennas Wirel. Propag. Lett.*, vol. 2, no. 14, pp. 254–258, 2003.

[21] S. M. Rao, D. R. Wilton, and A. W. Glisson, "Electromagnetic scattering by surfaces of arbitrary shape," *IEEE Trans. Antennas Propag.*, vol. 30, no. 3, pp. 409-417, May 1982.

[22] B. Hu, W. C. Chew, and S. Velamparambil, "Fast inhomogeneous plane wave algorithm for the analysis of electromagnetic scattering," *Radio Sci.*, vol. 36, no. 6, pp. 1327–1340, 2001

[23] B. Hu, W. C. Chew, E. Michielssen and J. S. Zhao, "Fast inhomogeneous plane wave algorithm for the fast analysis of two-dimensional scattering problems," *Radio Sci.*, vol. 34, pp. 759–772, Jul.–Aug. 1999.

[24] B. Hu and W. C. Chew, "Fast inhomogeneous plane wave algorithm for electromagnetic solutions in layered medium structures - 2D case," *Radio Sci.*, vol. 35, pp. 31–43, Jan.–Feb. 2000.

[25] B. Hu and W. C. Chew, "Fast inhomogeneous plane wave algorithm for scattering from objects above the multi-layered medium," *IEEE Geosci. Remote Sens.*, vol. 39, no. 5, pp. 1028–1038, May 2001.

[26] W. C. Chew, "Field due to a point source in a layered medium—symmetrized formulation," *Private Notes*, Mar. 25, 2003.

CHAPTER 7

Fast Inhomogeneous Plane Wave Algorithm for Layered Media

7.1 Introduction

Although new developments in fast algorithm result in the ability to solve large-scale electromagnetic simulation problem with presently available computational resources, all of them are targeted to problems with homogeneous background, also known as free-space (homogeneous space) problems. However, there are numerous applications of practical interests, where the free-space approximation is not sufficient. For example, correct modeling of electromagnetic phenomena in areas such as remote sensing, interconnect characterization, microstrip antennas and monolithic microwave integrated circuits (MMIC), calls for the capability to analyze structures in the vicinity of stratified medium containing an arbitrary number of dielectric layers. The casting of this problem into integral based form was systematically addressed by Michalski and Zheng [1], where several versions of dyadic Green's function for layered medium problem were derived. Later, Chew *et al.* developed an alternative formulation [2–4] and successfully applied to the development of fast method for layered medium [5,6]. As shown later, this formulation is more easily adapted to fast methods developed here and results in similar efficiency as that for free-space problems.

Although theoretical development for solving layered medium problem using integral equation based solvers was considered complete, two obstacles exist for solving such problems efficiently. The first difficulty comes from the necessity to fill the impedance matrix when using the traditional solution method. The core part is to evaluate the spatial Green's functions, which are expressed in the Sommerfeld-type integral forms. Because no general analytic closed forms are available, the numerical integration along the Sommerfeld integration path (SIP) is necessary and it is quite time consuming. Many techniques have been proposed to reduce this computation time. The majority of those techniques can be categorized into two main approaches. The first approach is represented by well known discrete complex image method (DCIM) [7–10]. The main idea is to fit the spectrum of dyadic Green's function encountered in Sommerfeld integrals using a set of special functions, for which closed-form Sommerfeld integration results exist. This approach garners significant attention in recent years because of its simplicity and ease of implementation. The second approach is to perform Sommerfeld integration through a cleverly designed integration path so that fast convergence can be achieved in numerical integration. For example, Cui and Chew [6] have demonstrated the efficiency to perform the numerical integration along the fast converging steepest descent path (SDP). By using these techniques, the impedance matrix filling time can be reduced to be in the same order of magnitude as its free-space counterpart. However, the total computational complexity to fill the impedance matrix remains $O(N^2)$.

Although the matrix-filling issue for layered medium problems can be mitigated using aforementioned techniques, the second difficulty, that is, solving the resultant matrix equation within a reasonable amount of time, poses a major hurdle for solving the practical problems. Similar to free-space problems, the impedance matrix resulted from MOM discretization for layer medium problems is a dense, complex one in general. Because the straightforward fast multipole method (FMM) implementation for layered medium problem is not possible, there are multiple efforts to develop efficient methods for large-scale layered medium problems. For example, the adaptive integral method (AIM) [11] coupled with DCIM [12], thin-stratified medium fast-multipole algorithm (TSM-FMA) developed by Zhao *et al.* to solve microstrip type of problems [5], steepest descent fast multipole method (SDFMM) developed by Jandhyala *et al.* [13], MLFMA-based implementatation using the asymptotic image term developed by Geng et. al [14], and FMM-PML-MPIE developed by Vande Ginste *et al.* [15,16].

In this chapter, we will review the recent progress on fast inhomogeneous plane wave algorithm (FIPWA) and its application to layered medium problems.

7.2 Integral Equations for Layered Medium

In this section, without loss of generality of describing the algorithm involved, the electromagnetic scattering from a three-dimensional perfect electric conducting (PEC) surface scatterer S, in the presence of a layered medium, is studied. The stratified medium is characterized by N layers with different dielectric properties as shown in Figure 7.1. The air-dielectric interface is assumed to be at $z = 0$, and the scatterer is above all planar dielectric layers. To compute the scattering from the target, the electric field integral equation (EFIE) for inhomogeneous medium is formulated as

$$i\omega\hat{n} \times \int_S \overline{\mathbf{G}}_e(\mathbf{r}, \mathbf{r}')\mu(\mathbf{r}') \cdot \mathbf{J}(\mathbf{r}')d\mathbf{r}' = -\hat{n} \times \mathbf{E}^{inc}(\mathbf{r}), \qquad \forall \mathbf{r} \in S, \qquad (7.1)$$

and the magnetic field integral equation (MFIE) for inhomogeneous medium is

$$\mu(\mathbf{r})^{-1}\hat{n} \times \nabla \times \int_S \overline{\mathbf{G}}_e(\mathbf{r}, \mathbf{r}')\mu(\mathbf{r}') \cdot \mathbf{J}(\mathbf{r}')dS' - \frac{1}{2}\mathbf{J}(\mathbf{r}) = -\hat{n} \times \mathbf{H}^{inc}(\mathbf{r}), \qquad (7.2)$$

where $\hat{n}$ is the normal outward vector at point $\mathbf{r}$, $\overline{\mathbf{G}}_e(\mathbf{r}, \mathbf{r}')$ is the dyadic Green's function, $\mathbf{J}(\mathbf{r}')$ is the surface current distribution, and $\mathbf{E}^{inc}(\mathbf{r})$ and $\mathbf{H}^{inc}(\mathbf{r})$ are the fields of the impinging plane wave in the presence of the layered medium.

To reduce the effect of the internal resonances (see Chapters 3 and 4), combined field integral equation (CFIE) for closed conducting objects is utilized and it is simply a linear

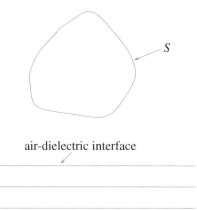

Figure 7.1: The PEC scatterer S above a layered medium. The outermost layers are Layer 1 and Layer N.

combination of EFIE and MFIE, that is,

$$\text{CFIE} = \alpha \text{EFIE} + \eta(1 - \alpha)\text{MFIE} \tag{7.3}$$

with $\alpha \in [0, 1]$ and normally $\alpha = 0.2 \sim 0.5$ is chosen.

Before discretizing Eqs. (7.1)–(7.3), the symmetric dyadic Green's function can be derived as follows [3] (see also Chapter 6):

$$
\begin{aligned}
i\omega\hat{\alpha} \cdot \overline{\mathbf{G}}(\mathbf{r}, \mathbf{r}')\mu(\mathbf{r}') \cdot \hat{\alpha}' &= (\hat{\alpha}_s \cdot \hat{\alpha}_s')(g_0 - g^{TE}) + \alpha_z \alpha_z'(g_0 + g^{TM}) \\
&\quad + \frac{1}{k^2}\hat{\alpha} \cdot \nabla\nabla \cdot \hat{\alpha}' g_0 + \frac{1}{k^2}\hat{\alpha} \cdot \nabla\nabla \cdot \hat{\alpha}'' g^{TM} \\
&\quad + \hat{\alpha} \cdot \nabla_s \nabla_s \cdot \hat{\alpha}'' g^{EM},
\end{aligned}
\tag{7.4}
$$

where

$$g_0(\mathbf{r}, \mathbf{r}') = \frac{i\omega\mu}{4\pi}\frac{e^{ik_1|\mathbf{r}-\mathbf{r}'|}}{|\mathbf{r} - \mathbf{r}'|}, \tag{7.5}$$

$$g^{\beta}(\mathbf{r}, \mathbf{r}') = \int_{SIP} dk_{\rho} \tilde{g}^{\beta}(k_{\rho}, \mathbf{r}, \mathbf{r}'), \tag{7.6}$$

$$\beta = (TE), (TM), EM, \tag{7.7}$$

$$\hat{\alpha} = \hat{\alpha}_s + \hat{\alpha}_z, \tag{7.8}$$

$$\hat{\alpha}' = \hat{\alpha}_s' + \hat{\alpha}_z', \tag{7.9}$$

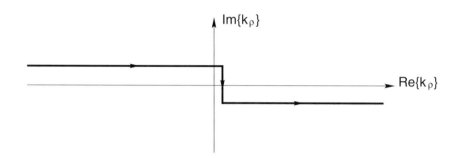

Figure 7.2: Illustration of the Sommerfeld integration path (SIP). The shift of the path is to avoid the singularities at $k_\rho = 0$ and the surface wave poles.

$$\hat{\alpha}'' = -\hat{\alpha}'_s + \hat{\alpha}'_z, \tag{7.10}$$

$$\tilde{g}^{TM}(k_\rho, \mathbf{r}, \mathbf{r}') = -\frac{\omega\mu}{8\pi}\frac{k_\rho}{k_{1z}}\tilde{R}^{TM}(k_\rho)H_0^{(1)}(k_\rho|\boldsymbol{\rho} - \boldsymbol{\rho}'|)e^{ik_{1z}(z+z')}, \tag{7.11}$$

$$\tilde{g}^{TE}(k_\rho, \mathbf{r}, \mathbf{r}') = \frac{\omega\mu}{8\pi}\frac{k_\rho}{k_{1z}}\tilde{R}^{TE}(k_\rho)H_0^{(1)}(k_\rho|\boldsymbol{\rho} - \boldsymbol{\rho}'|)e^{ik_{1z}(z+z')}, \tag{7.12}$$

$$\tilde{g}^{EM}(k_\rho, \mathbf{r}, \mathbf{r}') = \frac{1}{k_\rho^2}\left(\tilde{g}^{TE}(k_\rho, \mathbf{r}, \mathbf{r}') - \tilde{g}^{TM}(k_\rho, \mathbf{r}, \mathbf{r}')\right), \tag{7.13}$$

and k_1 is the wave number for the residing layer of the scatterer, $k_{1z} = \sqrt{k_1^2 - k_\rho^2}$, and $\tilde{R}^{TE}$ and $\tilde{R}^{TM}$ represent the generalized reflection coefficients for the layered medium, as defined in [17]. The spectral domain Green's function in terms of cylindrical waves can be derived from those of plane waves in Chapter 6 by using the relationship that

$$\frac{1}{2\pi}\iint_{-\infty}^{\infty} d\mathbf{k}_s F(k_\rho)e^{i\mathbf{k}_s\cdot(\mathbf{r}_s-\mathbf{r}'_s)} = \int_0^{\infty} dk_\rho k_\rho J_0(k_\rho\rho)F(k_\rho)$$

$$= \frac{1}{2}\int_{-\infty}^{\infty} dk_\rho k_\rho H_0^{(1)}(k_\rho\rho)F(k_\rho) \tag{7.14}$$

It should be noted that all information related to the layered medium, such as the location and the dielectric property of each layer, is embedded within the generalized reflection coefficients. Therefore, the Green's function presented in Eq. (7.4) is applicable to arbitrary layered structures. In the above equations, g_0 represents the free-space (or homogeneous medium) Green's function and the g^{TE} and g^{TM} denote the reflected wave (secondary field) parts, which are introduced to account for the effect of the layered medium. The integration in Eq. (7.6) is performed along the SIP in the k_ρ plane, which is shown in Figure 7.2.

Each component of the dyadic Green's function can be cast in the following generic form:

$$g_0(\mathbf{r}_j, \mathbf{r}_i) = \int_{SIP} dk_\rho \frac{k_\rho}{k_{1z}} H_0^{(1)}(k_\rho|\boldsymbol{\rho}_j - \boldsymbol{\rho}_i|)e^{ik_{1z}(z_j-z_i)}, \tag{7.15a}$$

$$g^R(\mathbf{r}_j, \mathbf{r}_i) = \int_{SIP} dk_\rho W(k_\rho)H_0^{(1)}(k_\rho|\boldsymbol{\rho}_j - \boldsymbol{\rho}_i|)e^{ik_{1z}(z_j+z_i)}, \tag{7.15b}$$

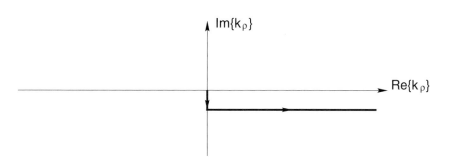

Figure 7.3: The folded SIP, which is used for the integral with the Bessel function kernel.

where $W(k_\rho)$ denote the weight function because it can be considered as the weights applied to the inhomogeneous plane waves pointing to different directions. For the sake of clarity, Eqs. (7.15a) and (7.15b) are used to describe the algorithm.

When $|\boldsymbol{\rho}_j - \boldsymbol{\rho}_i| \to 0$, it is better to work with the following formula:

$$g_0(\mathbf{r}_j, \mathbf{r}_i) = 2 \int_{FSIP} dk_\rho \frac{k_\rho}{k_{1z}} J_0(k_\rho |\boldsymbol{\rho}_j - \boldsymbol{\rho}_i|) e^{ik_{1z}(z_j - z_i)}, \quad (7.16a)$$

$$g^R(\mathbf{r}_j, \mathbf{r}_i) = 2 \int_{FSIP} dk_\rho W(k_\rho) J_0(k_\rho |\boldsymbol{\rho}_j - \boldsymbol{\rho}_i|) e^{ik_{1z}(z_j + z_i)}, \quad (7.16b)$$

to avoid the divergence of the Hankel function. Here, FSIP stands for the folded SIP, as shown in Figure 7.3.

Observation 1. It is easy to observe the similarity between g_0 and g^R in their spectral integral representations. Because g_0 represents the free-space Green's function and FIPWA has been developed to accelerate the matrix-vector multiplication for this kind of kernel, it is suggested that the same techniques be applied to g^R.

7.3   FIPWA for Free Space

FIPWA was originally developed as an alternative to FMM to accelerate the matrix vector multiplication encountered in the iterative solution of the electromagnetic scattering problems in free-space. It is briefly reviewed here for the sake of completeness. Readers are referred to [18] for more details.

To simplify the discussion, the kernel of the free-space Green's function, that is, e^{ikr}/r, is used to illustrate the algorithm. By using the Weyl identity [17] and using the symmetric property of the free-space Green's function, its integral representation can be formulated as

$$\frac{e^{ikr_{ji}}}{r_{ji}} = \frac{ik}{2\pi} \int_0^{2\pi} d\phi \int_{FSIP} d\theta \sin\theta e^{i\mathbf{k}(\theta,\phi)\cdot\mathbf{r}_{ji}}, \quad (7.17)$$

where $\mathbf{k}(\theta, \phi) = k\hat{k}(\theta, \phi) = k(\hat{x} \sin\theta \cos\phi + \hat{y} \sin\theta \sin\phi + \hat{z} \cos\theta)$ and $\mathbf{r}_{ji} = \mathbf{r}_j - \mathbf{r}_i$.

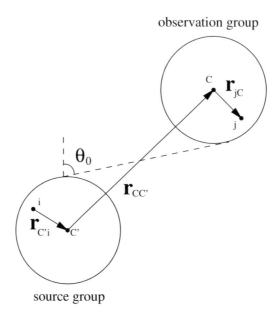

Figure 7.4: Illustration of the source and observation groups. The source group is centered at $\mathbf{r}_{C'}$ and the observation group is centered at $\mathbf{r}_C$.

Denoting $\mathbf{r}_{C'}$ and $\mathbf{r}_C$ as the centers of the groups containing the elements i and j, respectively, and using $\mathbf{r}_{ji} = \mathbf{r}_{jC} + \mathbf{r}_{CC'} + \mathbf{r}_{C'i}$, as shown in Figure 7.4, the above integral can be written as

$$\frac{e^{ikr_{ji}}}{r_{ji}} = \frac{ik}{2\pi} \int_0^{2\pi} d\phi \int_{FSIP} d\theta \sin\theta e^{i\mathbf{k}(\theta,\phi)\cdot\mathbf{r}_{jC}} \cdot e^{i\mathbf{k}(\theta,\phi)\cdot\mathbf{r}_{CC'}} \cdot e^{i\mathbf{k}(\theta,\phi)\cdot\mathbf{r}_{C'i}}. \tag{7.18}$$

To evaluate the integral efficiently, the SDP can be defined. When the source and observation groups are aligned along the z axis, that is, $\mathbf{r}_{CC'} = \hat{z}z_{CC'}$, the SDP on the θ plane is chosen and shown in Figure 7.5. The integration on ϕ is within $[0, 2\pi]$ and can be performed fairly easily.

We rewrite Eq. (7.18) as

$$\frac{e^{ikr_{ji}}}{r_{ji}} = \int_0^{2\pi} d\phi \int_\Gamma d\theta f(\theta)\mathcal{B}(\theta,\phi)e^{ikz_{CC'}\cos\theta}, \tag{7.19}$$

where

$$f(\theta) = \frac{ik}{2\pi}\sin\theta, \quad \mathcal{B}(\theta,\phi) = \mathcal{B}_{jC}\cdot\mathcal{B}_{C'i} = e^{i\mathbf{k}(\theta,\phi)\cdot\mathbf{r}_{jC}} \cdot e^{i\mathbf{k}(\theta,\phi)\cdot\mathbf{r}_{C'i}}, \tag{7.20}$$

and $\mathcal{B}(\theta,\phi)$ is the combination of the radiation and receiving pattern function for the source and observation groups, respectively, and $f(\theta)$ is considered to be the weight function. Denoting $\boldsymbol{\Omega}_\mathbf{q} = (\theta_{q_1}, \phi_{q_2})$, $\mathbf{W}_\mathbf{q} = (w_{q_1}, w_{q_2})$, and $\mathbf{q} = (q_1, q_2)$ as the appropriate abscissas and

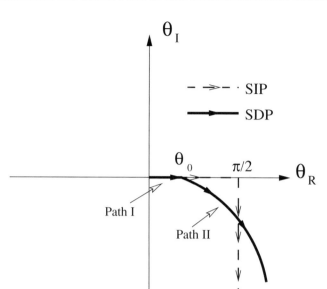

Figure 7.5: Illustration of the steepest descent path (SDP) in the θ plane for the free-space problems with the SIP in the background. The θ_0 is defined in Figure 7.4.

weights for the corresponding numerical quadrature methods for θ and ϕ along the SDP, Eq. (7.19) can be evaluated as

$$\frac{e^{ikr_{ji}}}{r_{ji}} = \sum_{\mathbf{q}} f(\mathbf{\Omega_q})\mathcal{B}(\mathbf{\Omega_q})e^{ikz_{CC'}\cos\theta_{q_1}}. \tag{7.21}$$

To diagonalize the translation operator, the value of the pattern function along the SDP, that is, $\mathcal{B}(\mathbf{\Omega}_q)$, can be obtained using the interpolation and extrapolation techniques as follows:

$$\mathcal{B}(\mathbf{\Omega_q}) = \sum_{\mathbf{s}} \mathcal{B}(\mathbf{\Omega_s})\mathcal{I}(\mathbf{\Omega_q}, \mathbf{\Omega_s}), \tag{7.22}$$

where $\mathcal{I}(\mathbf{\Omega_q}, \mathbf{\Omega_s})$ is the interpolation and extrapolation kernel and $\mathbf{\Omega_s}$ denotes the samples on the real θ and ϕ axis. After substituting Eq. (7.22) into (7.21), it can be derived as

$$\begin{aligned}
\frac{e^{ikr_{ji}}}{r_{ji}} &= \sum_{\mathbf{q}} f(\mathbf{\Omega_q})\mathcal{B}(\mathbf{\Omega_q})e^{ikz_{CC'}\cos\theta_{q_1}} \\
&= \sum_{\mathbf{q}} f(\mathbf{\Omega_q})\left[\sum_{\mathbf{s}} \mathcal{B}(\mathbf{\Omega_s})\mathcal{I}(\mathbf{\Omega_q}, \mathbf{\Omega_s})\right]e^{ikz_{CC'}\cos\theta_{q_1}} \\
&= \sum_{\mathbf{s}} \mathcal{B}(\mathbf{\Omega_s})\sum_{\mathbf{q}} f(\mathbf{\Omega_q})\mathcal{I}(\mathbf{\Omega_q}, \mathbf{\Omega_s})e^{ikz_{CC'}\cos\theta_{q_1}}
\end{aligned}$$

$$= \sum_{\mathbf{s}} \mathcal{B}(\mathbf{\Omega_s}) \cdot \mathcal{T}_{CC'}(\mathbf{\Omega_s}). \tag{7.23}$$

Here, $\mathcal{T}_{CC'}(\mathbf{\Omega}_s)$ is the diagonal translator and Eq. (7.23) can be used to construct an FMM-like fast algorithm to accelerate the matrix-vector multiplication. As detailed in [18], when source and observation groups are not aligned in the z axis, the diagonal translation operator can be obtained by performing the coordinate system rotation with the help of interpolation.

It has been observed that FIPWA achieves the diagonal translator using a different method from that used by FMM. Except for the different diagonalization procedures, these two algorithms are quite similar. Therefore, the computational complexity can be analyzed in a similar fashion, which has been discussed by many researchers, such as Song *et al.* [19]. In summary, the memory requirement for the multilevel implementation is $O(N)$ and the CPU time per iteration scales as $O(N \log N)$.

7.4 FIPWA for Layered Medium

In this section, FIPWA is extended to accelerate the matrix-vector multiplication for the problems involving a layered medium. As mentioned before, the Green's function for the layered medium can be separated into a direct (primary) part, which is the same as the free-space Green's function, and a reflected-wave part. In the previous section, FIPWA has been developed to handle the free-space problems. Therefore, in this section, effort focuses on the reflected-wave part.

7.4.1 Groups not aligned in $\hat{z}$ axis

Before describing the algorithm, the coordinate transform is performed by using

$$k_\rho = k \sin \theta, \quad k_z = k \cos \theta \tag{7.24}$$

and Eq. (7.15b) becomes

$$g^R(\mathbf{r}_j, \mathbf{r}_i) = \int_{SIP} d\theta W'(\theta) H_0^{(1)}(k \sin \theta |\boldsymbol{\rho}_j - \boldsymbol{\rho}_i|) e^{ik \cos \theta (z_j + z_i)}, \tag{7.25}$$

where $W'(\theta) = k \cos \theta W(\theta)$, and $\mathbf{r}_i$ and $\mathbf{r}_j$ denote the source and observation points, respectively. As discussed in the previous section, the source points and observation points are grouped, as shown in Figure 7.6, where the source and observation groups are centered at $\mathbf{r}_{C'}$ and $\mathbf{r}_C$, respectively. Comparing Eq. (7.25) with its free-space counterpart, the following observations can be made.

Observation 2. The $g^R(\mathbf{r}_j, \mathbf{r}_i)$ can be interpreted as the field originated from the source at $\mathbf{r}_i$, reflected by the dielectric layers and received by observation points located at $\mathbf{r}_j$. Equivalently, this effect can also be viewed as the field radiated from the mirror image of $\mathbf{r}_i$ with respect to the air-dielectric interface, denoted as $\mathbf{r}_{iI}$. Thus, the mirror images of the source points can also be combined to form a group, whose center, denoted as $\mathbf{r}_{C'I}$, is also the mirror image of the center for source group located at $\mathbf{r}_{C'}$. This grouping scheme is illustrated in Figure 7.6. Without introducing ambiguity, the mirror images of the source

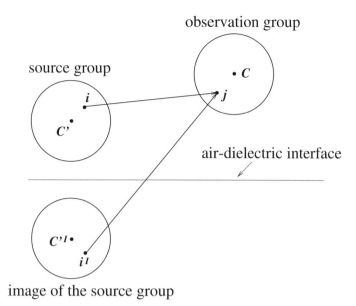

Figure 7.6: The grouping of the source and observation points. The mirror image of the source points can also be combined to form a mirror image group, both with respect to the air-dielectric interface.

point and source group are referred to as the source point and source group in the following discussion. It is to be cautioned that this is not the same as invoking the image theorem.

Observation 3. The basic expression of $g^R(\mathbf{r}_j, \mathbf{r}_i)$ is similar to the free-space one. The only difference lies in the weight function $W'(\theta)$. Therefore, all techniques developed for the FIPWA in free-space can be utilized for this kernel.

In the following, several important issues of the proposed algorithm, such as the proper choice of the SDP, the inclusion of the pole and branch point contributions, and the diagonalization of the translation matrices, are discussed.

Steepest descent path

By using the asymptotic expansion of the Hankel function, that is,

$$H_0^{(1)}(x) \sim \sqrt{\frac{2}{x\pi}} e^{i(x - \frac{\pi}{4})}, \quad x \to \infty, \tag{7.26}$$

the integrand in Eq. (7.25) behaves asymptotically as $e^{ik\rho \sin\theta + ikz \cos\theta}$. Thus, an SDP, along which the integrand decays exponentially away from its saddle point, can be defined in the θ plane. As the proper choice of the SDP has been discussed extensively in [20] and [21], the whole procedure is not repeated here. One possible choice of the SDP consists of three parts. They are:

$$\text{Path 1}: \qquad \theta = \theta_{left} - \theta_R + i\theta_I,$$

$$\begin{aligned}
&\cos\theta_R\cosh\theta_I = 1, \\
&\theta_R > 0, \ \ \theta_I > 0;
\end{aligned}$$

$$\text{Path 2}: \quad \theta_{left} \le \theta_R \le \theta_{right};$$
$$\text{Path 3}: \quad \theta = \theta_{right} + \theta_R - i\theta_I;$$
$$\cos\theta_R\cosh\theta_I = 1;$$
$$\theta_R > 0, \ \ \theta_I > 0; \tag{7.27}$$

where θ_{left} and θ_{right} represent the smallest and largest angles that the vectors, directed from the source points to the observation points located in the respective groups, make with the z axis, and they are shown in Figure 7.7. The SDP path is illustrated with the SIP in the background in Figure 7.8.

Two-dimensional fast multipole method

Before proceeding to discuss the diagonalization, the Hankel function kernel should be expanded using the plane wave. Here, the 2D FMM is chosen to perform this task.

From [22], the Hankel function can be expanded using the plane waves as follows:

$$H_0^{(1)}(k|\boldsymbol{\rho}_j - \boldsymbol{\rho}_i|) = \frac{1}{N}\sum_{q_2=1}^{N}\beta_{jC}(k,\phi_{q_2})\cdot\tau(k,\phi_{q_2},\boldsymbol{\rho}_{CC'I})\cdot\beta_{C'Ii^I}(k,\phi_{q_2}), \tag{7.28}$$

where τ is the diagonal translation coefficient of 2D FMM and is calculated using

$$\tau(k,\phi_{q_2},\boldsymbol{\rho}_{CC'I}) = \sum_{q=-N}^{N} H_n^{(1)}(k\rho_{CC'I})e^{-in\left[\phi_{CC'I}-\phi_{q_2}-\frac{\pi}{2}\right]}, \tag{7.29}$$

and

$$\beta_{jC}(k,\phi_{q_2}) = e^{ik(x_{jC}\cos\phi_{q_2}+y_{jC}\sin\phi_{q_2})}, \tag{7.30}$$

$$\beta_{C'Ii^I}(k,\phi_{q_2}) = e^{ik(x_{C'Ii^I}\cos\phi_{q_2}+y_{C'Ii^I}\sin\phi_{q_2})}, \tag{7.31}$$

where $x_{jC} = x_j - x_C$, and so on. The $\beta_{C'Ii^I}$ and β_{jC} are the 2D radiation and receiving patterns for the source and observation groups, respectively. Because k is a complex number, they are both inhomogeneous plane waves.

In principle, the 2D FIPWA [20] can also be used to obtain the plane wave expansion of the Hankel function, as long as the proper SDP for complex wave numbers is defined (remember the integration is now performed along the SDP and $k\sin\theta$ is a complex number in general).

Integration along the steepest descent path

The property of the integrand along the SDP has been studied in detail in [20] and [21]. On Paths 1 and 3, the integrand decays exponentially, whereas fast oscillation is observed on Path 2. Therefore, proper quadrature rules, such as the Gauss-Laguerre and Gauss-Legendre method, are chosen. For simplicity, (θ_{q_1}, w_{q_1}) denote the proper quadrature sample and weight

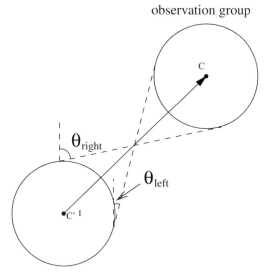

Figure 7.7: Illustration of observation group and the image of the source group. The image of the source group is centered at $\mathbf{r}_{C''}$.

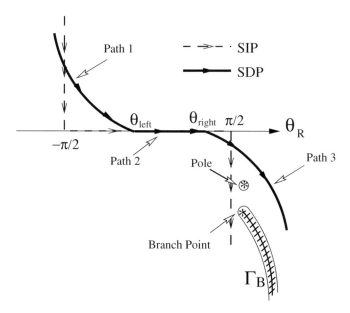

Figure 7.8: The SDP is shown, with the SIP in the background. One contributing pole and the branch point with the constant phase path, that is, Γ_B, passing through it are also shown.

along the SDP in the θ plane and (ϕ_{q_2}, w_{q_2}) is for the ϕ integration. Thus, the numerical integration is evaluated as

$$g^{SDP}(\mathbf{r}_j, \mathbf{r}_i) = \sum_{q_1} w_{q_1} W'(\theta_{q_1}) H_0^{(1)}(k \sin\theta_{q_1} |\boldsymbol{\rho}_j - \boldsymbol{\rho}_i|) e^{ik \cos\theta_{q_1}(z_j + z_i)}$$

$$= \frac{1}{N} \sum_{q_1} w_{q_1} W'(\theta_{q_1}) e^{ik \cos\theta_{q_1}(z_j + z_i)}$$

$$\times \sum_{q_2=1}^{N} \beta_{jC}(k \sin\theta_{q_1}, \phi_{q_2}) \cdot \tau(k \sin\theta_{q_1}, \phi_{q_2}, \boldsymbol{\rho}_{CC''}) \cdot \beta_{C''iI}(k \sin\theta_{q_1}, \phi_{q_2})$$

$$= \sum_{q_1} \sum_{q_2=1}^{N} \mathcal{B}_{jC}(\boldsymbol{\Omega}_{\mathbf{q}}) \cdot \mathcal{T}^{SDP}(\boldsymbol{\Omega}_{\mathbf{q}}, \mathbf{r}_{CC''}) \cdot \mathcal{B}_{C''iI}(\boldsymbol{\Omega}_{\mathbf{q}}), \tag{7.32}$$

where

$$\boldsymbol{\Omega}_{\mathbf{q}} = (\theta_{q_1}, \phi_{q_2}), \qquad \mathbf{q} = (q_1, q_2), \tag{7.33}$$

$$\mathcal{B}_{jC}(\boldsymbol{\Omega}_{\mathbf{q}}) = e^{i\mathbf{k}(\boldsymbol{\Omega}_{\mathbf{q}}) \cdot \mathbf{r}_{jC}}, \tag{7.34}$$

$$\mathcal{B}_{C''iI}(\boldsymbol{\Omega}_{\mathbf{q}}) = e^{i\mathbf{k}(\boldsymbol{\Omega}_{\mathbf{q}}) \cdot \mathbf{r}_{C''iI}}, \tag{7.35}$$

$$\mathbf{k}(\boldsymbol{\Omega}_{\mathbf{q}}) = k \left(\hat{x} \sin\theta_{q_1} \cos\phi_{q_2} + \hat{y} \sin\theta_{q_1} \sin\phi_{q_2} + \hat{z} \cos\theta_{q_1}\right), \tag{7.36}$$

$$\mathcal{T}^{SDP}(\boldsymbol{\Omega}_{\mathbf{q}}, \mathbf{r}_{CC''}) = \frac{w_{q_1}}{N} W'(\theta_{q_1}) \tau(k \sin\theta_{q_1}, \phi_{q_2}, \boldsymbol{\rho}_{CC''}) e^{ik \cos\theta_{q_1} z_{CC''}}, \tag{7.37}$$

and $\mathbf{r}_{ji} = \mathbf{r}_j - \mathbf{r}_i$.

In Eq. (7.32), the numerical integration is performed through the translation of the inhomogeneous plane waves from the source group to the observation group using the diagonal translation operator $\mathcal{T}^{SDP}$. The diagonal translator in this context means that there are no mutual interactions among the inhomogeneous plane waves pointing in different directions. Although the $\mathcal{T}^{SDP}$ is diagonal, this translation scheme is not desirable for two reasons. First, the number of quadrature samples, especially on Path 2, is directly proportional to the distance between two groups. Therefore, a large number of $\boldsymbol{\Omega}_{\mathbf{q}}$ is expected for distant groups. Second, as the integration path, that is, the SDP, is dependent on the relative position of the source group with respect to the observation groups, different sets of quadrature samples are required if this relative position changes. As is well known, the efficiency of the FMM-like algorithm is based on the precomputing and storing of the far-field pattern samples and the translation operator, that is, $\mathcal{B}_{C''iI}$, $\mathcal{B}_{jC}$ and $\mathcal{T}^{SDP}$, before the actual matrix-vector multiplication. To account for all possible source-observation group pairs, Eq. (7.32) requires one to precompute and store the samples for all possible SDPs. Obviously, it is inefficient in view of both the computation time and memory storage required. Previously developed algorithms [5], [13] store the samples along the integration path and thus are only suitable for the quasi-planar objects.

To solve this problem, the interpolation and extrapolation techniques are used. First, for the sake of clarity, Eq. (7.32) is rewritten as

$$g^{SDP}(\mathbf{r}_j, \mathbf{r}_i) = \sum_{\mathbf{q}} \mathcal{B}_{jC}(\boldsymbol{\Omega}_{\mathbf{q}}) \cdot \mathcal{T}_{CC''}^{SDP}(\boldsymbol{\Omega}_{\mathbf{q}}) \cdot \mathcal{B}_{C''iI}(\boldsymbol{\Omega}_{\mathbf{q}}). \tag{7.38}$$

Denoting $\mathbf{\Omega_s} = (\theta_{s_1}, \phi_{s_2})$ and $\mathbf{s} = (s_1, s_2)$ as the samples for the homogeneous plane waves, which means that θ_{s_1} and ϕ_{s_2} are real, and $\mathcal{I}(\mathbf{\Omega_q}, \mathbf{\Omega_s})$ as the interpolation/extrapolation kernel, the samples of the far-field patterns along the SDP can be obtained by

$$\mathcal{B}_{jC}(\mathbf{\Omega_q}) \cdot \mathcal{B}_{C''_I i_I}(\mathbf{\Omega_q}) = \sum_{\mathbf{s}} \mathcal{B}_{jC}(\mathbf{\Omega_s}) \cdot \mathcal{B}_{C''_I i_I}(\mathbf{\Omega_s})\mathcal{I}(\mathbf{\Omega_q}, \mathbf{\Omega_s}). \tag{7.39}$$

Then, Eq. (7.38) can be manipulated in the same fashion as in Eq. (7.23), that is,

$$
\begin{aligned}
g^{SDP}(\mathbf{r}_j, \mathbf{r}_i) &= \sum_{\mathbf{q}} \mathcal{B}_{jC}(\mathbf{\Omega_q}) \cdot \mathcal{T}_{CC''_I}^{SDP}(\mathbf{\Omega_q}) \cdot \mathcal{B}_{C''_I i_I}(\mathbf{\Omega_q}) \\
&= \sum_{\mathbf{q}} \mathcal{T}_{CC''_I}^{SDP}(\mathbf{\Omega_q}) \cdot \sum_{\mathbf{s}} \mathcal{B}_{jC}(\mathbf{\Omega_s}) \cdot \mathcal{B}_{C''_I i_I}(\mathbf{\Omega_s}) \cdot \mathcal{I}(\mathbf{\Omega_q}, \mathbf{\Omega_s}) \\
&= \sum_{\mathbf{s}} \mathcal{B}_{jC}(\mathbf{\Omega_s}) \cdot \tilde{\mathcal{T}}_{CC''_I}^{SDP}(\mathbf{\Omega_s}) \cdot \mathcal{B}_{C''_I i_I}(\mathbf{\Omega_s}),
\end{aligned}
\tag{7.40}
$$

where

$$\tilde{\mathcal{T}}_{CC''_I}^{SDP}(\mathbf{\Omega_s}) = \sum_{\mathbf{q}} \mathcal{T}_{CC''_I}^{SDP}(\mathbf{\Omega_q}) \cdot \mathcal{I}(\mathbf{\Omega_q}, \mathbf{\Omega_s}), \tag{7.41}$$

and $\tilde{\mathcal{T}}_{CC''_I}^{SDP}$ is the desired diagonal translation coefficients. Consequently, the integration along the SDP is performed by translating the homogeneous plane waves from the source group to the observation group, with the help of $\tilde{\mathcal{T}}_{CC''_I}^{SDP}$. The advantage of this manipulation is that the far-field patterns are sampled on the real axis, that is, $\mathbf{\Omega_s}$ in Eq. (7.40), and the sampling criteria are only dependent on the requirement of accurate reconstruction in Eq. (7.39). In other words, the way of the sampling, such as the number of the samples, is only determined by the group properties, such as the size of the groups, and it is independent of the source-observation group pairs involved in the numerical integration. Thus, only one set of samples of the far-field patterns is necessary for each group.

Observation 4. The radiation pattern function from the image source group, that is, $\mathcal{B}_{C''_I i_I}(\mathbf{\Omega_s})$, can be constructed from the source group, that is, $\mathcal{B}_{C'_i i}(\mathbf{\Omega_s})$, in Eq. (7.20) after exploiting the symmetric properties between them. Therefore, it is not necessary to store $\mathcal{B}_{C''_I i_I}(\mathbf{\Omega_s})$ when implementing the algorithm.

Pole contribution

In contrast to the free-space version, the weight function $W'(\theta)$, or essentially the generalized reflection coefficients embedded within it, has singularities, manifested as the pole and branch point in the complex θ plane. As a result, the deformation of the integration path from the SIP to SDP inevitably encompasses some of them, and the effects of these singularities to the final integration result must be included. As an illustration, a contributing pole is shown in Figure 7.8.

Using the same technique proposed in [21], all the poles in the complex θ plane can be located by using a recursive method. Assuming that θ_p is one of the contributing poles, the effect, denoted as g^P, can be computed using

$$g^P(\mathbf{r}_j, \mathbf{r}_i) = \Re es\{\theta_p\} H_0^{(1)}(k \sin \theta_p |\boldsymbol{\rho}_j - \boldsymbol{\rho}_i|) e^{ik \cos \theta_p(z_j + z_i)}$$

$$= \frac{1}{N} \Re es\{\theta_p\} e^{ik \cos \theta_p(z_j + z_i)}$$

$$\times \sum_{q_2=1}^{N} \beta_{jC}(k \sin \theta_p, \phi_{q_2}) \cdot \tau(k \sin \theta_p, \phi_{q_2}, \boldsymbol{\rho}_{CC''I}) \cdot \beta_{C''I i^I}(k \sin \theta_p, \phi_{q_2})$$

$$= \sum_{q_2=1}^{N} \mathcal{B}_{jC}(\theta_p, \phi_{q_2}) \cdot \mathcal{T}^P(\theta_p, \phi_{q_2}, \mathbf{r}_{CC''I}) \cdot \mathcal{B}_{C''I i^I}(\theta_p, \phi_{q_2}), \qquad (7.42)$$

where $\Re es\{\theta_p\}$ is the residue of the weight function $W'(\theta)$ at $\theta = \theta_p$, and $\mathcal{T}^P$ is the translation operator for the pole contribution and is calculated using

$$\mathcal{T}^P(\theta_p, \phi_{q_2}, \mathbf{r}_{CC''I}) = \frac{\Re es\{\theta_p\}}{N} \tau(k \sin \theta_p, \phi_{q_2}, \boldsymbol{\rho}_{CC''I}) e^{ik \cos \theta_p z_{CC''I}}. \qquad (7.43)$$

Observation 5. As the locations of the poles are fixed in the θ plane, the pole contribution can be accounted for by using a 2D FMM, as manifested in Eq. (7.42).

Pole determination

Although it is quite straightforward to include the pole contribution in FIPWA, a major problem is to find the accurate locations of all poles in a specific region. A handful of root-searching routines for nonlinear equations are available, but almost all of them can only locate one root in the given region. As the locations and the number of poles are not known beforehand, existing root solvers miss the locations of some poles. As a remedy, a new recursive method is introduced here to accurately determine arbitrary number of poles for a given region.

By using the Cauchy theorem in complex variables, which states that the contour integral of an analytic function is zero, this problem is solved successfully. The basic procedure is as follows. Given a specific region, normally a rectangular box, the contour integral is performed around it. A zero result implies that no pole exists inside. If the integration result is nontrivial, the box is further divided into smaller boxes around which the contour integral is performed. Obviously, only boxes with nontrivial contour integration results are further subdivided. In this manner, the approximate locations of the poles can be determined. Then, some root-searching routines can be used to refine them.

This approach has two advantages. First, this method is systematic and accurate. The method is mathematically sound. By subdividing the box in a multilevel manner, very small cells can be achieved in only a few levels, implying that the poles can be located with very high accuracy. Secondly, it is very efficient, because only the reflection coefficient, which has a closed form, possesses the poles, it is easy to perform the contour integral. As mentioned before, in each level, only several contour integrations are actually performed. Therefore, little time is used to locate the poles. A by-product of this method is that the residue is obtained directly from the result of contour integral. This is useful when the structure under

Table 7.1: Locations of Poles

Number	real part (ϕ_R)	imag part (ϕ_I)	$|\tilde{R}^{TM}|$ at this point
1	2.04328364451110	-0.164791069566742	1.1157E+12
2	2.04822034059814	-1.26312918284433	2.0315E+10
3	2.40029093077478	-1.11651541355518	3.1057E+14
4	2.91654903716033	-1.49117966962597	6.2335E+13

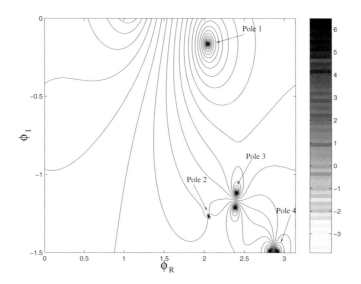

Figure 7.9: Contour plot of $\tilde{R}^{TM}$ in the ϕ plane. The region plotted is $\phi_R \in (0, \pi)$ and $\phi_I \in (-\frac{\pi}{2}, 0)$.

study is complex and the derivation of an analytic expression for the residue is tedious, if not impossible.

As an illustration of the effectiveness of the proposed pole-locating algorithm, the following example is shown. The working frequency is $f = 3$ GHz. The permittivity of the substrate is $\varepsilon_r = 2.56 + i2.56$, the thickness is $t = 5$ cm, and it is backed with the PEC ground plane. This structure is not typical in microstrip analysis and is chosen so that multiple poles exist. Figure 7.9 is the contour plot of the reflection coefficient $\tilde{R}^{TM}$ in the ϕ plane. It is shown that four poles are located on this plane. The poles found by this routine are listed in Table 7.1. Comparing this table with Figure 7.9, it is obvious that the program finds all four poles. For all other root-searching routines tested so far, only one pole (No. 1 in the table) can be located.

Because the poles are fixed in the ϕ plane, the computation time in the order of $O(N)$ is required to evaluate Eq. (7.42) in a multilevel implementation.

Branch point contribution

For layered medium problems, two branch points exist in the k_ρ plane, which are located at $k_\rho = k_1$ and $k_\rho = k_N$, and k_1 and k_N are the wave numbers of the outermost layers. As discussed in [21], the branch point at k_1 is eliminated under the coordinate transform. Thus, the only branch point in the θ plane corresponds to $k_\rho = k_N$, which is shown in Figure 7.8. A constant phase path passing through the branch point, denoted as Γ_B, can be designed so that the integrand decays exponentially away from the branch point. The branch point contribution, denoted as g^B, can be evaluated by performing the numerical integration along the constant phase path as follows:

$$
\begin{aligned}
g^B(\mathbf{r}_j, \mathbf{r}_i) &= \int_{\Gamma_B} d\theta \left[W'^+(\theta) - W'^-(\theta) \right] H_0^{(1)}(k \sin\theta |\boldsymbol{\rho}_j - \boldsymbol{\rho}_i|) e^{ik\cos\theta(z_j + z_i)} \\
&= \sum_n w_n \left[W'^+(\theta_n) - W'^-(\theta_n) \right] H_0^{(1)}(k \sin\theta_n |\boldsymbol{\rho}_j - \boldsymbol{\rho}_i|) e^{ik\cos\theta_n(z_j + z_i)} \\
&= \frac{1}{N} \sum_n w_n \mathcal{W}(\theta_n) e^{ik\cos\theta_n(z_j + z_i)} \\
&\quad \times \sum_{q_2=1}^{N} \beta_{jC}(k\sin\theta_n, \phi_{q_2}) \cdot \tau(k\sin\theta_p, \phi_{q_2}, \boldsymbol{\rho}_{CC''I}) \cdot \beta_{C''I\,iI}(k\sin\theta_p, \phi_{q_2}) \\
&= \sum_n \sum_{q_2=1}^{N} \mathcal{B}_{jC}(\theta_n, \phi_{q_2}) \cdot \mathcal{T}^B(\theta_n, \phi_{q_2}, \mathbf{r}_{CC''I}) \cdot \mathcal{B}_{C''I\,iI}(\theta_n, \phi_{q_2}),
\end{aligned}
\tag{7.44}
$$

where (θ_n, w_n) is the numerical quadrature coefficient and the diagonal translator $\mathcal{T}^B$ is expressed as

$$
\mathcal{T}^B(\theta_n, \phi_{q_2}, \mathbf{r}_{CC''I}) = \frac{w_n}{N} \mathcal{W}(\theta_n) \tau(k\sin\theta_n, \phi_{q_2}, \boldsymbol{\rho}_{CC''I}) e^{ik\cos\theta_n z_{CC''I}},
\tag{7.45}
$$

with $\mathcal{W}(\theta_n) = W'^+(\theta_n) - W'^-(\theta_n)$. Here, $W'^+(\theta_n)$ is the value of the weight function where k_ρ lies in the upper Riemann sheet, that is, $\Re\{k_{Nz}\} > 0$, whereas $W'^-(\theta_n)$ corresponds to the lower Riemann sheet, that is, $\Re\{k_{Nz}\} < 0$.

7.4.2   Groups aligned in $\hat{z}$ axis

If two groups are adjacent to each other in the $\boldsymbol{\rho}$ direction, that is, $\boldsymbol{\rho} \to 0$ for the source and observation points within the groups, the algorithm described in the previous section is no longer valid because of the divergence of the Hankel function. Therefore, another approach is proposed, which uses Eq. (7.16b) with Bessel function as the kernel.

Using the identity of the Bessel function,

$$
J_0(k_\rho \rho) = \frac{1}{2\pi} \int_0^{2\pi} d\phi \, e^{\mathbf{k}_\rho(\phi) \cdot \boldsymbol{\rho}},
\tag{7.46}
$$

and performing the same coordinate transform in Eq. (7.24), Eq. (7.16b) can be rewritten as

$$g^R(\mathbf{r}_j, \mathbf{r}_i) = \int_0^{2\pi} d\phi \int_{FSIP} d\theta W'(\theta) e^{\mathbf{k}(\theta,\phi)\cdot(\mathbf{r}_j - \mathbf{r}_i)}, \tag{7.47}$$

where $W'(\theta) = \frac{1}{\pi}\sin\theta W(\theta)$. The FSIP is shown in Figure 7.3. As mentioned before, the only difference between this Green's function and the free-space one, that is, Eq. (7.17), is the weight function. Therefore, the choice of the proper SDP is the same as the free-space case, which is shown in Figure 7.4. The diagonalization of the translation matrix is similar to the free-space one, as detailed in [18]. The pole and branch point contributions have to be included in the final integration result, which can be treated similarly as in the previous sections.

7.4.3   Observations

In the previous section, the application of the FIPWA to the layered medium problem is discussed in detail. After carefully going through the derivation, several observations are in order:

Observation 6. As mentioned at the start of this chapter, this approach treats the reflected part of the Green's function in the same way as the free-space one. The Sommerfeld-type integral presented in the reflected part is evaluated rigorously using the method of steepest descent. In addition, the numerical integration is performed by translating the homogeneous plane waves between groups, with the help of a diagonal translator. Furthermore, the radiation pattern and the receiving pattern for the free-space problem can be reused for the reflected-wave part. It is necessary to store additional far-field pattern samples for the reflected part.

Observation 7. Because the layered medium information is embedded in the generalized reflection coefficients, that is, $\tilde{R}^{TM}$ and $\tilde{R}^{TE}$, this approach, when applied to solve the problems with the objects above the layered medium, can handle arbitrary number of layers, with arbitrary parameters such as the dielectric constant and the thickness for each layer. By using this scheme, only one image term is actually required for arbitrary layered medium structure.

Observation 8. There are several methods previously designed for the layered medium problems using the SDP idea, such as the steepest descent fast multipole algorithm (SDFMA) [13] and the TSM-FMA [5]. Both methods make some assumptions, either about the structure under study or about the layered medium, and both limit their abilities to handle general problems. This approach can be considered a generalized one in which those restrictions are lifted.

Observation 9. The analysis of the computational complexity of this approach is easy. In this algorithm, the reflected wave part is treated similarly to the free-space part. As mentioned previously, the radiation and receiving patterns for the direct radiation term can be reused for the reflected-wave part. Therefore, the added computational cost for the layered medium problems is limited to the translation from the image source group to the observation group. The pole and branch point contributions do not affect the final computational complexity because these singularities are fixed in the θ plane. Following the similar procedure discussed in [18], the CPU time per iteration for the multilevel implementation is $O(N \log N)$, where N is the number of unknowns.

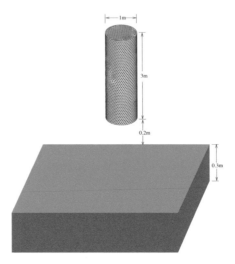

Figure 7.10: Illustration of the circular cylinder above a two-layer medium. The permittivity of the first dielectric layer is $\varepsilon_{r,2} = 2.56$ and the bottom one is $\varepsilon_{r,3} = 6.5 + i0.6$.

Observation 10. It is well known that the translation coefficients for the free-space problems are only the function of the relative position of the source and observation groups. On the contrary, in the layered medium problems, the translation coefficients for the reflected-wave part are also dependent on the location of the source group with respect to the air-dielectric interface. Therefore, additional storage of the translation coefficients is required. From the implementation point of view, the major memory consumption of the algorithm is the storage for the samples of the radiation patterns. Therefore, this additional storage does not affect the total memory requirement much, and the scaling property of the memory usage with respect to the number of unknowns is unchanged.

7.5   Numerical Results

In this section, some numerical results are presented to show the validity of this approach. The multilevel version of this algorithm is implemented on top of the ScaleME (scaleable multipole engine), a portable implementation of the dynamic MLFMA code [23]. The scattering from two objects, including a sphere, a circular cylinder, and a realistic tank, is calculated. The plane wave excitation is always assumed and all scattered fields are computed in the far-field zone. All examples are solved using a 600 MHz DEC Alpha workstation with 2 GB RAM, except the ones involving more than one million unknowns.

The first example is a circular cylinder above a two-layer medium, with permittivities of the two layers as $\varepsilon_{r,2} = 2.56$ and $\varepsilon_{r,3} = 6.5 + i0.6$. The thickness of the first dielectric layer is $t = 0.3$ m. The cylinder is 3 m long and has a diameter of 1 m. The distance between the lower end of the cylinder and the interface is 0.2 m, as shown in Figure 7.10. The object is discretized by 6,472 small triangle patches at $f = 600$ MHz and the number

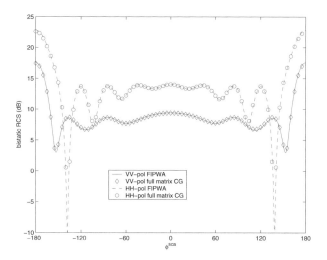

Figure 7.11: The bistatic radar cross section for a circular cylinder above a two-layer medium. The incident plane wave is at $(\theta^{inc}, \phi^{inc}) = (60^o, 0^o)$. The scattered field is calculated at $\theta^{sca} = 60^o$ and $\phi^{sca} \in [-180^o, 180^o]$. The cylinder is discretized with 6,472 small triangle patches and the number of unknowns is $N = 9,708$. The results are compared with those computed using the traditional CG program with full matrix-vector multiplication.

of unknowns is $N = 9,708$. The incident plane wave direction is $\theta^{inc} = 60^o$ and $\phi^{inc} = 0^o$ and the bistatic RCS is computed at $\theta^{scat} = 60^o$ and $\phi^{scat} \in [-180^o, 180^o]$. A five-level ML-FIPWA (multilevel FIPWA) program is used in the example. The results are shown in Figure 7.11 and are compared to those calculated using a conventional MOM code with the iterative solver using full matrix-vector multiplication. Very close agreement between the two codes is observed.

The next example is a tank model sitting on the ground, with permittivity of the ground material as $\varepsilon_r = 6.5 + i0.6$. The dimensions of the tank model are $8.5 \times 3.74 \times 2.19$ m. A patch model of the tank is shown in Figure 7.12. The incident plane wave direction is fixed at $\theta^{inc} = 60^o$ and $\phi^{inc} = 0^o$. The bistatic RCS is computed at $\theta^{scat} = 60^o$ and $\phi^{scat} \in [0^o, 360^o]$. In Figure 7.13, the bistatic RCS at $f = 100$ MHz is computed using a five-level ML-FIPWA code and compared with the one calculated using the conventional MOM code. The tank is discretized using 2,816 small triangle patches and the number of unknowns is $N = 8,334$.

The scaling properties of the memory requirement and the CPU time per matrix-vector multiplication are shown in Figure 7.14. They are compared with the ones from the free-space codes, including a free-space ML-FIPWA code and the Fast Illinois Solver Code (FISC) [19, 24]. To make a fair comparison of the computational complexity between this algorithm and the free-space ones, the parameters for the fast algorithms, such as the number of samples, the order of interpolation scheme, and the number of levels, are chosen to be the same for both the free-space and the layered medium codes. To generate this figure, three frequency

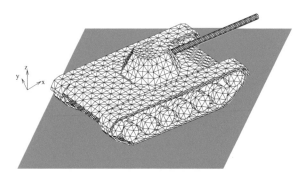

Figure 7.12: Illustration of the tank model sitting on the ground. The dimensions of the tank are $8.5 \times 3.74 \times 2.19$ m.

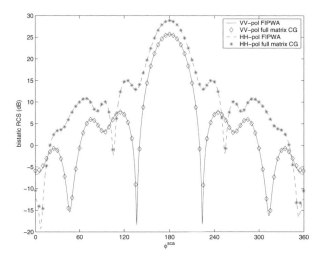

Figure 7.13: The bistatic radar cross section for the tank model at $f = 100$ MHz. The incident plane wave is at $(\theta^{inc}, \phi^{inc}) = (60^o, 0^o)$. The scattered field is calculated at $\theta^{sca} = 60^o$ and $\phi^{sca} \in [0^o, 360^o]$. The tank is discretized with 5,556 small triangle patches and the number of unknowns is $N = 8,334$.

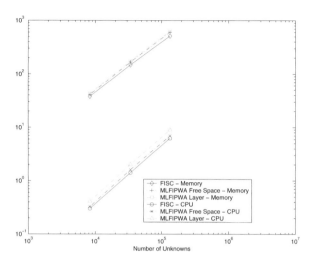

Figure 7.14: The scaling properties of the memory requirement and the CPU time for one matrix-vector multiplication with respect to the number of unknowns.

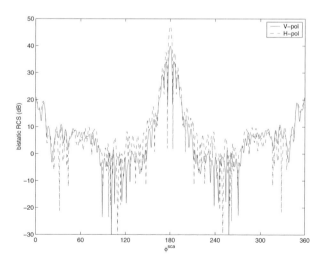

Figure 7.15: The bistatic RCS of the tank at $f = 1,200$ MHz. The solid line is the VV response and the dashed line is the HH response.

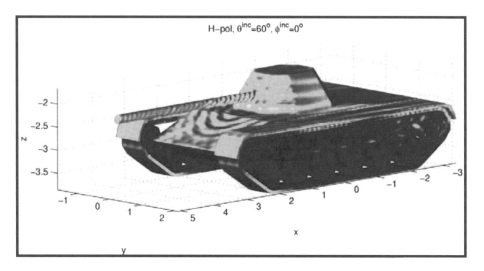

Figure 7.16: The current distribution of the tank residing in free-space. The incident plane wave direction is $(\theta^{inc}, \phi^{inc}) = (60^o, 0^o)$ and it is H-polarized. The working frequency is $f = 1,200$ MHz and the number of unknowns involved is $N = 1,210,458$.

points are chosen. They are 100 MHz, 200 MHz, and 400 MHz, corresponding to the number of unknowns $N = 8,334$, $N = 33,921$, and $N = 133,578$, respectively. From this figure, it is shown that the memory increases by merely 5% and the CPU time per iteration increases by about 25% compared with the free-space ML-FIPWA code. Even compared with FISC, the increase is about 13% and 30%, respectively. This is a good performance as the MLFMA implemented in FISC is tightly coupled to the MOM code and several specialized optimizations for efficiency are possible. This result shows that the layered medium problems can be solved as efficiently as the free-space ones. The additional cost to account for the layered medium is fairly small.

To show the capability of this algorithm, the scattering of this tank at $f = 1,200$ MHz is solved with the number of unknowns as $N = 1,210,458$. An eight-level program is utilized and requires about 6.6 GB and about 65 hours of CPU time to achieve the residual error at 0.001. The problem was solved on one R10000/195-MHz processor of the SGI Origin 2000 array at the National Center for Supercomputing Applications (NCSA). In Figure 7.15, the bistatic RCS of the VV (vertical-vertical) and HH (horizontal-horizontal) responses is shown. If the conventional solver is used to solve the same size problem, the estimated memory requirement shall exceed 13 TB and will take about 30 years of CPU time to reach the convergence.

The comparison of the current distribution on the tank, when residing in free-space and sitting on the ground, is also shown in Figures 7.16 and 7.17. The H-polarization (horizontal polarization) results are presented and the effect of the layered medium is clearly shown.

However, there is an open issue on solving the ill-conditioned problems using the iterative solver. For example, the tank model sits on top of a two-layer medium, which is the same medium as in Figure 7.10. The working frequency is $f = 100$ MHz. The scattering results of

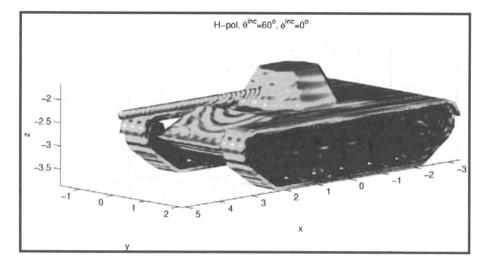

Figure 7.17: The current distribution of the tank sitting on the ground. The incident plane wave direction is $(\theta^{inc}, \phi^{inc}) = (60^o, 0^o)$ and it is H-polarized. The working frequency is $f = 1,200$ MHz and the number of unknowns involved is $N = 1,210,458$.

this problem are compared in Figure 7.18. It is obvious that the results obtained using FIPWA is a little distorted compared to the ones obtained using the full matrix solver, especially in the back-scattering region. It turns out the resultant impedance matrix is an ill-conditioned one and the number of iteration used is insufficient; hence a preconditioning procedure is required.

The same implementation can be readily applied to the buried object problem, which is a contra-posed problem to the one with objects above layered medium. All of the examples shown here are the PEC targets buried in the ground, with lossy and lossless ground materials presented. The plane wave excitation is always assumed and all scattered fields are computed in the far-field zone. In the following, the scattering from three types of targets is computed. They include a buried land mine, buried unexploded ordnance, and a buried bunker. Most of them are of interest to the remote sensing community.

The first example is a buried land mine, whose geometry is shown in Figure 7.19. The land mine is constructed with two circular cylinders. The diameter for the top one is 17.3 cm and 27 cm for the bottom one. The heights of the two cylinders are 1.62 cm and 7 cm, respectively. The working frequency is 600 MHz and the object is discretized using 1,764 small triangle patches. The mine is buried within a ground material with $\varepsilon_r = (3.3, 0.3)$ and the buried depth is 85 cm. The incident plane wave direction is fixed at $\theta^{inc} = 60^o$ and $\phi^{inc} = -90^o$. In Figure 7.20, the bistatic RCS, including the HV (horizontal-vertical) and HH (horizontal-horizontal) response, is computed using the three-level ML-FIPWA code and the conventional iterative solver with a direct matrix-vector multiplication. The figure shows close agreement between the ML-FIPWA program and the conventional solver.

The second example deals with an unexploded ordnance model, as shown in Figure 7.21.

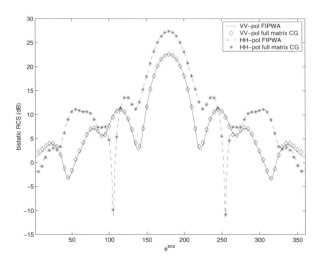

Figure 7.18: The bistatic RCS of the tank sitting on a two-layer medium at $f = 100$ MHz. The layered medium is described in Figure 7.9. The solid line is the VV response and the dashed line is the HH response. There is some distortion in the backscattering region, because of the ill-condition of the impedance matrix.

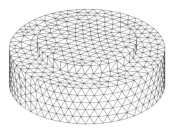

Figure 7.19: The geometry of the land mine. It has been discretized by using 1,764 small triangle patches, and the number of unknowns is $N = 2,646$.

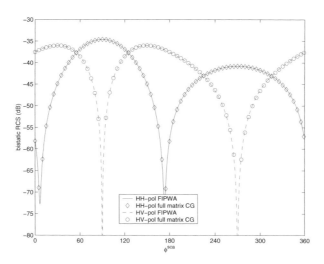

Figure 7.20: The bistatic RCS for the buried land mine problem.

Figure 7.21: The geometry model of the unexploded ordnance. It has been discretized by using 5,688 small triangle patches, and the number of unknowns is $N = 8,532$.

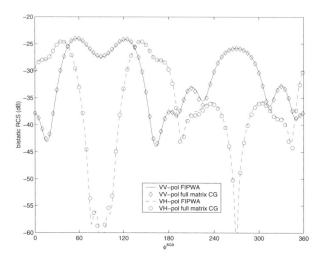

Figure 7.22: The bistatic RCS for the buried unexploded ordnance problem. The VV response and VH response are shown here.

The geometry is extracted from [14]. It is a cylinder with a hemispherical endcap with a diameter of 40.6 cm and a length of 153 cm. The object has a symmetry axis and it makes an angle of 30° relative to the z-axis. The working frequency is 600 MHz and the object is discretized using 5,688 small triangle patches. The object is buried within a ground material with $\varepsilon_r = 3.5$ and the buried depth is 85 cm. The plane wave is incident from $\theta^{inc} = 60^o$ and $\phi^{inc} = -90^o$. Again, the bistatic RCS is computed using a five-level ML-FIPWA code and the conventional iterative solver, which is used as a benchmark. The results are shown in Figure 7.22 and the close agreement between two solvers is shown.

As a more practical large-scale buried object problem, the scattering from a underground bunker model is solved. As shown in Figure 7.23, the dimensions of the bunker are 5 m by 5 m with the height of 2 m. The object is buried within a ground material with $\varepsilon_r = (3.3, 0.3)$ and the buried depth is 2.5 m. The scaling properties of the CPU time per iteration and memory requirement are shown in Figures 7.24 and 7.25, respectively. They are compared with the free-space ML-FIPWA code for the same structure. To have a fair comparison, the background material and the parameters for FIPWA are chosen to be the same for both programs. From these figures, it is shown that compared to the free-space code, the CPU time increases by merely 10% and the memory required to solve the same size problem increases by about 13%. These timing results are similar to what have been observed for objects above layered medium, that is, the buried object problems can be solved almost as efficiently as the free-space problem. This excellent scaling property is again mainly due to the plane wave expansion of the dyadic Green's function for the buried object problems. To generate these figures, four frequencies are chosen. They are 75 MHz, 150 MHz, 300 MHz, and 600 MHz, corresponding to the number of unknowns as $N = 7{,}674$, $N = 30{,}090$, $N = 120{,}180$, and $N = 480{,}696$.

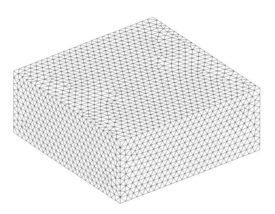

Figure 7.23: The geometry of the underground bunker. The one shown here is discretized by using $5,116$ small triangle patches.

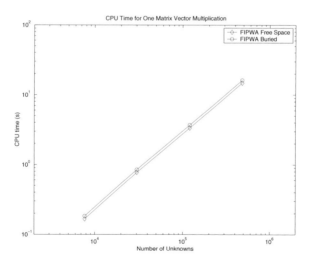

Figure 7.24: The scaling property of the CPU time per iteration for the different sizes of the underground bunker problem. It is compared with the free-space ML-FIPWA code, and CPU time required for the buried object problem increases by less than 10%.

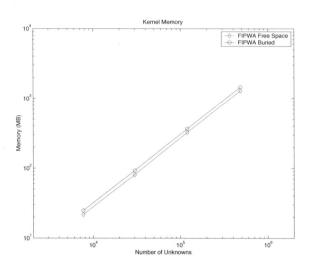

Figure 7.25: The scaling property of the required memory for different sizes of the underground bunker problems. It is compared with the free-space ML-FIPWA code, and the memory required for the buried object problem increases by about than 13%.

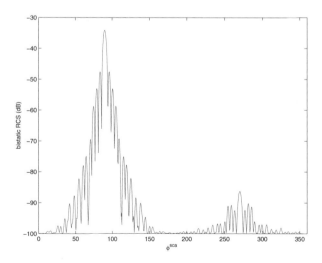

Figure 7.26: The bistatic RCS (VV-pol) from the underground bunker at $f = 900$ MHz. The number of unknowns is N=1,074,588 and it is solved by using one SGI R10000/195-MHz processor in the Origin 2000 array at NCSA. It uses about 3.94 GB memory and takes about 18 h of CPU time to achieve the residual error of 0.001.

To show the capability of this algorithm, the bunker problem has been solved at $f = 900$ MHz. The number of unknowns for this frequency is $N =$ 1,074,588 and an eight-level ML-FIPWA program is used. To solve this problem requires 3.94 GB memory and about 18 h of CPU time on one R10000/195-MHz processor of the SGI Origin 2000 array at NCSA, and the residual error is set at 0.001. In Figure 7.26, the bistatic RCS of the VV response is shown. If a conventional solver is used to solve the same size problem, the estimated memory requirement will exceed 9.3 TB and takes about 11 years of CPU time, which is unrealistic. This also underscores the importance of the fast method in solving large-scale problems.

7.6   Conclusions

In this chapter, the FIPWA has been successfully applied to solve the layered-medium problems, by exploiting the similarity of the Green's function encountered in a layered medium with its free-space counterpart. This approach has several merits, such as being efficient, accurate, and simple. Compared to the previously developed methods for solving the problems in this category, this method is more general and versatile. The multilevel algorithm is implemented and the CPU time per iteration is shown to be $O(N \log N)$, and the memory requirement scales as $O(N)$. It is interesting to note that the computational labor of this algorithm increases marginally over its free-space counterpart; typically the increase is less than 20%. As a result, we conclude that the problems in the presence of the layered medium can be solved as efficiently as the free-space ones. A large-scale problem with more than one million unknowns has been solved using this multilevel program.

Bibliography

[1] K. A. Michalski, and D. Zheng, "Electromagnetic scattering and radiation by surfaces of arbitrary shape in layered media, Parts I and II," *IEEE Trans. Antennas Propag.*, vol. 38, pp. 335–352, Mar 1990.

[2] W. C. Chew, "Field due to a point source in a layered medium - symmetrized formulation," *Private Notes*, March 25, 2003.

[3] W. C. Chew, J. S. Zhao, and T. J. Cui, "The layered medium Green's function - A new look," *Microwave Opt. Tech. Lett.*, vol. 31, no. 4, pp. 252–255, 2001.

[4] W. C. Chew, J. L. Xiong, and M. A. Saville, "A Matrix-Friendly Formulation of Layered Medium Green's Function," *Antennas Wirel. Propag. Lett.*, vol. 5, no. 1, pp. 490–494, Dec 2006.

[5] J. S. Zhao, W. C. Chew, C. C. Lu, E. Michielssen, and J. M. Song, "Thin-stratified medium fast-multipole algorithm for solving microstrip structures," *IEEE Trans. Microwave Theory Tech.*, vol. 46, no. 4, pp. 395–403, Apr 1998.

[6] T. J. Cui, and W. C. Chew, "Fast evaluation of Sommerfeld integrals for EM scattering and radiation by three-dimensional buried objects," *IEEE Geosci. Remote Sens.*, vol. 37, no. 2, pp.887–900, Mar 1999.

[7] D. G. Fang, J. J. Yang, and G. Y. Delisle, "Discrete image theory for horizontal electric dipole in a multilayer medium," *IEE Proc.-H*, vol. 135, pp. 295–303, Oct 1988.

[8] Y. L. Chow, J. J. Yang, D. G. Fang, and G. E. Howard, "A closed-form spatial Green's function for the thick microstrip substrate," *IEEE Trans. Microwave Theory Tech.*, vol. 39, pp. 588–592, Mar 1991.

[9] M. I. Aksun, "A robust approach for the derivation of closed-form Green's functions," *IEEE Trans. Microwave Theory Tech.*, vol. 44, pp. 651–658, May 1996.

[10] V. I. Okhmatovski, and A. C. Cangellaris, "Evaluation of layered media Green's function via rational function fitting," *IEEE Microwave Compon. Lett.*, vol. 14, pp. 22–24, Jan 2004.

[11] E. Bleszynski, M. Bleszynski, and T. Jaroszewicz, "AIM: Adaptive integral method compression algorithm for solving large scale electromagnetic scattering and radiation problems," *Radio Sci.*, vol 31, pp. 1225–1251, Sep–Oct 1996.

[12] F. Ling, and J. M. Jin, "Discrete complex image method for Green's functions of general multilayer media," *IEEE Microwave Guided Wave Lett.*, vol. 10, no. 10, pp. 400–402, Oct 2000.

[13] V. Jandhyala, E. Michielssen, B. Shanker, and W. C. Chew, "A combined steepest descent fast multipole algorithm for the fast analysis of three-dimensional scattering by rough surfaces," *IEEE Geosci. Remote Sens.*, vol. 36, pp. 738–748, May 1998.

[14] N. Geng, A. Sullivan, and L. Carin, "Multilevel fast-multipole algorithm for scattering from conducting targets above or embedded in a lossy half space," *IEEE Geosci. Remote Sens.*, vol. 38, pp. 1561–1573, Jul 2000.

[15] D. Vande Ginste, H. Rogier, D. De Zutter, and F. Olyslager, "A fast multipole method for layered media based on the application of perfectly matched layers - the 2-D case," *IEEE Trans. Antennas Propag.*, vol. 52, no. 10, pp. 2631–2640, Oct 2004.

[16] D. Vande Ginste, E. Michielssen, F. Olyslager, D. De Zutter, "An efficient perfectly matched layer based multilevel fast multipole algorithm for large planar microwave structures," *IEEE Trans. Antennas Propag.*, vol. 54, no. 5, pp. 1538–1548, May 2006.

[17] W. C. Chew, *Waves and Fields in Inhomogeneous Media*. New York: Van Nostrand Reinhold, 1990. Reprinted by Piscataway, NJ: IEEE Press, 1995.

[18] B. Hu, W. C. Chew, and S. Velamparambil, "Fast inhomogeneous plane wave algorithm for the analysis of electromagnetic scattering," *Radio Sci..*, Volume 36, Issue 6, pp. 1327–1340, 2001

[19] J. M. Song, C. C. Lu, and W. C. Chew, "Multilevel fast multipole algorithm for electromagnetic scattering by large complex objects," *IEEE Trans. Antennas Propag.*, vol 45, pp. 1488–1493, Oct. 1997.

[20] B. Hu, W. C. Chew, E. Michielssen and J. S. Zhao, "Fast inhomogeneous plane wave algorithm for the fast analysis of two-dimensional scattering problems," *Radio Sci.*, vol. 34, pp. 759–772, Jul–Aug 1999.

[21] B. Hu and W. C. Chew, "Fast inhomogeneous plane wave algorithm for electromagnetic solutions in layered medium structures - 2D case," *Radio Sci.*, vol. 35, pp. 31–43, Jan–Feb 2000.

[22] C.C. Lu and W.C. Chew, "A fast algorithm for solving hybrid integral equation," *IEE Proc.-H*, vol. 140, no. 6, pp.44455-460, Dec. 1993.

[23] S. Velamparambil, J. M. Song, and W. C. Chew, "ScaleME: A portable, distributed memory multilevel fast multipole kernel for electromagnetic and acoustic integral equation solvers," *Ctr. Comput. Electromag., Dept. Elect. Comp. Eng., U. Illinois, Urbana-Champaign*, Urbana, IL, Tech. Rep. CCEM-23-99, 1999.

[24] J. M. Song and W. C. Chew, "Large scale computations using FISC," in *IEEE APS Int. Symp. Dig.*, 2000, pp. 1856–1859.

CHAPTER 8

Electromagnetic Wave versus Elastic Wave

8.1 Introduction

Students of electromagnetics eventually become expert in wave physics. However, many disciplines in engineering and science involve wave physics. For instance, elastic wave is used in nondestructive testing, ultrasound imaging, seismic exploration, well logging, etc. The subject of elastic wave is within the grasp of students of electromagnetics easily. Even though this book is mainly about electromagnetic wave, we will introduce the elastic wave concepts here so that students can easily learn about this subject when they need this knowledge in the real world. Other areas that can benefit from the understanding of wave physics are the areas of quantum physics, ocean acoustics, and marine engineering.

This chapter is written for advanced students of electromagnetics who have had some advanced level courses in the area. Some mathematical sophistication is assumed in the reader. For some preliminary background in electromagnetics, one can consult [1, Chapter 1]. This chapter is also developed from a set of lecture notes written for advanced students in electromagnetics [2].

8.2 Derivation of the Elastic Wave Equation

The field of elastodynamics is broad, but in this chapter, we will consider small perturbations to the equations of elastodynamics. In this manner, the equations can be greatly simplified by linearization, giving rise to the elastic wave equation. By so doing, many wave phenomena in elastic solids can be analyzed using the linearized equation. In elasticity, where strain and stress analysis is involved, tensors of higher rank are involved. Hence, it is more expedient to use indicial notation (or index notation) to describe the equation sometimes.

The elastic wave equation governs the propagation of waves in solids. We shall illustrate its derivation as follows: the wave in a solid causes small perturbation of the particles in the solid. The particles are displaced from their equilibrium positions. The elasticity of the solid will provide the restoring forces for the displaced particles. Hence, the study of the balance of these forces will lead to the elastic wave equation [3–5].

The displacement of the particles in a solid from their equilibrium position causes a small displacement field $\mathbf{u}(\mathbf{x}, t)$ where $\mathbf{u}$ is the displacement of the particle at position $\mathbf{x}$ at time t. Here, $\mathbf{x}$ is a position vector in three dimensions. (Usually, in electromagnetics and many physics literature, we use $\mathbf{r}$ for position vector, but we use $\mathbf{x}$ here so that indicial notation can be used conveniently. In indicial notation, x_1, x_2, and x_3 refer to x, y, and z, respectively.) The displacement field $\mathbf{u}(\mathbf{x}, t)$ will stretch and compress distances between particles. For instance, particles at $\mathbf{x}$ and $\mathbf{x} + \delta\mathbf{x}$ are $\delta\mathbf{x}$ apart at equilibrium. But under a perturbation by $\mathbf{u}(\mathbf{x}, t)$, the change in their separation is given by

$$\delta\mathbf{u}(\mathbf{x}, t) = \mathbf{u}(\mathbf{x} + \delta\mathbf{x}, t) - \mathbf{u}(\mathbf{x}, t) \tag{8.1}$$

By using Taylor's series expansion, the above becomes

$$\delta\mathbf{u}(\mathbf{x}, t) \simeq \delta\mathbf{x} \cdot \nabla\mathbf{u}(\mathbf{x}, t) + O(\delta\mathbf{x}^2) \tag{8.2}$$

or in indicial notation,

$$\delta u_i \simeq \partial_j u_i \delta x_j \tag{8.3}$$

where $\partial_j u_i = \frac{\partial u_i}{\partial x_j}$. This change in separation $\delta\mathbf{u}$ can be decomposed into a symmetric and an antisymmetric parts as follows:

$$\delta u_i = \frac{1}{2} \overbrace{(\partial_j u_i + \partial_i u_j)}^{\text{symmetric}} \delta x_j + \frac{1}{2} \overbrace{(\partial_j u_i - \partial_i u_j)}^{\text{antisymmetric}} \delta x_j \tag{8.4}$$

Using indicial notation, it can be shown that

$$\begin{aligned}
[(\nabla \times \mathbf{u}) \times \delta\mathbf{x}]_i &= \epsilon_{ijk}(\nabla \times \mathbf{u})_j \delta x_k \\
&= \epsilon_{ijk}\epsilon_{jlm}\partial_l u_m \delta x_k
\end{aligned} \tag{8.5}$$

where ϵ_{ijk} is a Levi-Civita alternating tensor. From the identity that [1, Appendix B]

$$\epsilon_{ijk}\epsilon_{jlm} = -\epsilon_{jik}\epsilon_{jlm} = \delta_{im}\delta_{kl} - \delta_{il}\delta_{km} \tag{8.6}$$

we deduce that

$$[(\nabla \times \mathbf{u}) \times \delta\mathbf{x}]_i = (\partial_k u_i - \partial_i u_k)\delta x_k \tag{8.7}$$

Hence,

$$\delta u_i = \overbrace{e_{ij}\delta x_j}^{\text{stretch}} + \frac{1}{2} \overbrace{[(\nabla \times \mathbf{u}) \times \delta\mathbf{x}]_i}^{\text{rotation}} = (\overline{\mathbf{e}} \cdot \delta\mathbf{x})_i + \frac{1}{2}[\nabla \times \mathbf{u}) \times \delta\mathbf{x}]_i \tag{8.8}$$

where we have used an overline over a boldface to denote a second rank tensor, and we have defined the second rank tensor $\overline{\mathbf{e}}$ as

$$e_{ij} = \frac{1}{2}(\partial_j u_i + \partial_i u_j) \tag{8.9}$$

The first term in (8.8) results in a change in distance between the particles, whereas the second term, which corresponds to a rotation, has a higher order effect. This can be shown easily as follows: The perturbed distance between the particles at $\mathbf{x}$ and $\mathbf{x}+\delta\mathbf{x}$ is now $\delta\mathbf{x}+\delta\mathbf{u}$. The length square of this distance is (using (8.8))

$$\begin{aligned}
(\delta\mathbf{x}+\delta\mathbf{u})^2 &\cong \delta\mathbf{x}\cdot\delta\mathbf{x}+2\delta\mathbf{x}\cdot\delta\mathbf{u}+O(\delta\mathbf{u}^2)\\
&= |\delta\mathbf{x}|^2+2\delta\mathbf{x}\cdot\overline{\mathbf{e}}\cdot\delta\mathbf{x}+O(\delta\mathbf{u}^2)
\end{aligned} \tag{8.10}$$

assuming that $\delta\mathbf{u}\ll\delta\mathbf{x}$, because $\mathbf{u}$ is small, and $\delta\mathbf{u}$ is even smaller. Hence, the last term in (8.10) can be ignored. The second term in (8.8) vanishes in (8.10) because it is orthogonal to $\delta\mathbf{x}$.

The above analysis shows that the stretch in the distance between the particles is determined to first order by the first term in (8.8). The tensor $\overline{\mathbf{e}}$ describes how the particles in a solid are stretched in the presence of a displacement field: it is called the strain tensor. This strain produced by the displacement field will produce stresses in the solid.

Stress in a solid is described by a stress tensor $\overline{\boldsymbol{\mathcal{T}}}$. Given a surface $\triangle S$ in the body of solid with a unit normal $\hat{n}$, the stress in the solid will exert a force on this surface $\triangle S$. This force acting on a surface, known as traction, is given by (see Figure 8.1)

$$\mathbf{T}=\hat{n}\cdot\overline{\boldsymbol{\mathcal{T}}}\triangle S \tag{8.11}$$

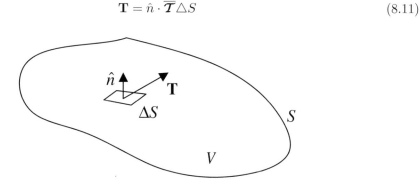

Figure 8.1: The balance of forces on an elastic body.

Hence, if we know the traction on the entire closed surface S of a volume V, the total force acting on the body is given by

$$\oint_S \hat{n}\cdot\overline{\boldsymbol{\mathcal{T}}}dS=\int_V \nabla\cdot\overline{\boldsymbol{\mathcal{T}}}dV \tag{8.12}$$

where the second equality follows from Gauss' theorem, assuming that $\overline{\boldsymbol{\mathcal{T}}}$ is defined as a continuous function of space. This force caused by stresses in the solid, must be balanced by other forces acting on the body, for example, the inertial force and body forces. Hence, by Newton's law,

$$\int_V \rho\frac{\partial^2\mathbf{u}}{\partial t^2}dV=\int_V \nabla\cdot\overline{\boldsymbol{\mathcal{T}}}dV+\int_V \mathbf{f}dV \tag{8.13}$$

where ρ is the mass density and $\mathbf{f}$ is a force density, for example, due to some externally applied sources in the body. The left-hand side is the inertial force (mass times acceleration) while the right-hand side is the total applied force on the body. Because (8.13) holds true for an arbitrary volume V, we conclude that

$$\rho \frac{\partial^2 \mathbf{u}}{\partial t^2} = \nabla \cdot \overline{\boldsymbol{\mathcal{T}}} + \mathbf{f} \tag{8.14}$$

as our equation of motion.

Because the stress force, the first term on the right hand side of (8.14), is caused by strains in the solid, $\overline{\boldsymbol{\mathcal{T}}}$ should be a function of $\overline{\mathbf{e}}$. Under the assumption of small perturbation, $\overline{\boldsymbol{\mathcal{T}}}$ should be linearly dependent on $\overline{\mathbf{e}}$ (generalized Hooke's law). The most general linear relationship between the second rank tensors is [3]

$$\mathcal{T}_{ij} = C_{ijkl} e_{kl} \tag{8.15}$$

The above is the constitutive relation for a solid, where C_{ijkl} is a fourth rank tensor.

For an isotropic medium, C_{ijkl} should be independent of any coordinate rotation. The most general form for a fourth rank tensor that has such independence is

$$C_{ijkl} = \lambda \delta_{ij} \delta_{kl} + \mu_1 \delta_{ik} \delta_{jl} + \mu_2 \delta_{il} \delta_{jk} \tag{8.16}$$

Furthermore, because e_{kl} is symmetric, $C_{ijkl} = C_{ijlk}$. Therefore, the simplified form of C_{ijkl} is

$$C_{ijkl} = \lambda \delta_{ij} \delta_{kl} + \mu (\delta_{jk} \delta_{il} + \delta_{il} \delta_{ik}) \tag{8.17}$$

Consequently, in an isotropic medium, the constitutive relation is characterized by two constants λ and μ known as Lamé constant. Note that as a consequence of (8.17), $\mathcal{T}_{ij} = \mathcal{T}_{ji}$ in (8.15). Hence, both the strain and the stress tensors are symmetric tensors in an isotropic medium. The above arguments hold true for inhomogeneous media as well. Reciprocity theorem similar to that in electromagnetics exists in linear elastodynamics according to Chapter 2.

Using (8.17) in (8.15), after using (8.9), we have that

$$\begin{aligned} \mathcal{T}_{ij} &= \lambda \delta_{ij} e_{ll} + \mu (e_{ij} + e_{ji}) \\ &= \lambda \delta_{ij} \partial_l u_l + \mu (\partial_j u_i + \partial_i u_j), \end{aligned} \tag{8.18}$$

which is true for inhomogeneous and isotropic media. Then

$$\begin{aligned} (\nabla \cdot \overline{\boldsymbol{\mathcal{T}}})_j = \partial_i \mathcal{T}_{ij} &= \partial_j (\lambda \partial_l u_l) + \partial_i (\mu \partial_j u_i) + \partial_i (\mu \partial_i u_j) \\ &= \lambda \partial_j (\partial_l u_l) + (\partial_l u_l) \partial_j \lambda + \mu \partial_j (\partial_i u_i) + (\partial_j u_i) \partial_i \mu + \partial_i \mu \partial_i u_j \\ &= (\lambda + \mu)[\nabla \nabla \cdot \mathbf{u}]_j + (\nabla \cdot \mu \nabla \mathbf{u})_j + (\nabla \cdot \mathbf{u})(\nabla \lambda)_j + [(\nabla \mathbf{u}) \cdot \nabla \mu]_j \end{aligned} \tag{8.19}$$

If μ and λ are constants of positions, that is, for a homogeneous medium, the above becomes

$$\nabla \cdot \overline{\boldsymbol{\mathcal{T}}} = (\lambda + \mu) \nabla \nabla \cdot \mathbf{u} + \mu \nabla^2 \mathbf{u} \tag{8.20}$$

Using (8.20) in (8.14), we have

$$\rho \frac{\partial^2 \mathbf{u}}{\partial t^2} = (\lambda + \mu) \nabla \nabla \cdot \mathbf{u} + \mu \nabla^2 \mathbf{u} + \mathbf{f} \tag{8.21}$$

which is the elastic wave equation for homogeneous and isotropic media. By using the identity that $\nabla^2 \mathbf{u} = \nabla\nabla\mathbf{u} - \nabla \times \nabla\mathbf{u}$, the above equation can be rewritten as

$$(\lambda + 2\mu)\nabla\nabla \cdot \mathbf{u} - \mu\nabla \times (\nabla \times \mathbf{u}) - \rho\ddot{\mathbf{u}} = -\mathbf{f}, \tag{8.22}$$

where the double over dots implies second derivative with respect to time.

8.3   Solution of the Elastic Wave Equation—A Succinct Derivation

By Fourier transform,

$$\mathbf{u}(\mathbf{x},t) = \frac{1}{2\pi} \int_{-\infty}^{\infty} d\omega e^{-i\omega t}\mathbf{u}(\mathbf{x},\omega) \tag{8.23}$$

the elastic wave equation (8.22) becomes

$$(\lambda + 2\mu)\nabla\nabla \cdot \mathbf{u} - \mu\nabla \times (\nabla \times \mathbf{u}) + \omega^2\rho\mathbf{u} = -\mathbf{f} \tag{8.24}$$

where $\mathbf{u} = \mathbf{u}(\mathbf{x},\omega)$, and $\mathbf{f} = \mathbf{f}(\mathbf{x},\omega)$ now.

The above equation supports transverse waves as well as longitudinal waves. The transverse wave means that if the displacement $\mathbf{u}$ is a plane wave, it represents motion that is transverse to the $\mathbf{k}$ vector of the plane wave. For a longitudinal wave, the $\mathbf{u}$ is in the direction of the $\mathbf{k}$ vector if $\mathbf{u}$ is a plane wave. Hence, the displacement $\mathbf{u}$ can be decomposed into the transverse wave part plus the longitudinal wave part. The transverse waves are known as shear waves (S waves), whereas the longitudinal waves are known as compressional waves. In acoustics, the longitudinal waves are the same as the pressure waves (P waves). Hence, S and P waves are also used to denote these two types of waves.

To see this decomposition, one takes $\nabla\times$ of the above equation, and defines

$$\boldsymbol{\Omega} = \nabla \times \mathbf{u}, \tag{8.25}$$

the rotation of $\mathbf{u}$, we have

$$\mu\nabla \times (\nabla \times \boldsymbol{\Omega}) - \omega^2\rho\boldsymbol{\Omega} = \nabla \times \mathbf{f} \tag{8.26}$$

The $\boldsymbol{\Omega}$ represents the field of the S wave. Because $\nabla \times (\nabla \times \boldsymbol{\Omega}) = \nabla\nabla \cdot \boldsymbol{\Omega} - \nabla^2\boldsymbol{\Omega}$, and that $\nabla \cdot \boldsymbol{\Omega} = \nabla \cdot \nabla \times \mathbf{u} = 0$, the above is just

$$\nabla^2\boldsymbol{\Omega} + k_s^2\boldsymbol{\Omega} = -\frac{1}{\mu}\nabla \times \mathbf{f} \tag{8.27}$$

where $k_s^2 = \omega^2\rho/\mu = \omega^2/c_s^2$ and $c_s = \sqrt{\frac{\mu}{\rho}}$. Hence, this wave propagates with the velocity c_s which is the shear wave velocity.

The solution to the above equation is

$$\boldsymbol{\Omega}(\mathbf{x},\omega) = \frac{1}{\mu} \int d\mathbf{x}' g_s(\mathbf{x} - \mathbf{x}')\nabla' \times \mathbf{f}(\mathbf{x}',\omega) \tag{8.28}$$

where $g_s(\mathbf{x} - \mathbf{x}') = e^{ik_s|\mathbf{x}-\mathbf{x}'|}/4\pi|\mathbf{x} - \mathbf{x}'|$ is the scalar Green's function for the shear wave.

Taking the divergence of (8.24), and defining

$$\theta = \nabla \cdot \mathbf{u}, \tag{8.29}$$

the dilatational part of $\mathbf{u}$, we have

$$(\lambda + 2\mu)\nabla^2\theta + \omega^2\rho\theta = -\nabla \cdot \mathbf{f}. \tag{8.30}$$

In the above, θ is related to the P wave. The solution to the above is

$$\theta(\mathbf{x}, \omega) = \frac{1}{\lambda + 2\mu} \int d\mathbf{x}' g_c(\mathbf{x} - \mathbf{x}')\nabla' \cdot \mathbf{f}(\mathbf{x}', \omega) \tag{8.31}$$

where $k_c^2 = \omega^2\rho/(\lambda + 2\mu) = \omega^2/c_c^2$, $c_c = \sqrt{(\lambda + 2\mu)/\rho}$, and $g_c(\mathbf{x} - \mathbf{x}') = e^{ik_c|\mathbf{x}-\mathbf{x}'|}/4\pi|\mathbf{x} - \mathbf{x}'|$ is the scalar Green's function for the compressional wave. Notice that the P wave or the compressional wave travels with a velocity of c_c.

From (8.24), after using (8.25) and (8.29), we deduce that

$$\mathbf{u}(\mathbf{x}, \omega) = -\frac{\mathbf{f}}{\mu k_s^2} + \frac{1}{k_s^2}\nabla \times \boldsymbol{\Omega} - \frac{1}{k_c^2}\nabla\theta \tag{8.32}$$

In a source-free region where $\mathbf{f} = 0$, the above is the Helmholtz decomposition of the displacement field $\mathbf{u}$ into a divergence-free part, the S wave, and a curl-free part, the P wave.

Notice that when $\mu = 0$, from (8.26), $\boldsymbol{\Omega}$ is nonzero only in the source region, and is zero outside the source. Hence, only compressional wave exists when $\mu = 0$, and this can be used to model a fluid medium which is an acoustic medium. An acoustic medium is a special case of an elastic medium with $\mu = 0$ where shear waves cannot exist.

To further manipulate the above, using (8.28) and (8.31), we have

$$\mathbf{u}(\mathbf{x}, \omega) = -\frac{\mathbf{f}}{\mu k_s^2} + \frac{1}{\mu k_s^2}\nabla \times \int d\mathbf{x}' g_s(\mathbf{x} - \mathbf{x}')\nabla' \times \mathbf{f}(\mathbf{x}', \omega)$$
$$- \frac{1}{(\lambda + 2\mu)k_c^2}\nabla \int d\mathbf{x}' g_c(\mathbf{x} - \mathbf{x}')\nabla' \cdot \mathbf{f}(\mathbf{x}', \omega) \tag{8.33}$$

Using integration by parts, and the fact that $\nabla'g(\mathbf{x} - \mathbf{x}') = -\nabla g(\mathbf{x} - \mathbf{x}')$, we obtain

$$\mathbf{u}(\mathbf{x}, \omega) = -\frac{\mathbf{f}}{\mu k_s^2} + \frac{1}{\mu k_s^2}\nabla \times \nabla \times \int d\mathbf{x}' g_s(\mathbf{x} - \mathbf{x}')\mathbf{f}(\mathbf{x}', \omega)$$
$$- \frac{1}{(\lambda + 2\mu)k_c^2}\nabla\nabla \int d\mathbf{x}' g_c(\mathbf{x} - \mathbf{x}')\mathbf{f}(\mathbf{x}', \omega) \tag{8.34}$$

Using the usual identity to simplify $\nabla \times \nabla \times \mathbf{A}$, and the fact that

$$\nabla^2 g_s(\mathbf{x} - \mathbf{x}') + k_s^2 g_s(\mathbf{x} - \mathbf{x}') = -\delta(\mathbf{x} - \mathbf{x}'),$$

we arrive at

$$\mathbf{u}(\mathbf{x}, \omega) = \left(\overline{\mathbf{I}} + \frac{\nabla\nabla}{k_s^2} \right) \cdot \frac{1}{\mu} \int d\mathbf{x}' g_s(\mathbf{x} - \mathbf{x}') \mathbf{f}(\mathbf{x}', \omega)$$

$$- \frac{\nabla\nabla}{k_c^2} \cdot \frac{1}{\lambda + 2\mu} \int d\mathbf{x}' g_c(\mathbf{x} - \mathbf{x}') \mathbf{f}(\mathbf{x}', \omega) \tag{8.35}$$

We can write the above more concisely as

$$\mathbf{u}(\mathbf{x}) = \int d\mathbf{x}' \overline{\mathbf{G}}(\mathbf{x}, \mathbf{x}') \cdot \mathbf{f}(\mathbf{x}') \tag{8.36}$$

with the ω dependence suppressed, and the dyadic Green's function $\overline{\mathbf{G}}(\mathbf{x}, \mathbf{x}')$ is defined as

$$\overline{\mathbf{G}}(\mathbf{x}, \mathbf{x}') = \left(\overline{\mathbf{I}} + \frac{\nabla\nabla}{k_s^2} \right) \frac{1}{\mu} g_s(\mathbf{x} - \mathbf{x}') - \frac{\nabla\nabla}{k_c^2} \frac{1}{\lambda + 2\mu} g_c(\mathbf{x} - \mathbf{x}'). \tag{8.37}$$

The first term of the dyadic Green's function produces the transverse wave, and is similar to the electromagnetic dyadic Green's function [1, Chapter 1]. The second term produces the longitudinal wave, and does not exist in a simple electromagnetic medium. It is to be noted that the scalar Green's functions have $1/|\mathbf{x} - \mathbf{x}'|$ singularity, and the double ∇ operator will accentuate this singularity in both terms above. This singularity is also known as the hypersingularity in the computational electromagnetics parlance. However, it can be shown that these hypersingularities from both terms above cancel each other when $\mathbf{x} \to \mathbf{x}'$. Since the electromagnetic dyadic Green's function only has the first term (transverse wave), the hypersingularity persists.

The hypersingularity, often studied in electromagnetics as the singularity of the dyadic Green's function [1, Chapter 7], is both a curse and a blessing! It is a curse because it makes the numerical evaluation of many integrals involving the dyadic Green's function difficult. It is a blessing because it implies that there is rich physics in this hypersingularity. It is within this hypersingularity that dwells the world of circuit physics that we use to create a plethora of new technologies. Because within the proximity of this singularity, where dimensions of structures are much smaller than the wavelength, lie the world of magnetoquasistatics (the world of inductors), and the world of the electro-quasistatics (the world of the capacitors) (see Chapter 5).

8.3.1 Time-Domain Solution

If $\mathbf{f}(\mathbf{x}, t) = \hat{x}_j \delta(\mathbf{x}) \delta(t)$, then $\mathbf{f}(\mathbf{x}, \omega) = \hat{x}_j \delta(\mathbf{x})$ and

$$\mathbf{u}(\mathbf{x}, \omega) = \left(\overline{\mathbf{I}} + \frac{\nabla\nabla}{k_s^2} \right) \cdot \frac{e^{ik_s r}}{4\pi\mu r} \hat{x}_j - \frac{\nabla\nabla}{k_c^2} \cdot \frac{e^{ik_c r}}{4\pi(\lambda + 2\mu)r} \hat{x}_j \tag{8.38}$$

where $r = |\mathbf{x}|$, or in indicial notation

$$u_i(\mathbf{x}, \omega) = \delta_{ij} \frac{e^{ik_s r}}{4\pi\mu r} + \partial_i \partial_j \frac{1}{4\pi r} \left(\frac{e^{ik_s r}}{\mu k_s^2} - \frac{e^{ik_c r}}{(\lambda + 2\mu) k_c^2} \right) \tag{8.39}$$

Since $k_s = \omega/c_s = \omega\sqrt{\rho/\mu}$, $k_c = \omega/c_c = \omega\sqrt{\rho/(\lambda+2\mu)}$, the above can be inverse Fourier transformed to yield

$$u_i(\mathbf{x}, t) = \delta_{ij}\frac{\delta(t - r/c_s)}{4\pi\mu r}$$
$$- \partial_i\partial_j\frac{1}{4\pi r\rho}\left[\left(t - \frac{r}{c_s}\right)u\left(t - \frac{r}{c_s}\right) - \left(t - \frac{r}{c_c}\right)u\left(t - \frac{r}{c_c}\right)\right] \quad (8.40)$$

Writing $\partial_i\partial_j = (\partial_i r)(\partial_j r)\frac{\partial^2}{\partial r^2}$, the above becomes

$$u_i(\mathbf{x}, t) = [\delta_{ij} - (\partial_i r)(\partial_j r)]\frac{\delta(t - r/c_s)}{4\pi\mu r}$$
$$- (\partial_i r)(\partial_j r)\frac{1}{2\pi\rho r^3}\left[tu\left(t - \frac{r}{c_s}\right) - tu\left(t - \frac{r}{c_c}\right)\right]$$
$$+ (\partial_i r)(\partial_j r)\frac{\delta(t - r/c_c)}{4\pi(\lambda+2\mu)r}. \quad (8.41)$$

Consequently, if $\mathbf{f}(\mathbf{x}, t)$ is a general body force, by convolutional theorem,

$$u_i(\mathbf{x}, t) = [\delta_{ij} - (\partial_i r)(\partial_j r)]\frac{f_j(\mathbf{x}, t - \frac{r}{c_s})}{4\pi\mu r}$$
$$+ (\partial_i r)(\partial_j r)\frac{1}{2\pi\rho r^3}\int_{r/c_c}^{r/c_s}\tau f_j(\mathbf{x}, t - \tau)d\tau$$
$$+ (\partial_i r)(\partial_j r)\frac{f_j(\mathbf{x}, t - \frac{r}{c_c})}{4\pi(\lambda+2\mu)r} \quad (8.42)$$

8.4 Alternative Solution of the Elastic Wave Equation via Fourier-Laplace Transform

Alternatively, the elastic wave equation

$$(\lambda+\mu)\nabla\nabla\cdot\mathbf{u} + \mu\nabla^2\mathbf{u} - \rho\ddot{\mathbf{u}} = -\mathbf{f} \quad (8.43)$$

can be solved by Fourier-Laplace transform. We let

$$\mathbf{u}(\mathbf{x}, t) = \frac{1}{(2\pi)^4}\int_{-\infty}^{\infty}d\omega e^{-i\omega t}\int_{-\infty}^{\infty}d\mathbf{k}e^{i\mathbf{k}\cdot\mathbf{x}}\mathbf{u}(\mathbf{k}, \omega), \quad (8.44)$$

then (8.43) becomes

$$-(\lambda+\mu)\mathbf{k}\mathbf{k}\cdot\mathbf{u} - \mu k^2\mathbf{u} + \omega^2\rho\mathbf{u} = -\mathbf{f} \quad (8.45)$$

where $\mathbf{u} = \mathbf{u}(\mathbf{k}, \omega)$, $\mathbf{f} = \mathbf{f}(\mathbf{k}, \omega)$ now, and $k^2 = \mathbf{k}\cdot\mathbf{k}$. The above can be formally solved to yield

$$\mathbf{u}(\mathbf{k}, \omega) = \left[(\lambda+\mu)\mathbf{k}\mathbf{k} + \mu k^2\overline{\mathbf{I}} - \omega^2\rho\overline{\mathbf{I}}\right]^{-1}\cdot\mathbf{f}. \quad (8.46)$$

The inverse of the above tensor commutes with itself, so that the inverse must be of the form $\alpha\bar{\mathbf{I}} + \beta\mathbf{k}\mathbf{k}$, that is,

$$\left[(\lambda+\mu)\mathbf{k}\mathbf{k} + \mu k^2\bar{\mathbf{I}} - \omega^2\rho\bar{\mathbf{I}}\right] \cdot \left[\alpha\bar{\mathbf{I}} + \beta\mathbf{k}\mathbf{k}\right] = \bar{\mathbf{I}} \tag{8.47}$$

Solving the above yields that

$$\alpha = \frac{1}{(\mu k^2 - \omega^2\rho)} \tag{8.48}$$

$$
\begin{aligned}
\beta &= \frac{-(\lambda+\mu)}{\left[\mu k^2 - \rho\omega^2\right]\left[(\lambda+2\mu)k^2 - \rho\omega^2\right]} \\
&= -\frac{\mu}{\rho\omega^2\left[\mu k^2 - \rho\omega^2\right]} + \frac{(\lambda+2\mu)}{\rho\omega^2\left[(\lambda+2\mu)k^2 - \rho\omega^2\right]} \\
&= -\frac{1}{k_s^2\mu\left[k^2 - k_s^2\right]} + \frac{1}{k_c^2(\lambda+2\mu)\left[k^2 - k_c^2\right]}
\end{aligned}
\tag{8.49}
$$

where $k_s^2 = \omega^2\rho/\mu$, $k_c^2 = \omega^2\rho/(\lambda+2\mu)$. Consequently,

$$\mathbf{u}(\mathbf{k},\omega) = \left[\bar{\mathbf{I}} - \frac{\mathbf{k}\mathbf{k}}{k_s^2}\right] \cdot \frac{\mathbf{f}(\mathbf{k},\omega)}{\mu(k^2 - k_s^2)} + \frac{\mathbf{k}\cdot\mathbf{f}(\mathbf{k},\omega)}{k_c^2(\lambda+2\mu)(k^2 - k_c^2)} \tag{8.50}$$

To obtain $\mathbf{u}(\mathbf{x},\omega)$, we use the relation that

$$\mathbf{u}(\mathbf{x},\omega) = \frac{1}{(2\pi)^3}\int_{-\infty}^{\infty} d\mathbf{k}\, e^{i\mathbf{k}\cdot\mathbf{x}}\mathbf{u}(\mathbf{k},\omega). \tag{8.51}$$

First, we can perform the integral assuming that we have a point source for $\mathbf{f}$. It can be shown by contour integration technique [1, Chapters 2 and 7] that

$$\frac{1}{(2\pi)^3}\int d\mathbf{k}\, e^{i\mathbf{k}\cdot\mathbf{x}}\frac{1}{(k^2 - k_0^2)} = \frac{e^{ik_0 r}}{4\pi r} \tag{8.52}$$

where $r = |\mathbf{x}|$. Hence, it can be shown that

$$\frac{1}{(2\pi)^3}\int d\mathbf{k}\, e^{i\mathbf{k}\cdot\mathbf{x}}\left(\bar{\mathbf{I}} - \frac{\mathbf{k}\mathbf{k}}{k_s^2}\right)\frac{1}{\mu(k^2 - k_s^2)} = \left(\bar{\mathbf{I}} + \frac{\nabla\nabla}{k_s^2}\right)\frac{e^{ik_s r}}{4\pi\mu r} \tag{8.53}$$

where $k_s = \omega/c_s$, and $c_s = \sqrt{\mu/\rho}$ is the shear wave velocity. Similarly,

$$\frac{1}{(2\pi)^3}\int d\mathbf{k}\, e^{i\mathbf{k}\cdot\mathbf{x}}\frac{\mathbf{k}\mathbf{k}}{k_c^2(\lambda+2\mu)(k^2 - k_c^2)} = -\frac{\nabla\nabla}{k_c^2}\frac{e^{ik_c r}}{4\pi(\lambda+2\mu)r} \tag{8.54}$$

where $k_c = \omega/c_c$, $c_c = \sqrt{(\lambda+2\mu)/\rho}$ is the compressional wave velocity.

By the convolutional theorem,

$$
\begin{aligned}
\mathbf{u}(\mathbf{x},\omega) = {} & \left(\bar{\mathbf{I}} + \frac{\nabla\nabla}{k_s^2}\right) \cdot \int d\mathbf{x}'\frac{e^{ik_s|\mathbf{x}-\mathbf{x}'|}}{4\pi\mu|\mathbf{x}-\mathbf{x}'|}\mathbf{f}(\mathbf{x}',\omega) \\
& - \frac{\nabla\nabla}{k_c^2} \cdot \int d\mathbf{x}'\frac{e^{ik_c|\mathbf{x}-\mathbf{x}'|}}{4\pi(\lambda+2\mu)|\mathbf{x}-\mathbf{x}'|}\mathbf{f}(\mathbf{x}',\omega)
\end{aligned}
\tag{8.55}
$$

The above is exactly the same as (8.35), the result obtained in the previous Section 8.3.

8.5   Boundary Conditions for Elastic Wave Equation

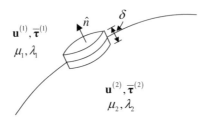

Figure 8.2: Boundary condition at the interface of two elastic media.

The equation of motion for elastic waves is

$$\nabla \cdot \overline{\boldsymbol{T}} + \mathbf{f} = -\omega^2 \rho \mathbf{u} \tag{8.56}$$

By integrating this over a pillbox whose thickness is infinitesimally small at the interface between two regions (see Figure 8.2), and assuming that $\mathbf{f}$ is not singular at the interface, it can be shown that

$$\hat{n} \cdot \overline{\boldsymbol{T}}^{(1)} = \hat{n} \cdot \overline{\boldsymbol{T}}^{(2)} \tag{8.57}$$

The above corresponds to three equations for the boundary conditions at an interface.

If (8.57) is written in terms of a local Cartesian coordinates with $\hat{z}$ being the unit normal $\hat{n}$, then (8.57) is equivalent to the continuity of $\mathcal{T}_{zz}$, $\mathcal{T}_{zx}$, and $\mathcal{T}_{zy}$ across an interface. More explicitly, one can express

$$\mathcal{T}_{zz} = \lambda \nabla \cdot \mathbf{u} + 2\mu \partial_z u_z = \lambda(\partial_x u_x + \partial_y u_y) + (\lambda + 2\mu)\partial_z u_z \tag{8.58}$$

In the above λ and μ are functions of z in the local coordinates, since the medium parameters are assumed to vary across the interface. The z derivative is assumed to be taken across the interface as well. Furthermore, $\mathcal{T}_{zz}$ cannot be singular (does not contain Dirac delta functions), then sum of the terms on the right-hand side involving $\partial_x u_x$, $\partial_y u_y$, and $\partial_z u_z$ must be regular. Since ∂_x and ∂_y are tangential derivatives, and u_x and u_y are smooth functions of x and y, $\partial_x u_x$ and $\partial_y u_y$ are regular. Therefore $\partial_z u_z$ must be regular. Requiring $\partial_z u_z$ to be regular implies that at the interface

$$u_z^{(1)} = u_z^{(2)} \tag{8.59}$$

Furthermore,

$$\mathcal{T}_{zx} = \mu \partial_x u_z + \mu \partial_z u_x, \tag{8.60}$$

by the same token, the continuity of $\mathcal{T}_{zx}$ implies that $\partial_z u_x$ must be regular. This induces the boundary condition that

$$u_x^{(1)} = u_x^{(2)} \tag{8.61}$$

By the same argument from $\mathcal{T}_{zy}$, we have

$$u_y^{(1)} = u_y^{(2)} \tag{8.62}$$

Hence, in addition to (8.57), we have boundary conditions (8.59), (8.61), and (8.62) which form a total of six boundary conditions at a solid-solid interface.

At a fluid-solid interface, $\mu_1 = 0$ in the fluid region assuming that it is region 1, then from (8.60), $\mathcal{T}_{zx}$ and $\mathcal{T}_{zy}$ are zero at the interface, in order for them to be continuous across an interface. Furthermore, u_x and u_y need not be continuous anymore. The boundary conditions are (8.57) and (8.59), a total of four boundary conditions.

At a fluid-fluid interface, only $\mathcal{T}_{zz}$ is nonzero. Its continuity implies $\lambda \nabla \cdot \mathbf{u} = p$ is continuous or the pressure is continuous. Furthermore, it induces the boundary condition (8.59). Hence there are only two boundary conditions.

8.6   Decomposition of Elastic Wave into SH, SV and P Waves for Layered Media

In a planar layered medium where the interfaces are horizontal, the transverse waves, or the shear waves, can be classified as shear horizontal (SH) where the particle motion is parallel to the interface, and shear vertical (SV) where the particle motion is in a plane orthogonal to the horizontal interface. The SH and SV waves are analogous to the TE_z and TM_z waves in electromagnetics [1, Chapter 2]. The P wave is the longitudinal wave, and in the case when the shear wave disappears, as in a purely acoustic medium, it is the pressure wave.

From the boundary conditions, it can be seen that the SH waves propagate through a planar layered medium independent of the SV and P waves. Moreover, the SV and P waves are coupled together at the planar interfaces by the boundary condition. Hence, given an arbitrary source, we can use the Weyl or Sommerfeld identity to expand the waves into plane waves. If these plane waves can be further decomposed into SH, SV and P waves, then the transmission and reflection of these waves through a planar layered medium can be easily found using the method of Chapter 2 in reference [1].

It has been shown previously in Section 8.3 that an arbitrary source produces a displacement field in a homogeneous isotropic medium given by

$$\mathbf{u}(\mathbf{x}) = -\frac{\mathbf{f}}{\mu k_s^2} + \underbrace{\frac{1}{k_s^2} \nabla \times \boldsymbol{\Omega}}_{\mathbf{u}^s} - \underbrace{\frac{1}{k_c^2} \nabla \theta}_{\mathbf{u}^p} \tag{8.63}$$

Outside the source region the first term is zero. The second term corresponds to S waves while the third term corresponds to P waves. Hence, in a source-free region, the S component of (8.63) is

$$\mathbf{u}^s(\mathbf{x}) = \frac{1}{k_s^2} \nabla \times \boldsymbol{\Omega} \tag{8.64}$$

From the definition of $\boldsymbol{\Omega}$, we have

$$\boldsymbol{\Omega} = \nabla \times \mathbf{u}^s(\mathbf{x}). \tag{8.65}$$

In the above, $\boldsymbol{\Omega}$ was previously derived in the Section 8.3 to be

$$\boldsymbol{\Omega}(\mathbf{x}) = \frac{1}{\mu} \nabla \times \int d\mathbf{r}' \frac{e^{ik_s|\mathbf{x}-\mathbf{x}'|}}{4\pi|\mathbf{x}-\mathbf{x}'|} \mathbf{f}(\mathbf{x}') \tag{8.66}$$

Using the Sommerfeld-Weyl identities, (8.66) can be expressed as a linear superposition of plane waves. Assuming that $\mathbf{\Omega}$ and $\mathbf{u}^s$ are plane waves in (8.64) and (8.65), replacing ∇ by $i\mathbf{k}_s$, when we assume the vertical axis to be the z axis, we note that only the SV waves have $u_z^s \neq 0$, and only the SH waves have $\Omega_z \neq 0$. Hence, we can use u_z^s, namely, the z component of $\mathbf{u}^s$, to characterize SV waves; and we can use Ω_z, namely, the z component of $\mathbf{\Omega}$, to characterize SH wave.

Assuming that $\mathbf{f}(\mathbf{x}) = \hat{a}A\delta(\mathbf{x})$, that is, a point excitation polarized in the $\hat{a}$ direction, then (8.66) becomes

$$\mathbf{\Omega}(\mathbf{x}) = \frac{A}{\mu}(\nabla \times \hat{a})\frac{e^{ik_s r}}{4\pi r} \tag{8.67}$$

where $r = |\mathbf{x}|$.

In the electromagnetic case, the z components of the $\mathbf{H}$ and $\mathbf{E}$ fields are used to characterize TE_z and TM_z waves, respectively. We can adopt the same method here. Consequently, using the fact that $\nabla^2 g_s = -k_s^2 g_s$ outside the point source (where g_s is the scalar Green's function for the shear wave), extracting the z component of the S-wave field, $\mathbf{u}_z^s = \hat{z}u_s$ follows from (8.64) to be

$$\mathbf{u}_z^s(\mathbf{x}) = \frac{A}{\mu k_s^2}(\hat{z}\hat{z}k_s^2 + \nabla_z\nabla)\cdot\hat{a}\frac{e^{ik_s r}}{4\pi r} \tag{8.68}$$

The z component of (8.67) characterizes an SH wave while (8.68) characterizes an SV wave. For a student of electromagnetics, one can think that $\mathbf{\Omega}$ is analogous to the $\mathbf{H}$ field, while $\mathbf{u}$ is analogous to the $\mathbf{E}$ field.

The P wave can be characterized by θ which has been previously derived to be (see (8.31))

$$\begin{aligned}
\theta &= \frac{\nabla\cdot}{\lambda + 2\mu}\int d\mathbf{r}'\frac{e^{ik_c|\mathbf{x}-\mathbf{x}'|}}{4\pi|\mathbf{x}-\mathbf{x}'|}\mathbf{f}(\mathbf{x}') \\
&= \frac{A\nabla\cdot\hat{a}}{\lambda + 2\mu}\frac{e^{ik_c r}}{4\pi r} \tag{8.69}
\end{aligned}$$

for this particular point source. Alternatively, P waves can be characterized by $u_z^p = -(1/k_c^2)\partial_z\theta$

Using the Sommerfeld identity [1, Chapter 2],

$$\frac{e^{ikr}}{r} = i\int_0^\infty dk_\rho\frac{k_\rho}{k_z}e^{ik_z|z|}J_0(k_\rho\rho), \tag{8.70}$$

where $k_\rho^2 + k_z^2 = k^2$, $\rho = \sqrt{x^2 + y^2}$, $r = \sqrt{\rho^2 + z^2}$, the above can be expressed as

$$\Omega_z(\mathbf{r}) = \frac{iA\hat{z}\cdot(\nabla\times\hat{a})}{4\pi\mu}\int_0^\infty dk_\rho\frac{k_\rho}{k_{sz}}e^{ik_{sz}|z|}J_0(k_\rho\rho), \quad \text{SH}, \tag{8.71}$$

$$u_z^s(\mathbf{r}) = \frac{iA(\hat{z}\cdot\hat{a}k_s^2 + \partial z\nabla\cdot\hat{a})}{4\pi\mu k_s^2}\int_0^\infty dk_\rho\frac{k_\rho}{k_{sz}}e^{ik_{sz}|z|}J_0(k_\rho\rho), \quad \text{SV}, \tag{8.72}$$

$$u_z^p(\mathbf{r}) = \frac{\pm A\nabla\cdot\hat{a}}{4\pi(\lambda + 2\mu)k_c^2}\int_0^\infty dk_\rho k_\rho e^{ik_{cz}|z|}J_0(k_\rho\rho), \quad \text{P} \tag{8.73}$$

where $k_\rho^2 + k_{sz}^2 = k_s^2$, $k_\rho^2 + k_{cz}^2 = k_c^2$. The above have expressed the field from a point source in terms of a linear superposition of plane waves. We have used u_z^p to characterize P waves so that it has the same dimension as u_z^s.

When the point source is placed above a planar layered medium, the SH wave characterized by Ω_z will propagate through the layered medium independently of the other waves. Hence, the SH wave for a point source on top of a layered medium can be expressed as

$$\Omega_z(\mathbf{r}) = \frac{iA\hat{z} \cdot (\nabla \times \hat{a})}{4\pi\mu} \int_0^\infty dk_\rho \frac{k_\rho}{k_{sz}} J_0(k_\rho\rho) \left[e^{ik_{sz}|z|} + R_{HH} e^{ik_{sz}(z+2d_1)} \right] \tag{8.74}$$

Since the SV waves and the P waves are always coupled together in a planar layered medium, we need to write them as a couplet:

$$\phi = \begin{bmatrix} u_z^s \\ u_z^p \end{bmatrix} = \int_0^\infty dk_\rho k_\rho e^{i\bar{\mathbf{k}}_z|z|} \begin{bmatrix} \widetilde{u}_{z\pm}^s \\ \widetilde{u}_{z\pm}^p \end{bmatrix} \quad \begin{matrix} z > 0 \\ z < 0 \end{matrix} \tag{8.75}$$

where

$$\widetilde{u}_{z\pm}^s = \frac{iA(\hat{z} \cdot \hat{a}k_s^2 \pm ik_{sz}\nabla_\pm^s \cdot \hat{a})}{4\pi\mu k_s^2 k_{sz}} J_0(k_\rho\rho), \quad \widetilde{u}_{z\pm}^p = \frac{\pm A\nabla_\pm^p \cdot \hat{a}}{4\pi(\lambda + 2\mu)k_c^2} J_0(k_\rho\rho) \tag{8.76}$$

$$\bar{\mathbf{k}}_z = \begin{bmatrix} k_{sz} & 0 \\ 0 & k_{cz} \end{bmatrix}, \quad \nabla_\pm^s = \nabla_s \pm \hat{z}ik_{sz}, \quad \nabla_\pm^p = \nabla_s \pm \hat{z}ik_{cz}. \tag{8.77}$$

In the above $\nabla_s = \hat{x}\partial_x + \hat{y}\partial_y$. When the point excitation is placed on top of a layered medium, we have

$$\phi = \int_0^\infty dk_\rho k_\rho \left[e^{i\bar{\mathbf{k}}_z|z|} \cdot \widetilde{\mathbf{u}}_\pm + e^{i\bar{\mathbf{k}}_z(z+d_1)} \cdot \overline{\mathbf{R}} \cdot e^{i\bar{\mathbf{k}}_z d_1} \cdot \widetilde{\mathbf{u}}_- \right] J_0(k_\rho\rho) \tag{8.78}$$

where $\widetilde{\mathbf{u}}_\pm^t = \left[\widetilde{u}_{z\pm}^s, \widetilde{u}_{z\pm}^p \right]$ and $\overline{\mathbf{R}}$ is the appropriate reflection matrix describing the reflection and cross-coupling between the SV and P waves.

The above derivation could be repeated with the Weyl identity if we so wish. The reflection coefficient R_{HH} and the reflection matrix $\overline{\mathbf{R}}$ can be found by matching boundary conditions. Then the field due to an elastic point source in or on top of a layered medium can be derived similar to the formalism described in Chapter 2 in [1].

8.7   Elastic Wave Equation for Planar Layered Media

The elastic wave equation for isotropic inhomogeneous media is

$$\partial_j(\lambda\partial_l u_l) + \partial_i(\mu\partial_i u_j) + \partial_i(\mu\partial_j u_i) + \omega^2\rho u_j = 0 \tag{8.79}$$

In vector notation, this may be written as

$$\nabla(\lambda\nabla \cdot \mathbf{u}) + \nabla \cdot (\mu\nabla\mathbf{u}) + (\mu\nabla\mathbf{u}) \cdot \overleftarrow{\nabla} + \omega^2\rho\mathbf{u} = 0 \tag{8.80}$$

where $\overleftarrow{\nabla}$ operates on terms to its left.

If λ and μ are functions of z only, and $\frac{\partial}{\partial x} = 0$ and $\mathbf{u} = \hat{x}u_x$, that is, we have an SH wave only, then extracting the $\hat{x}$ component of (8.80) leads to

$$\nabla_s \cdot \mu \nabla_s u_x + \omega^2 \rho u_x = 0 \tag{8.81}$$

where for this example, $\nabla_s = \hat{y}\frac{\partial}{\partial y} + \hat{z}\frac{\partial}{\partial z}$. Hence for this problem, a displacement field

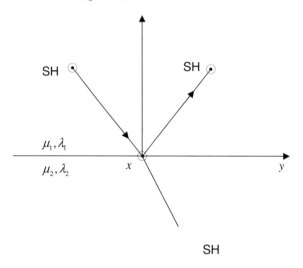

Figure 8.3: Reflected and transmitted SH plane wave at the interface of two-layer media

polarized in x with no variation in x is a pure shear wave with no coupling to the P wave. Even when μ and λ are discontinuous in z, only SH waves will be reflected and transmitted, as shown in Figure 8.3.

However, if the incident plane wave is an SV (shear vertical) wave, the displacement of the particles at an interface will induce both P (compressional) and SV reflected and transmitted waves. To see this, we let $\mathbf{u}_s = \hat{y}u_y + \hat{z}u_z$. Then (8.80) becomes

$$\nabla_s(\lambda \nabla_s \cdot \mathbf{u}_s) + \nabla_s \cdot (\mu \nabla_s \mathbf{u}_s) + (\mu \nabla_s \mathbf{u}_s) \cdot \overleftarrow{\nabla}_s + \omega^2 \rho \mathbf{u}_s = 0. \tag{8.82}$$

The above is the equation that governs the shear and compressional waves in a one-dimensional inhomogeneity where $\frac{\partial}{\partial x} = 0$.

Since λ, μ and $\mathbf{u}_s$ are smooth functions of y, $\frac{\partial}{\partial y}$ is smooth. Since μ and λ functions which are discontinuous with a step jump in the z direction, the z derivative may be discontinuous unless boundary conditions are satisfied by the field. To this end, we extract the z component of the terms in (8.82) that contain z derivative. They are

$$\frac{\partial}{\partial z}\left(\lambda \nabla_s \cdot \mathbf{u}_s + 2\mu\frac{\partial}{\partial z}u_z\right) \tag{8.83}$$

In order for $\frac{\partial}{\partial z}$ to be nonsingular in (8.82), we require

$$\lambda \nabla_s \cdot \mathbf{u}_s + 2\mu\frac{\partial}{\partial z}u_z \tag{8.84}$$

to be continuous. This is the same as requiring $\mathcal{T}_{zz}$ to be continuous.

Similarly, taking the y component of (8.82) that contains z derivative, we require

$$\mu\frac{\partial}{\partial z}u_y + \mu\frac{\partial}{\partial y}u_z \tag{8.85}$$

to be continuous. This is the same as requiring $\mathcal{T}_{zy}$ to be continuous. In order for (8.84) and (8.85) to be regular, u_z and u_y have to be continuous functions of z. Hence, the boundary conditions at a solid-solid interface are

$$u_z^{(1)} = u_z^{(2)} \tag{8.86}$$

$$u_y^{(1)} = u_y^{(2)} \tag{8.87}$$

$$\lambda_1 \nabla_s \cdot \mathbf{u}_s^{(1)} + 2\mu_1\frac{\partial}{\partial z}u_z^{(1)} = \lambda_2 \nabla_s \cdot \mathbf{u}_s^{(2)} + 2\mu_2\frac{\partial}{\partial z}u_z^{(2)} \tag{8.88}$$

$$\mu_1\left(\frac{\partial}{\partial z}u_y^{(1)} + \frac{\partial}{\partial y}u_z^{(1)}\right) = \mu_2\left(\frac{\partial}{\partial z}u_y^{(2)} + \frac{\partial}{\partial y}u_z^{(2)}\right) \tag{8.89}$$

The reflection of SH waves by a plane interface is purely a scalar problem. However, the reflection of a P wave or an SV wave by a plane boundary is a vector problem. In this case, $\mathbf{u}_s$ can always be decomposed into two components $\mathbf{u}_s = \hat{v}u_v + \hat{p}u_p$ where $\hat{p}$ is a unit vector in the direction of wave propagation, and $\hat{v}$ is a unit vector in the yz-plane orthogonal to $\hat{p}$.

For $z > 0$, the incident wave can be written as

$$\begin{aligned}
\mathbf{u}_s^{inc} &= \begin{bmatrix} u_v^{inc} \\ u_p^{inc} \end{bmatrix} = \begin{bmatrix} v_0 e^{-ik_{1vz}z} \\ p_0 e^{-ik_{1pz}z} \end{bmatrix} e^{ik_y y} \\
&= \begin{bmatrix} e^{-ik_{1vz}z} & 0 \\ 0 & e^{-ik_{1pz}z} \end{bmatrix} \begin{bmatrix} v_0 \\ p_0 \end{bmatrix} e^{ik_y y} \\
&= e^{-i\overline{\mathbf{k}}_{1z}z} \cdot \mathbf{u}_0 e^{ik_y y} \tag{8.90}
\end{aligned}$$

where

$$\mathbf{u}_0 = \begin{bmatrix} v_0 \\ p_0 \end{bmatrix}, \quad \overline{\mathbf{k}}_{1z} = \begin{bmatrix} k_{1vz} & 0 \\ 0 & k_{1pz} \end{bmatrix} \tag{8.91}$$

In the presence of a boundary, the reflected wave can be written as

$$\mathbf{u}_s^{ref} = \begin{bmatrix} u_v^{ref} \\ u_p^{ref} \end{bmatrix} = e^{i\overline{\mathbf{k}}_{1z}z} \cdot \mathbf{u}_r e^{ik_y y} \tag{8.92}$$

The most general relation between $\mathbf{u}_r$ and $\mathbf{u}_0$ is that

$$\mathbf{u}_r = \overline{\mathbf{R}} \cdot \mathbf{u}_0 \tag{8.93}$$

where

$$\overline{\mathbf{R}} = \begin{bmatrix} R_{vv} & R_{vp} \\ R_{pv} & R_{pp} \end{bmatrix} \tag{8.94}$$

By the same token, the transmitted wave is

$$\mathbf{u}_s^{tra} = e^{-i\overline{\mathbf{k}}_{2z}z} \cdot \mathbf{u}_t e^{ik_y y} \tag{8.95}$$

where

$$\mathbf{u}_t = \overline{\mathbf{T}} \cdot \mathbf{u}_0 \tag{8.96}$$

and

$$\overline{\mathbf{T}} = \left[\begin{array}{cc} T_{vv} & T_{vp} \\ T_{pv} & T_{pp} \end{array} \right] \tag{8.97}$$

There are four unknowns in $\mathbf{u}_r$ and $\mathbf{u}_t$ which can be found from four equations as a consequence of the boundary conditions (8.86) to (8.89).

8.7.1 Three-Layer Medium Case

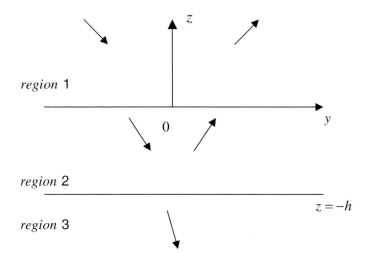

Figure 8.4: Reflected and transmitted waves in three-layer media

The three-layer problem can be solved using the constraint boundary conditions introduced in Chapter 2 of [1]. Then a recursive formula can be derived from which the reflection matrix for an N-layer problem can be easily found.

When three layers are present as shown in Figure 8.4, the field in Region 1 can be written as

$$\begin{aligned} \mathbf{u}_1 &= e^{-i\overline{\mathbf{k}}_{1z}z} \cdot \mathbf{a}_1 + e^{i\overline{\mathbf{k}}_{1z}z} \cdot \mathbf{b}_1, \\ &= \left[e^{-i\overline{\mathbf{k}}_{1z}z} + e^{i\overline{\mathbf{k}}_{1z}z} \cdot \widetilde{\overline{\mathbf{R}}}_{12} \right] \cdot \mathbf{a}_1 \end{aligned} \tag{8.98}$$

where we have defined $\mathbf{b}_1 = \widetilde{\overline{\mathbf{R}}}_{12} \cdot \mathbf{a}_1$. The $e^{ik_y y}$ dependence is dropped assuming that it is implicit.

In Region 2, we have

$$
\begin{aligned}
\mathbf{u}_2 &= e^{-i\overline{\mathbf{k}}_{2z}z} \cdot \mathbf{a}_2 + e^{i\overline{\mathbf{k}}_{2z}z} \cdot \mathbf{b}_2 \\
&= \left[e^{-i\overline{\mathbf{k}}_{2z}z} + e^{i\overline{\mathbf{k}}_{2z}(z+h)} \cdot \overline{\mathbf{R}}_{23} \cdot e^{i\overline{\mathbf{k}}_{2z}h} \right] \cdot \mathbf{a}_2
\end{aligned}
\tag{8.99}
$$

where we have defined

$$
e^{-i\overline{\mathbf{k}}_{2z}h} \cdot \mathbf{b}_2 = \overline{\mathbf{R}}_{23} \cdot e^{i\overline{\mathbf{k}}_{2z}h} \cdot \mathbf{a}_2
\tag{8.100}
$$

and $\overline{\mathbf{R}}_{23}$ is just the one-interface reflection coefficient previously defined.

The amplitude $\mathbf{a}_2$ is determined by the transmission of the amplitude of the downgoing wave in Region 1 (which is a_1) plus the reflection of the upgoing wave in Region 2.

As a result, we have at $z = 0$,

$$
\mathbf{a}_2 = \overline{\mathbf{T}}_{12} \cdot \mathbf{a}_1 + \overline{\mathbf{R}}_{21} \cdot e^{i\overline{\mathbf{k}}_{2z}h} \cdot \overline{\mathbf{R}}_{23} \cdot e^{i\overline{\mathbf{k}}_{2z}h} \cdot \mathbf{a}_2
\tag{8.101}
$$

Solving the above yields

$$
\mathbf{a}_2 = \left[\overline{\mathbf{I}} - \overline{\mathbf{R}}_{21} \cdot e^{i\overline{\mathbf{k}}_{2z}h} \cdot \overline{\mathbf{R}}_{23} \cdot e^{i\overline{\mathbf{k}}_{2z}h} \right]^{-1} \cdot \overline{\mathbf{T}}_{12} \cdot \mathbf{a}_1
\tag{8.102}
$$

The amplitude $\mathbf{b}_1$ of the upgoing wave in Region 1 is the consequence of the reflection of the downgoing wave in Region 1 plus the transmission of the upgoing wave in Region 2. Hence, at $z = 0$,

$$
\mathbf{b}_1 = \widetilde{\overline{\mathbf{R}}}_{12} \cdot \mathbf{a}_1 = \overline{\mathbf{R}}_{12} \cdot \mathbf{a}_1 + \overline{\mathbf{T}}_{21} \cdot e^{i\overline{\mathbf{k}}_{2z}h} \cdot \overline{\mathbf{R}}_{23} \cdot e^{i\overline{\mathbf{k}}_{2z}h} \cdot \mathbf{a}_2
\tag{8.103}
$$

Using (8.101), the above can be solved for $\widetilde{\overline{\mathbf{R}}}_{12}$, yielding

$$
\begin{aligned}
\widetilde{\overline{\mathbf{R}}}_{12} = \overline{\mathbf{R}}_{12} \quad &+ \quad \overline{\mathbf{T}}_{12} \cdot e^{i\overline{\mathbf{k}}_{2z}h} \cdot \overline{\mathbf{R}}_{23} \cdot e^{i\overline{\mathbf{k}}_{2z}h} \\
&\cdot \left[\overline{\mathbf{I}} - \overline{\mathbf{R}}_{21} \cdot e^{i\overline{\mathbf{k}}_{2z}h} \cdot \overline{\mathbf{R}}_{23} \cdot e^{i\overline{\mathbf{k}}_{2z}h} \right]^{-1} \cdot \overline{\mathbf{T}}_{12},
\end{aligned}
\tag{8.104}
$$

where $\widetilde{\overline{\mathbf{R}}}_{12}$ is the generalized reflection operator for a layered medium. If a region is added beyond Region 3, we need only to change $\overline{\mathbf{R}}_{23}$ to $\widetilde{\overline{\mathbf{R}}}_{23}$ in the above to account for subsurface reflection.

The above is a recursive relation which in general, can be written as

$$
\begin{aligned}
\widetilde{\overline{\mathbf{R}}}_{i,i+1} = \quad &\overline{\mathbf{R}}_{i,i+1} + \overline{\mathbf{T}}_{i+1,i} \cdot e^{i\overline{\mathbf{k}}_{i+1,z}h_{i+1}} \cdot \widetilde{\overline{\mathbf{R}}}_{i+1,i+2} \cdot e^{i\overline{\mathbf{k}}_{i+1,z}h} \cdot \\
&\left[\overline{\mathbf{I}} - \overline{\mathbf{R}}_{i+1,i} \cdot e^{i\overline{\mathbf{k}}_{i+1,z}h_{i+1}} \cdot \widetilde{\overline{\mathbf{R}}}_{i+1,i+2} \cdot e^{i\overline{\mathbf{k}}_{i+1,z}h_{i+1}} \right]^{-1} \cdot \overline{\mathbf{T}}_{i,i+1}
\end{aligned}
\tag{8.105}
$$

where h_{i+1} is the thickness of the $(i + 1)$-th layer. Eq. (8.102) is then

$$
\mathbf{a}_{i+1} = \left[\overline{\mathbf{I}} - \overline{\mathbf{R}}_{i+1,i} \cdot e^{i\overline{\mathbf{k}}_{i+1,z}h} \cdot \widetilde{\overline{\mathbf{R}}}_{i+1,i+2} \cdot e^{i\overline{\mathbf{k}}_{i+1,z}h} \right]^{-1} \cdot \overline{\mathbf{T}}_{i,i+1} \cdot \mathbf{a}_i
\tag{8.106}
$$

8.8 Finite Difference Scheme for the Elastic Wave Equation

The equation of motion for elastic waves is given by

$$\rho \frac{\partial^2 u_x}{\partial t^2} = \frac{\partial \mathcal{T}_{xx}}{\partial x} + \frac{\partial \mathcal{T}_{xz}}{\partial z} \tag{8.107}$$

$$\rho \frac{\partial^2 u_z}{\partial t^2} = \frac{\partial \mathcal{T}_{xz}}{\partial x} + \frac{\partial \mathcal{T}_{zz}}{\partial z} \tag{8.108}$$

$$\mathcal{T}_{xx} = (\lambda + 2\mu) \frac{\partial u_x}{\partial x} + \lambda \frac{\partial u_z}{\partial z} \tag{8.109}$$

$$\mathcal{T}_{zz} = (\lambda + 2\mu) \frac{\partial u_z}{\partial z} + \lambda \frac{\partial u_x}{\partial x} \tag{8.110}$$

$$\mathcal{T}_{xz} = \mathcal{T}_{zx} = \mu \left(\frac{\partial u_x}{\partial z} + \frac{\partial u_z}{\partial x} \right) \tag{8.111}$$

Defining $v_i = \partial u_i / \partial t$, the above can be transformed into a first-order system, that is,

$$\frac{\partial v_x}{\partial t} = \rho^{-1} \left(\frac{\partial \mathcal{T}_{xx}}{\partial x} + \frac{\partial \mathcal{T}_{xz}}{\partial z} \right) \tag{8.112}$$

$$\frac{\partial u_z}{\partial t} = \rho^{-1} \left(\frac{\partial \mathcal{T}_{xz}}{\partial x} + \frac{\partial \mathcal{T}_{zz}}{\partial z} \right) \tag{8.113}$$

$$\frac{\partial \mathcal{T}_{xx}}{\partial t} = (\lambda + 2\mu) \frac{\partial v_x}{\partial x} + \lambda \frac{\partial v_z}{\partial z} \tag{8.114}$$

$$\frac{\partial \mathcal{T}_{zz}}{\partial t} = (\lambda + 2\mu) \frac{\partial v_z}{\partial z} + \lambda \frac{\partial v_x}{\partial x} \tag{8.115}$$

$$\frac{\partial \mathcal{T}_{xz}}{\partial t} = \mu \left(\frac{\partial v_x}{\partial z} + \frac{\partial v_z}{\partial x} \right) \tag{8.116}$$

Using a central difference scheme, the above can be written as

$$\begin{aligned}
v_{x,i,j}^{k+\frac{1}{2}} - v_{x,i,j}^{k-\frac{1}{2}} &= \rho_{ij}^{-1} \frac{\Delta t}{\Delta x} \left[\mathcal{T}_{xx,i+\frac{1}{2},j}^{k} - \mathcal{T}_{xx,i-\frac{1}{2},j}^{k} \right] \\
&+ \rho_{ij}^{-1} \frac{\Delta t}{\Delta z} \left[\mathcal{T}_{xz,i,j+\frac{1}{2}}^{k} - \mathcal{T}_{xz,i,j-\frac{1}{2}}^{k} \right]
\end{aligned} \tag{8.117}$$

$$\begin{aligned}
v_{z,i+\frac{1}{2},j+\frac{1}{2}}^{k+\frac{1}{2}} - v_{z,i+\frac{1}{2},j+\frac{1}{2}}^{k-\frac{1}{2}} &= \rho_{i+\frac{1}{2},j+\frac{1}{2}}^{-1} \frac{\Delta t}{\Delta x} \left[\mathcal{T}_{xz,i+1,j+\frac{1}{2}}^{k} - \mathcal{T}_{xz,i,j+\frac{1}{2}}^{k} \right] \\
&+ \rho_{i+\frac{1}{2},j+\frac{1}{2}}^{-1} \frac{\Delta t}{\Delta z} \left[\mathcal{T}_{zz,i+\frac{1}{2},j+1}^{k} - \mathcal{T}_{zz,i+\frac{1}{2},j}^{k} \right]
\end{aligned} \tag{8.118}$$

$$\mathcal{T}_{xx,i+\frac{1}{2},j}^{k+1} - \mathcal{T}_{xx,i+\frac{1}{2},j}^{k} = (\lambda + 2\mu)_{i+\frac{1}{2},j} \frac{\Delta t}{\Delta x} \left[v_{x,i+1,j}^{k+\frac{1}{2}} - v_{x,i,j}^{k+\frac{1}{2}} \right]$$
$$+ \lambda_{i+\frac{1}{2},j} \frac{\Delta t}{\Delta z} \left[v_{z,i+\frac{1}{2},j+\frac{1}{2}}^{k+\frac{1}{2}} - v_{z,i+\frac{1}{2},j-\frac{1}{2}}^{k+\frac{1}{2}} \right] \quad (8.119)$$

$$\mathcal{T}_{zz,i+\frac{1}{2},j}^{k+1} - \mathcal{T}_{zz,i+\frac{1}{2},j}^{k} = (\lambda + 2\mu)_{i+\frac{1}{2},j} \frac{\Delta t}{\Delta z} \cdot \left[v_{z,i,j+1}^{k+\frac{1}{2}} - v_{z,i,j}^{k+\frac{1}{2}} \right]$$
$$+ \lambda_{i+\frac{1}{2},j} \frac{\Delta t}{\Delta x} \left[v_{x,i+\frac{1}{2},j}^{k+\frac{1}{2}} - v_{x,i,j}^{k+\frac{1}{2}} \right] \quad (8.120)$$

$$\mathcal{T}_{xz,i,j+\frac{1}{2}}^{k+1} - \mathcal{T}_{xz,i,j+\frac{1}{2}}^{k} = \mu_{i,j+\frac{1}{2}} \frac{\Delta t}{\Delta z} \left[v_{x,i,j+1}^{k+\frac{1}{2}} - v_{x,i,j}^{k+\frac{1}{2}} \right]$$
$$+ \mu_{i,j+\frac{1}{2}} \frac{\Delta t}{\Delta x} \left[v_{z,i+\frac{1}{2},j+\frac{1}{2}}^{k+\frac{1}{2}} - v_{z,i-\frac{1}{2},j+\frac{1}{2}}^{k+\frac{1}{2}} \right] \quad (8.121)$$

For a homogenous medium, the stability criterion is

$$v_\rho \Delta t \sqrt{\frac{1}{(\Delta x)^2} + \frac{1}{(\Delta z)^2}} < 1 \quad (8.122)$$

The above finite difference formulas are easy to implement and has been used to solve for various practical problems [6–9].

8.9 Integral Equation for Elastic Wave Scattering

The integral equations for elastic wave scattering can be derived from its partial differential equation (PDE) counterpart by applying the equivalence principle and extinction theorem as described in Chapter 8 in reference [1]

The PDE of elastic wave has been previously derived in Section 8.2, and reproduced here under a slightly different notation as

$$\gamma \nabla \nabla \cdot \mathbf{u} - \mu \nabla \times \nabla \times \mathbf{u} + \omega^2 \rho \mathbf{u} = -\mathbf{f} \quad (8.123)$$

where $\gamma = \lambda + 2\mu$.

The dyadic Green's function for an elastic medium has been derived in Section 8.3 in Eq. (8.37), which is repeated here as

$$\overline{\mathbf{G}}(\mathbf{x}, \mathbf{x}') = \frac{1}{\mu} \left(\overline{\mathbf{I}} + \frac{\nabla \nabla}{k_s^2} \right) g_s(\mathbf{x} - \mathbf{x}') - \frac{1}{\gamma} \frac{\nabla \nabla}{k_c^2} g_c(\mathbf{x} - \mathbf{x}') \quad (8.124)$$

It satisfies

$$\gamma \nabla \nabla \cdot \overline{\mathbf{G}}(\mathbf{x}, \mathbf{x}') - \mu \nabla \times \nabla \times \overline{\mathbf{G}}(\mathbf{x}, \mathbf{x}') + \omega^2 \rho \overline{\mathbf{G}}(\mathbf{x}, \mathbf{x}') = \overline{\mathbf{I}} \delta(\mathbf{x} - \mathbf{x}') \quad (8.125)$$

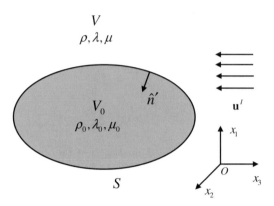

Figure 8.5: Elastic wave scattering by a homogeneous scatterer V_0 embedded in an infinite isotropic medium V. $\mathbf{u}^I$ is an incident wave.

Right multiplying (8.123) by $\overline{\mathbf{G}}(\mathbf{x}, \mathbf{x}')$ and left multiplying (8.125) by $\mathbf{u}(\mathbf{x})$, and subtracting the result, we have

$$
\begin{aligned}
&\gamma \left[\nabla(\nabla \cdot \mathbf{u}(\mathbf{x})) \cdot \overline{\mathbf{G}}(\mathbf{x}, \mathbf{x}') - \mathbf{u}(\mathbf{x}) \cdot \nabla\nabla \cdot \overline{\mathbf{G}}(\mathbf{x}, \mathbf{x}') \right] \\
&- \mu \left[\nabla \times \nabla \times \mathbf{u}(\mathbf{x}) \cdot \overline{\mathbf{G}}(\mathbf{x}, \mathbf{x}') - \mathbf{u}(\mathbf{x}) \cdot \nabla \times \nabla \times \overline{\mathbf{G}}(\mathbf{x}, \mathbf{x}') \right] \\
&= -\mathbf{f}(\mathbf{x}) \cdot \overline{\mathbf{G}}(\mathbf{x}, \mathbf{x}') - \delta(\mathbf{x} - \mathbf{x}')\mathbf{u}(\mathbf{x})
\end{aligned}
\tag{8.126}
$$

Using the vector identities

$$
\begin{aligned}
\nabla \cdot [(\nabla \cdot \mathbf{u})\mathbf{B} - \mathbf{u}\nabla \cdot \mathbf{B}] &= \nabla(\nabla \cdot \mathbf{u}) \cdot \mathbf{B} - \mathbf{u} \cdot (\nabla\nabla \cdot \mathbf{B}) \tag{8.127} \\
\nabla \cdot [(\nabla \times \mathbf{u}) \times \mathbf{B} + \mathbf{u} \times \nabla \times \mathbf{B}] &= (\nabla \times \nabla \times \mathbf{u}) \cdot \mathbf{B} - \mathbf{u} \cdot \nabla \times \nabla \times \mathbf{B}
\end{aligned}
$$

$$
\tag{8.128}
$$

we can integrate (8.126) over a volume V bounded by S, as shown in Figure 8.5, to result in

$$
\begin{aligned}
&\gamma \oint_S dS\hat{n} \cdot \left[(\nabla \cdot \mathbf{u}(\mathbf{x}))\overline{\mathbf{G}}(\mathbf{x}, \mathbf{x}') - \mathbf{u}(\mathbf{x})\nabla \cdot \overline{\mathbf{G}}(\mathbf{x}, \mathbf{x}') \right] \\
&- \mu \oint_S dS\hat{n} \cdot \left[(\nabla \times \mathbf{u}(\mathbf{x})) \times \overline{\mathbf{G}}(\mathbf{x}, \mathbf{x}') + \mathbf{u}(\mathbf{x}) \times \nabla \times \overline{\mathbf{G}}(\mathbf{x}, \mathbf{x}') \right] \\
&+ \int_V \mathbf{f}(\mathbf{x}) \cdot \overline{\mathbf{G}}(\mathbf{x}, \mathbf{x}')d\mathbf{x} = \begin{cases} -\mathbf{u}(\mathbf{x}'), & \mathbf{x}' \in V \\ 0, & \mathbf{x}' \notin V \end{cases}
\end{aligned}
\tag{8.129}
$$

The above can be rewritten as

$$\gamma \oint dS' \left[\nabla' \cdot \mathbf{u}(\mathbf{x}')\hat{n}' \cdot \overline{\mathbf{G}}(\mathbf{x}',\mathbf{x}) - \hat{n}' \cdot \mathbf{u}(\mathbf{x}')\nabla' \cdot \overline{\mathbf{G}}(\mathbf{x}',\mathbf{x})\right]$$

$$-\mu \oint dS' \left[\hat{n}' \times (\nabla' \times \mathbf{u}(\mathbf{x}')) \cdot \overline{\mathbf{G}}(\mathbf{x}',\mathbf{x}) + (\hat{n}' \times \mathbf{u}(\mathbf{x}')) \cdot \nabla' \times \overline{\mathbf{G}}(\mathbf{x}',\mathbf{x})\right]$$

$$+\mathbf{u}^I(\mathbf{x}) = \begin{cases} \mathbf{u}(\mathbf{x}), & \mathbf{x} \in V \\ 0, & \mathbf{x} \notin V \end{cases} \tag{8.130}$$

The extinction theorem described in Chapter 8 of reference [1] is also embedded in the above equation. We can use the above to formulate integral equations for elastic wave scattering. In the above, one can show that

$$\begin{aligned}
\nabla \cdot \overline{\mathbf{G}}(\mathbf{x},\mathbf{x}') &= \frac{1}{\mu}\left[\nabla g_s(\mathbf{x}-\mathbf{x}') + \frac{\nabla}{k_s^2}\nabla^2 g_s\right] - \frac{1}{\gamma}\frac{\nabla}{k_c^2}\nabla^2 g_c \\
&= \frac{1}{\mu}\left[\nabla g_s(\mathbf{x}-\mathbf{x}') + \frac{1}{k_s^2}\nabla(-\delta(\mathbf{x}-\mathbf{x}') - k_s^2 g_s)\right] \\
&\quad - \frac{1}{\gamma}\frac{1}{k_c^2}\nabla(-\delta(\mathbf{x}-\mathbf{x}') - k_c^2 g_c) \\
&= -\frac{1}{\mu}\frac{\nabla}{k_s^2}\delta(\mathbf{x}-\mathbf{x}') + \frac{1}{\gamma}\nabla g_c + \frac{1}{\gamma}\frac{1}{k_c^2}\nabla\delta(\mathbf{x}-\mathbf{x}') \\
&= \frac{1}{\gamma}\nabla g_c(\mathbf{x}-\mathbf{x}') \tag{8.131}
\end{aligned}$$

Using $\overline{\mathbf{G}}^T(\mathbf{x}',\mathbf{x}) = \overline{\mathbf{G}}(\mathbf{x},\mathbf{x}')$, $[\nabla' \times \overline{\mathbf{G}}(\mathbf{x}',\mathbf{x})]^T = \nabla \times \overline{\mathbf{G}}(\mathbf{x},\mathbf{x}')$, where the superscript T denotes a transpose, we have

$$\gamma \oint dS' \left\{[\overline{\mathbf{G}}(\mathbf{x},\mathbf{x}') \cdot \hat{n}'][\nabla' \cdot \mathbf{u}(\mathbf{x}')] - \frac{1}{\gamma}[\nabla' g_c(\mathbf{x}-\mathbf{x}')][\hat{n}' \cdot \mathbf{u}(\mathbf{x}')]\right\}$$

$$-\mu \oint dS' \left\{\overline{\mathbf{G}}(\mathbf{x},\mathbf{x}') \cdot [\hat{n}' \times \nabla' \times \mathbf{u}(\mathbf{x}')] + [\nabla' \times \overline{\mathbf{G}}(\mathbf{x},\mathbf{x}')] \cdot [\hat{n}' \times \mathbf{u}(\mathbf{x}')]\right\}$$

$$+\mathbf{u}^I(\mathbf{x}) = \begin{cases} \mathbf{u}(\mathbf{x}), & \mathbf{x} \in V \\ 0, & \mathbf{x} \notin V \end{cases} \tag{8.132}$$

The above formula was first given by Morse and Feshbach [10, (13.1.10)]. If we add a null quantity [11]

$$\begin{aligned}
-2\rho c_s^2 \int_S \hat{n}' \cdot [\nabla' \times (\mathbf{u} \times \overline{\mathbf{G}})]dS' &= -2\rho c_s^2 \int_V \nabla' \cdot [\nabla' \times (\mathbf{u} \times \overline{\mathbf{G}})]d\mathbf{x}' \\
&= 0 \tag{8.133}
\end{aligned}$$

to (8.132), then the formula becomes

$$\int_S \left\{\mathbf{t}(\mathbf{x}') \cdot \overline{\mathbf{G}}(\mathbf{x},\mathbf{x}') - \mathbf{u}(\mathbf{x}') \cdot [\hat{n}' \cdot \overline{\overline{\boldsymbol{\Sigma}}}(\mathbf{x},\mathbf{x}')]\right\} dS' + \mathbf{u}^I(\mathbf{x})$$

$$= \begin{cases} \mathbf{u}(\mathbf{x}), & \mathbf{x} \in V \\ 0, & \mathbf{x} \notin V \end{cases} \tag{8.134}$$

where $\mathbf{t}$ is the traction vector which can be related to $\mathbf{u}$ by Hooke's law, and $\overline{\overline{\boldsymbol{\Sigma}}}(\mathbf{x}, \mathbf{x}') = \lambda \overline{\mathbf{I}} \nabla \cdot \overline{\mathbf{G}} + \mu (\nabla \overline{\mathbf{G}} + \overline{\mathbf{G}} \nabla)$ is a third-rank Green's tensor introduced by Pao and Varatharajulu [11]. Eq. (8.134) is also known as the Somigliana's identity [12]

$$\overline{\mathbf{C}}^T(\mathbf{x}) \mathbf{u}(\mathbf{x}) = \int_S [\overline{\mathbf{U}}^T(\mathbf{x}, \mathbf{x}') \mathbf{t}(\mathbf{x}') - \overline{\mathbf{T}}^T(\mathbf{x}, \mathbf{x}') \mathbf{u}(\mathbf{x}')] \, dS' + \mathbf{u}^I(\mathbf{x}) \tag{8.135}$$

by identifying $\overline{\mathbf{U}} = \overline{\mathbf{G}}$ and $\overline{\mathbf{T}} = \hat{n}' \cdot \overline{\overline{\boldsymbol{\Sigma}}}$, where $\overline{\mathbf{U}}$ and $\overline{\mathbf{T}}$ are the Stokes' displacement and traction tensors, respectively. In Eq. (8.135), $\overline{\mathbf{C}}(\mathbf{x})$ is a tensor which takes the identity dyad for $\mathbf{x}$ in V, 0 for $\mathbf{x}$ in V_0, and a real function of the geometry of S in the vicinity of $\mathbf{x}$ for $\mathbf{x}$ on S. If the geometry is smooth at $\mathbf{x}$, then $\overline{\mathbf{C}}(\mathbf{x}) = \overline{\mathbf{I}}/2$. If we incorporate the boundary conditions at the surface of a scatterer, which are the continuity of $\mathbf{t}$ and $\mathbf{u}$ for a solid-solid interface as shown in Section 8.5, we have the integral equations

$$\frac{1}{2} \mathbf{u}(\mathbf{x}) + \int_S [\overline{\mathbf{T}}^T(\mathbf{x}, \mathbf{x}') \cdot \mathbf{u}(\mathbf{x}') - \overline{\mathbf{G}}^T(\mathbf{x}, \mathbf{x}') \cdot \mathbf{t}(\mathbf{x}')] \, dS'$$
$$= \mathbf{u}^I(\mathbf{x}), \quad \mathbf{x} \in S$$
$$\frac{1}{2} \mathbf{u}(\mathbf{x}) + \int_S [\overline{\mathbf{G}}_0^T(\mathbf{x}, \mathbf{x}') \cdot \mathbf{t}(\mathbf{x}') - \overline{\mathbf{T}}_0^T(\mathbf{x}, \mathbf{x}') \cdot \mathbf{u}(\mathbf{x}')] \, dS'$$
$$= 0, \quad \mathbf{x} \in S \tag{8.136}$$

The first equation above is obtained by letting the field point $\mathbf{x}$ approach the surface S from the exterior of the scatterer and the second equation is achieved when $\mathbf{x}$ approaches S from the interior of the scatterer. The kernels $\overline{\mathbf{G}}$ and $\overline{\mathbf{T}}$ are related to the medium in V while $\overline{\mathbf{G}}_0$ and $\overline{\mathbf{T}}_0$ are related to the medium in V_0. We have assumed that the geometry is smooth at the field point $\mathbf{x}$ on S so that the coefficient $1/2$ appears in the equations. From these boundary integral equations (BIE's)[1], the total displacement vector $\mathbf{u}$ and total traction vector $\mathbf{t}$ at the scatterer surface can be solved. If the scatterer is a traction-free cavity, then the total traction on the surface vanishes and the above equations reduce to

$$\frac{1}{2} \mathbf{u}(\mathbf{x}) + \int_S \overline{\mathbf{T}}^T(\mathbf{x}, \mathbf{x}') \cdot \mathbf{u}(\mathbf{x}') \, dS' = \mathbf{u}^I(\mathbf{x}), \quad \mathbf{x} \in S. \tag{8.137}$$

On the other hand, if the scatterer is a fixed rigid inclusion, then the total displacement on the surface vanishes and the above equations reduce to

$$\int_S \overline{\mathbf{G}}^T(\mathbf{x}, \mathbf{x}') \cdot \mathbf{t}(\mathbf{x}') \, dS' = -\mathbf{u}^I(\mathbf{x}), \quad \mathbf{x} \in S. \tag{8.138}$$

The above BIE's can be solved using different methods such as boundary element method (BEM) [12]– [20], method of fundamental solutions (MFS) [21], Nyström method [22], or method of moments (MOM) [23]. In the MOM implementation as described in Chapters 2–4, when applied to the elastic wave case, we need to decompose the unknown vectors into tangential and normal components. This is because the unknown vectors are three-dimensional

---

[1] In electromagnetic literature, these are also known as surface integral equations.

at the scatterer surface, which are different from the induced currents in electromagnetics. Taking Eq. (8.138) as an example, we decompose the unknown traction vector as

$$\mathbf{t}(\mathbf{x}') = \mathbf{t}_t(\mathbf{x}') + \mathbf{t}_n(\mathbf{x}') \tag{8.139}$$

where $\mathbf{t}_t(\mathbf{x}')$ is the tangential component and $\mathbf{t}_n(\mathbf{x}')$ is the normal component. We then expand the two components using Rao-Wilton-Glisson (RWG) basis and pulse basis, respectively

$$
\begin{aligned}
\mathbf{t}_t(\mathbf{x}') &= \sum_{n=1}^{N_t} \alpha_n \mathbf{f}_n(\mathbf{x}') \\
\mathbf{t}_n(\mathbf{x}') &= \sum_{n=1}^{N_n} \beta_n \hat{n}_n(\mathbf{x}').
\end{aligned}
\tag{8.140}
$$

In the above, $\mathbf{f}_n(\mathbf{x}')$ is the RWG basis as shown in Figure 8.6, $\hat{n}_n(\mathbf{x}')$ is the unit normal vector of the nth triangle patch used as a pulse basis, and α_n and β_n represent the unknown expansion coefficients to be solved. The RWG basis is redefined here as [24]

$$
\mathbf{f}_n(\mathbf{x}') = \begin{cases}
\frac{\ell_n}{2S_n^+} \mathbf{\Lambda}_n^+(\mathbf{x}'), & \mathbf{x}' \in S_n^+ \\
\frac{\ell_n}{2S_n^-} \mathbf{\Lambda}_n^-(\mathbf{x}'), & \mathbf{x}' \in S_n^- \\
0, & \text{otherwise.}
\end{cases}
\tag{8.141}
$$

where ℓ_n is the length of the common edge of two neighboring triangles, S_n^+ and S_n^- are the areas of the two triangles, and $\mathbf{\Lambda}_n^+(\mathbf{x}')$ and $\mathbf{\Lambda}_n^-(\mathbf{x}')$ are the distance vectors as indicated in Figure 8.6. We have in total N_t nonboundary edges connecting two neighboring triangles in which the RWG bases are defined and N_n triangles in which the pulse bases are defined. After using the above expansion, the BIE can be written as

$$
\sum_{n=1}^{N_t} \alpha_n \int_{S_n} \overline{\mathbf{G}}(\mathbf{x}, \mathbf{x}') \cdot \mathbf{f}_n(\mathbf{x}') \, dS' +
$$

$$
\sum_{n=1}^{N_n} \beta_n \int_{S_n} \overline{\mathbf{G}}(\mathbf{x}, \mathbf{x}') \cdot \hat{n}_n(\mathbf{x}') \, dS' = -\mathbf{u}^I(\mathbf{x}). \tag{8.142}
$$

where we omit the transpose T on $\overline{\mathbf{G}}(\mathbf{x}, \mathbf{x}')$ because $\overline{\mathbf{G}}(\mathbf{x}, \mathbf{x}')$ is symmetrical [12]. The next step is the testing using the basis functions as the weighting functions. By doing so, the following matrix equations are formed

$$
\begin{aligned}
\sum_{n=1}^{N_t} \alpha_n A_{mn} + \sum_{n=1}^{N_n} \beta_n B_{mn} &= E_{mn}, \quad m = 1, \cdots, N_t \\
\sum_{n=1}^{N_t} \alpha_n C_{mn} + \sum_{n=1}^{N_n} \beta_n D_{mn} &= F_{mn}, \quad m = 1, \cdots, N_n.
\end{aligned}
$$

$$\tag{8.143}$$

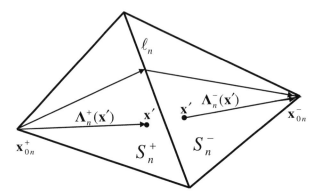

Figure 8.6: RWG basis $\mathbf{f}_n$ defined in two neighboring triangles S_n^+ and S_n^-. These two triangles share the nth nonboundary edge whose length is ℓ_n. Here S_n^+ and S_n^- also denote the corresponding areas of the two triangles.

where

$$
\begin{aligned}
A_{mn} &= \langle \mathbf{f}_m(\mathbf{x}), \overline{\mathbf{G}}(\mathbf{x}, \mathbf{x}'), \mathbf{f}_n(\mathbf{x}') \rangle \\
&= \frac{\ell_m \ell_n}{4 S_m^+ S_n^+} \langle \mathbf{\Lambda}_m^+(\mathbf{x}), \overline{\mathbf{G}}(\mathbf{x}, \mathbf{x}'), \mathbf{\Lambda}_n^+(\mathbf{x}') \rangle + \\
&\quad \frac{\ell_m \ell_n}{4 S_m^+ S_n^-} \langle \mathbf{\Lambda}_m^+(\mathbf{x}), \overline{\mathbf{G}}(\mathbf{x}, \mathbf{x}'), \mathbf{\Lambda}_n^-(\mathbf{x}') \rangle + \\
&\quad \frac{\ell_m \ell_n}{4 S_m^- S_n^+} \langle \mathbf{\Lambda}_m^-(\mathbf{x}), \overline{\mathbf{G}}(\mathbf{x}, \mathbf{x}'), \mathbf{\Lambda}_n^+(\mathbf{x}') \rangle + \\
&\quad \frac{\ell_m \ell_n}{4 S_m^- S_n^-} \langle \mathbf{\Lambda}_m^-(\mathbf{x}), \overline{\mathbf{G}}(\mathbf{x}, \mathbf{x}'), \mathbf{\Lambda}_n^-(\mathbf{x}') \rangle \\
B_{mn} &= \langle \mathbf{f}_m(\mathbf{x}), \overline{\mathbf{G}}(\mathbf{x}, \mathbf{x}'), \hat{n}_n(\mathbf{x}') \rangle \\
&= \frac{\ell_m}{2 S_m^+} \langle \mathbf{\Lambda}_m^+(\mathbf{x}), \overline{\mathbf{G}}(\mathbf{x}, \mathbf{x}'), \hat{n}_n(\mathbf{x}') \rangle + \\
&\quad \frac{\ell_m}{2 S_m^-} \langle \mathbf{\Lambda}_m^-(\mathbf{x}), \overline{\mathbf{G}}(\mathbf{x}, \mathbf{x}'), \hat{n}_n(\mathbf{x}') \rangle \\
C_{mn} &= \langle \hat{n}_m(\mathbf{x}), \overline{\mathbf{G}}(\mathbf{x}, \mathbf{x}'), \mathbf{f}_n(\mathbf{x}') \rangle \\
&= \frac{\ell_n}{2 S_n^+} \langle \hat{n}_m(\mathbf{x}), \overline{\mathbf{G}}(\mathbf{x}, \mathbf{x}'), \mathbf{\Lambda}_n^+(\mathbf{x}') \rangle + \\
&\quad \frac{\ell_n}{2 S_n^-} \langle \hat{n}_m(\mathbf{x}), \overline{\mathbf{G}}(\mathbf{x}, \mathbf{x}'), \mathbf{\Lambda}_n^-(\mathbf{x}') \rangle \\
D_{mn} &= \langle \hat{n}_m(\mathbf{x}), \overline{\mathbf{G}}(\mathbf{x}, \mathbf{x}'), \hat{n}_n(\mathbf{x}') \rangle \\
E_{mn} &= \langle \mathbf{f}_m(\mathbf{x}), \mathbf{u}^I(\mathbf{x}) \rangle \\
&= -\frac{\ell_m}{2 S_m^+} \langle \mathbf{\Lambda}_m^+(\mathbf{x}), \mathbf{u}^I(\mathbf{x}) \rangle - \frac{\ell_m}{2 S_m^-} \langle \mathbf{\Lambda}_m^-(\mathbf{x}), \mathbf{u}^I(\mathbf{x}) \rangle \\
F_{mn} &= -\langle \hat{n}_m(\mathbf{x}), \mathbf{u}^I(\mathbf{x}) \rangle
\end{aligned}
\tag{8.144}
$$

with

$$\Lambda_n^+(\mathbf{x}') = \mathbf{x}' - \mathbf{x}_{0n}^+$$
$$\Lambda_n^-(\mathbf{x}') = \mathbf{x}_{0n}^- - \mathbf{x}'. \tag{8.145}$$

The involved inner products above are defined exemplarily as

$$\langle \mathbf{f}_m(\mathbf{x}), \overline{\mathbf{G}}(\mathbf{x}, \mathbf{x}'), \mathbf{f}_n(\mathbf{x}') \rangle =$$
$$\int_{S_m} dS \mathbf{f}_m(\mathbf{x}) \cdot \int_{S_n} \overline{\mathbf{G}}(\mathbf{x}, \mathbf{x}') \cdot \mathbf{f}_n(\mathbf{x}') \, dS'$$
$$\langle \mathbf{f}_m(\mathbf{x}), \mathbf{u}^I(\mathbf{x}) \rangle = \int_{S_m} \mathbf{f}_m(\mathbf{x}) \cdot \mathbf{u}^I(\mathbf{x}) \, dS. \tag{8.146}$$

To illustrate the MOM solutions for the above BIE's, we first consider the scattering by a fixed rigid sphere with a radius of $a = 1.0$. The host medium has Poisson's ratio $\nu = 0.25$ and mass density $\rho = 1.0$. The incident wave is a time-harmonic dilatational plane wave propagating along $-x_3$ direction with a unit circular frequency ($\omega = 1.0$) and normalized wave number of $\kappa_c a = \pi$. Figure 8.7 shows the radial and tangential (elevated) components of total traction along the principal cut ($\phi = 0°$ and $\theta = 0° \sim 180°$) at the sphere surface. It can be seen that the solutions agree with the analytical solutions very well. The analytical solutions for general elastic wave scattering by a sphere can be found in [25].

We then consider the scattering in an elastic medium by a traction-free spherical cavity. The cavity has a radius of $a = 1.0$ and the host medium is characterized by Poisson's ratio $\nu = 1/3$, Young's modulus $E = 2/3$ and mass density $\rho = 1.0$. The incident wave is the same as before but with the normalized wave number of $\kappa_c a = 0.913$. Figure 8.8 plots the radial and tangential (elevated) components of total displacement along the principal cut at the surface. The solutions are also very close to the analytical solutions.

For the generalized case with both the host medium and scatterer being elastic, we select $\lambda = 0.53486$, $\mu = 0.23077$ and $\rho = 1.0$ for the host medium, and $\lambda_0 = 0.23716$, $\mu_0 = 0.52641$ and $\rho_0 = 1.9852$ for the elastic spherical inclusion with a unit radius. The incident wave is the same as the one for scattering by the spherical cavity. Figure 8.9 depicts the radial and tangential (elevated) components of total displacement at the surface along the principal cut. These results are also in excellent agreement with the analytical solutions.

8.10   Conclusions

In this chapter, we start with the motion of particles under a small perturbation in a homogeneous elastic medium and derive the elastic wave equation in a PDE form. The equation is then solved by using different techniques, that is, Helmholtz decomposition, time-domain approach, Fourier-Laplace transformation, and finite difference scheme. For inhomogeneous media, we derive the boundary conditions for different interfaces, that is, solid-solid, solid-fluid and fluid-fluid interfaces, and address the elastic wave physics in planar layered media. Moreover, we derive the corresponding integral equations from the PDE counterpart for elastic wave scattering by using the equivalence principle and extinction theorem. The integral equations are solved by using MOM which is first implemented in elastodynamics. Since the

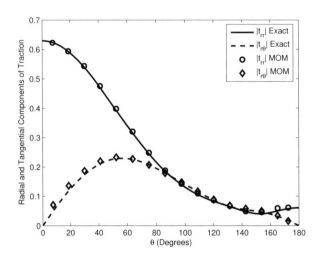

Figure 8.7: Radial and tangential (elevated) components of total traction along the principal cut at the surface of a rigid sphere, $k_c a = \pi$.

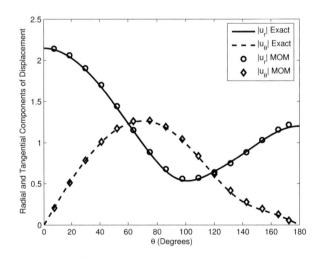

Figure 8.8: Radial and tangential (elevated) components of total displacement along the principal cut at the surface of a spherical cavity, $k_c a = 0.913$.

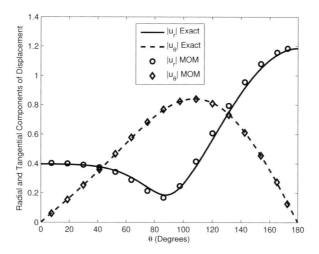

Figure 8.9: Radial and tangential (elevated) components of total displacement along the principal cut at the surface of an elastic sphere, $k_c a = 0.913$.

unknown functions are three-dimensional over a surface in the BIE's, the implementation of MOM is different from that in electromagnetics. Several numerical examples are used to demonstrate the effectiveness of the solution technique.

Bibliography

[1] W. C. Chew, *Waves and Fields in Inhomogeneous Media*, New York: Van Nostrand Reinhold, 1990. Reprinted by Piscataway, NJ: IEEE Press, 1995.

[2] W. C. Chew, *Elastic Wave Class Notes*, U. Illinois, Urbana-Champaign, Fall, 1991.

[3] K. Aki and P. G. Richards, *Quantitative Seismology: theory and methods*, Chapter 1, San Francisco: Freeman, 1980.

[4] J. A. Hudson, *The Excitation and Propagation of Elastic Waves*, pp. 15–24, Cambridge, MA: Combridge University Press, 1980.

[5] J. D. Archenbach, *Wave Propagation in Elastic Solids*, Chapter 1, Amsterdam: North Holland, 1973.

[6] Y.-H. Chen, W. C. Chew, and Q. H. Liu, "A three-dimensional finite difference code for the modeling of sonic logging tools," *J. Acoust. Soc. Am.*, vol. 93, pp. 702–712, 1998.

[7] W. C. Chew and Q. H. Liu, "Perfectly matched layers for elastodynamics: A new absorbing boundary condition," *J. Comput. Acoust.*, vol. 4, pp. 341–359, 1996.

[8] C. J. Randall, "Absorbing boundary condition for the elastic wave equation: velocity-stress formulation," *Geophys.*, vol. 54, pp. 1141–1152, 1989.

[9] Q.-H. Liu, E. Schoen, F. Daube, C. Randall, H.-L. Liu, and P. Lee, "A three-dimensional finite difference simulation of sonic logging," *J. Acoust. Soc. Am.*, vol. 100, pp. 72–79, 1996.

[10] P. M. Morse and H. Feshbach, *Methods of Theoretical Physics*, pp. 1770, New York: McGraw-Hill, 1953.

[11] Y.-H. Pao and V. Varatharajulu, "Huygens' principle, radiation conditions, and integral formulas for the scattering of elastic waves," *J. Acoust. Soc. Am.*, vol. 59, pp. 1361–1371, 1976.

[12] F. J. Rizzo, D. J. Shippy, and M. Rezayat, "A boundary integral equation method for radiation and scattering of elastic waves in three dimensions," *Int. J. Numer. Methods. Eng.*, vol. 21, pp. 115–129, 1985.

[13] G. D. Manolis and D. E. Beskos, *Boundary Element Methods in Elastodynamics*, London: Unwin Hyman, 1988.

[14] I. R. Gonsalves, D. J. Shippy, and F. J. Rizzo, "Direct boundary integral equations for elastodynamics in three-dimensional half-spaces," *Comput. Mech.*, vol. 6, pp. 279–292, 1990.

[15] F. J. Rizzo and D. J. Shippy, "An advanced boundary integral equation method for three-dimensional thermoelasticity," *Int. J. Numer. Methods. Eng.*, vol. 11, pp. 1753–1768, 1977.

[16] T. A. Cruse, "An improved boundary-integral equation method for three dimensional elastic stress analysis," *Comput. Struct.*, vol. 4, pp. 741–754, 1974.

[17] Y. Liu and F. J. Rizzo, "Hypersingular boundary integral equations for radiation and scattering of elastic waves in three dimensions," *Comput. Methods Appl. Mech. Eng.*, vol. 107, pp. 131–144, 1993.

[18] W. S. Hall, *The Boundary Element Method*, London: Kluwer Academic Publ., 1994.

[19] P. J. Schafbuch, R. B. Thompson, and F. J. Rizzo, "Application of the boundary element method to the elastic wave scattering by irregular defects," *J. Nondestr. Eval.*, vol. 2, pp. 113–127, 1990.

[20] P. S. Kondapalli, "Frequency response of elastic bodies of revolution by the boundary element method," *J. Acoust. Soc. Am.*, vol. 98, pp. 1558–1564, 1995.

[21] P. S. Kondapalli and D. J. Shippy, "The method of fundamental solutions for transmission and scattering of elastic waves," *Comput. Methods Appl. Mech. Eng.*, vol. 96, pp. 255–269, 1992.

[22] M. S. Tong and W. C. Chew, "Nyström method for elastic wave scattering by three-dimensional obstacles," *J. Comput. Phys.*, vol. 226, pp. 1845–1858, 2007.

[23] M. S. Tong and W. C. Chew, "Unified boundary integral equation for the scattering of elastic wave and acoustic wave: solution by the method of moments," *Waves Random Complex Media*, vol. 18, pp. 303–324, 2008.

[24] S. M. Rao, D. R. Wilton, and A. W. Glisson, "Electromagnetic scattering by surfaces of arbitrary shape," *IEEE Trans. Antennas Propag.*, vol. 30, pp. 409–418, 1982.

[25] Y.-H. Pao and C. C. Mow, "Scattering of plane compressional waves by a spherical obstacle," *J. Appl. Phys.*, vol. 34, pp. 493–499, 1963.

Glossary of Acronymns

AIM adaptive integral method;

BCG, BiCG biconjugate gradient;

BCGSTAB BCG stabilized;

BEM boundary element method;

BIE boundary integral equation;

CEM computational electromagnetics;

CFD computational fluid dynamics;

CFIE combined field integral equation;

CG conjugate gradient;

CPU central processing unit time;

CSIE combined source integral equation;

DCIM discrete complex image method;

DGLM dyadic Green's function for layered media;

EFIE electric field integral equation;

EPA equivalence principle algorithm;

FDM finite difference method;

FEM finite element method;

FERM finite element radiation method;

FIPWA fast inhomogeneous plane wave algorithm;

FISC Fast Illinois Solver Code;

FMM fast multipole method;

FMM-PML-MPIE FMM, perfectly matched layer, mixed potential integral equation;

FSIP folded Sommerfeld integration path;

GIBC generalized impedance boundary condition;

GMRES generalized minimal residual;

H polarization horizontal polarization;

HH response horizontal-horizontal response;

HV response horizontal-vertical response;

IBC impedance boundary condition;

IBC-MPW IBC modified plane wave;

IBC-WI IBC wave impedance;

LF-FMA low-frequency fast multipole algorithm;

LF-MLFMA low-frequency multilevel fast multipole algorithm;

LF-MOM low-frequency method of moments;

LPP long patch pair;

LUD lower-upper triangular decomposition;

MF-FMA mixed-form fast multipole algorithm;

MFIE magnetic field integral equation;

MFS method of fundamental solutions;

ML-FIPWA multilevel fast inhomogeneous plane wave algorithm;

MLFMA multilevel fast multipole algorithm;

MMIC monolithic microwave integrated circuit;

MOM method of moments;

NCSA National Center for Supercomputing Applications;

P wave pressure wave, compressional wave;

PDE partial differential equation;

PEC perfect electric conducting;

PMC perfect magnetic conductor;

PMCHWT formulation Poggio-Miller-Chang-Harrington-Wu-Tsai formulation;

QMR quasi-minimal residual;

RAM random access memory;

RCS radar cross section;

RFID radio frequency identification tag;

RWG Rao-Wilton-Glisson;

S wave shear wave;

SBI self-box inclusion;

ScaleME scaleable multipole engine;

SDFMA steepest descent fast multipole algorithm;

SDFMM steepest descent fast multipole method;

SDP steepest descent path;

SH wave shear horizontal wave;

SIE surface integral equation;

SIP Sommerfeld integration path;

SPP short patch pair;

SV wave shear vertical wave;

TDS thin dielectric sheet;

TE transverse electric;

TFQMR transpose free quasi-minimal residual;

TM transverse magnetic;

TSM-FMA thin-stratified medium fast-multipole algorithm;

VIE volume integral equation;

VV response vertical-vertical response;

About the Authors

Weng Cho CHEW received the B.S. (1976), M.S. (1978), Engineer's (1978), and Ph.D. (1980) degrees in EE from MIT. He is serving as the Dean of Engineering at The University of Hong Kong on leave of absence from the University of Illinois at Urbana-Champaign (UIUC). Previously, he was a professor and the director of the Electromagnetics Laboratory at the University of Illinois. Before that, he was a manager and program leader at Schlumberger-Doll Research. He was on IEEE Adcom for AP-S and GRSS, and was active with various journals and societies. He researches in wave physics for various applications and fast algorithms for solving wave problems, and originates several fast algorithms for solving scattering and inverse problems. A research group that he led, for the first time, solved dense matrix systems with tens of millions of unknowns for integral equations of scattering. He wrote the book *Waves and Fields in Inhomogeneous Media*, coauthored the book *Fast and Efficient Methods in Computational Electromagnetics*, as well as numerous journal, conference publications, and book chapters. He is an IEEE, OSA, and IOP Fellow, and was an NSF PYI (USA). He received the Schelkunoff Best Paper Award for TAP, the IEEE Graduate Teaching Award, UIUC Campus Wide Teaching Award, and IBM Faculty Awards. He was a Founder Professor and is currently, a Y.T. Lo Endowed Chair Professor at UIUC. Recently, he served as an IEEE Distinguished Lecturer, and Cheng Tsang Man Visiting Professor at Nanyang Technological University in Singapore. ISI Citation elected him to the category of Most-Highly Cited Authors (top 0.5%). This year, he has been chosen to receive the IEEE AP-S Distinguished Educator Award.

Mei Song TONG is a visiting research scientist at the Center for Computational Electromagnetics and Electromagnetics Laboratory (CCEML), Department of Electrical and Computer Engineering (ECE), University of Illinois at Urbana-Champaign (UIUC). He received the B.S. and M.S. degrees from Huazhong University of Science and Technology (China) in 1985 and 1988, respectively, and the Ph.D. degree from Arizona State University (ASU) in 2004, all in electrical engineering. His research interests include numerical methods in electromagnetics, acoustics and elastodynamics, simulation and design for antennas and RF/microwave circuits, and electronic packaging for digital devices. He is a senior member of the Institute of Electrical and Electronics Engineers (IEEE).

Bin HU received the B.S. degree in Electrical Engineering from Tsinghua University in 1995, the M.S. degree from Syracuse University in 1997, and Ph.D. degree from University of Illinois at Urbana-Champaign in 2001. Since 2001, he has been with Process Technology Modeling division at the Intel Corporation, currently as a Senior Staff Research Scientist. Dr. Hu leads the development of advanced lithography modeling techniques, a key enabling component of resolution enhancement solutions for Intel's next generation process technology. His research interests include all aspects of lithography modeling techniques and fast algorithms in computational electromagnetics. Dr. Hu is the co-author of 9 journal articles and book chapters, 15 conference papers, and 4 (pending) patents. At Intel, Dr. Hu received numerous awards for his work on lithography modeling, including the prestigious Intel Achievement Award. He was the recipient of Y.T. Lo Outstanding Research Award in 2000.

Index

Printed in the United States
by Baker & Taylor Publisher Services